Introduction to
Paint Chemistry

Introduction to
Paint Chemistry
and principles of paint technology

J. Bentley

and

G. P. A. Turner

Both formerly with ICI Paints
Slough, Berkshire, UK

Fourth edition

CHAPMAN & HALL

London · Weinheim · New York · Tokyo · Melbourne · Madras

Published by Chapman & Hall, 2–6 Boundary Row, London SE1 8HN, UK

Chapman & Hall, 2–6 Boundary Row, London SE1 8HN, UK

Chapman & Hall GmbH, Pappelallee 3, 69469 Weinheim, Germany

Chapman & Hall USA, 115 Fifth Avenue, New York, NY 10003, USA

Chapman & Hall Japan, ITP-Japan, Kyowa Building, 3F, 2-2-1 Hirakawacho, Chiyoda-ku, Tokyo 102, Japan

Chapman & Hall Australia, 102 Dodds Street, South Melbourne, Victoria 3205, Australia

Chapman & Hall India, R. Seshadri, 32 Second Main Road, CIT East, Madras 600 035, India

First edition 1967
Second edition 1980
Third edition 1988
Reprinted 1990, 1991, 1993, 1995
Fourth edition 1998

© 1967, 1980, 1988 G. P. A. Turner; 1998 J. Bentley and G. P. A. Turner

Typeset in 10/12pt Times by Academic & Technical Typesetting, Bristol, UK
Printed in Great Britain by St Edmundsbury Press, Bury St Edmunds, Suffolk

ISBN 0 412 72320 4 (HB)   0 412 72330 1 (PB)

A catalogue record for this book is available from the British Library

∞ Printed on acid-free text paper, manufactured in accordance with ANSI/ NISO Z39.48-1992 and ANSI/NISO Z39.48-1984 (Permanence of Paper).

Contents

Preface to the fourth edition   vii

Acknowledgements   ix

Names and units   xi

PART ONE: General science

1 Atoms to equations: the basis of chemical reactions   3
2 Inorganic chemistry: acids, bases and salts   17
3 Organic chemistry: alkanes to oils   30
4 Organic chemistry: ethers to isocyanates   47
5 Solid forms: polymers and solid structures   61
6 Colour: physics and chemistry   77

PART TWO: Applied science

7   Paint: first principles   91
8   Pigmentation: pigments and paint making   103
9   Solvents: chemistry and physics of solvents and diluents   118
10 Paint additives: modifying application, curing and appearance 136
11 Lacquers, emulsion paints and non-aqueous dispersions:
        paints drying by evaporation   150
12 Oil and alkyd paints: paints drying through oxidation   167
13 Thermosetting alkyd, polyester and acrylic paints: paints based
        on nitrogen resins   187
14 Epoxy coatings: paints based on epoxy resins   203
15 Polyurethanes: isocyanate-based coatings   220

16   Radiation curing and unsaturated polyesters: finishes curing
        through unsaturation   237
17   Chemical treatment of substrates   251
18   Health, safety and environment   259

        Appendix A: suggestions for further reading   263
        Appendix B: functional groups of organic chemistry in
        systematic nomenclature   266
        Index   267

Preface to the fourth edition

Introduction to Paint Chemistry was first published in 1967 with the intention of providing both a textbook for students and an introduction to the subject for those with little or no technical knowledge. This remains the objective. The book was revised in 1980 and again in 1988, but change continues. In this fourth edition, we have sought to bring it up to date with the newest developments in the technology and, with an additional chapter, to add a perspective on the importance now given to safety and also to environmental issues in paint system design, manufacture and use. An additional feature is the attention that has been given to the use of nomenclature and units, attempting to reconcile current usage in both education and in industry, providing a bridge where these differ.

The book is divided into two parts. Part One begins at the very basis of matter – its atomic structure – and works step by step through a sufficient selection of chemistry and physics to allow any interested reader to cope with the chemistry and the technology of paint in Part Two. The reader should absorb as much of Part One as he or she feels necessary. It is worth noting, however, that the topics in it are specially selected from a paint point of view and that, for example, detail on oils in Chapter 3, on polymers in Chapter 5 and on light and colour in Chapter 6 could well be missing in some chemistry degree courses.

Part Two begins with four chapters applicable to paints of every sort and then goes on to six particular paint systems, covering the greater part of paints and varnishes in current use. The classification of paints within these six chapters is largely by drying mechanism. Thus the important family of acrylic finishes does not get a chapter to itself, since the drying mechanisms of the many types of acrylic coating covered in this book are all different and not essentially acrylic mechanisms. The finishes are described in Chapters 11, 13, 14, 15 and 16. Again, there is no chapter on water-based paints, since water-based paints are made from a variety of chemically different water-soluble or water-dispersible resins and dry by a variety of mechanisms. The techniques for making resins soluble or dispersible in water are described in Chapters 9 and 11 and exemplified in other chapters.

As aids to understanding chemical names, an appendix gives a brief summary of chemical families, and the inner covers of the book cross-reference all systematic and trivial names used in the book. We aim to make conversions between names used in formal teaching and those used in industry easier. A second appendix contains suggestions for further reading and we hope that the index will be sufficiently full to allow quick and easy reference to any topic covered in the book.

J.B.
G.P.A.T.

Acknowledgements

We are grateful to the publishers for allowing us this space in which we thank the following for helping us to complete this book.

The Directors of ICI for leave to publish the original work.

Mr B.M. Letsky, Industrial Finish Consultant, for the nitrocellulose lacquer formula in Table 11.2. Rohm & Haas Company, Philadelphia, USA, for the acrylic lacquer formula in Table 11.2. Worlée Chemie GmbH for the water-borne air-drying formula in Table 12.1. DSM Resins (UK) Ltd for the emulsified alkyd paint formula in Table 12.5, the metallic car finish formula in Table 13.2 and the acrylic isocyanate formulae in Table 15.3. BIP Ltd for the woodfinish formula in Table 13.1. Amoco Chemicals UK Ltd for the high solids and water-based formula in Tables 13.4 and 13.5.

Eastman Chemical (UK) Ltd for the coil coating enamel formula in Table 13.3 and the polyester epoxy powder formulae in Table 14.4. Shell Chemicals for the can coating formula in Table 14.1 and the electro-deposition primer formula in Table 14.2. CIBA Polymers for the solventless epoxy formula in Table 14.3. Cargill Blagden for the one-pack polyurethane formulae in Tables 15.1 and 15.2. UCB sa for the UV-curing acrylic formula in Table 16.2.

Oxford University Press for permission to base Figs 5.3 and 5.4 on an illustration in *Chemical Crystallography*, by C.W. Bunn. Frederick J. Drake & Co., Chicago, USA, for permission to base Figs 6.9 and 6.10 on illustrations in *Color in Decoration and Design* by F.M. Crewdson. Eiger-Torrance for the details on which Figs 8.2 and 8.3 were based. Reinhold Publishing Corporation for permission to base Fig. 9.3 on an illustration in *Principles of Emulsion Technology* by P. Becher.

Dr C. Carr of the Paint Research Association for providing the Hansen solubility parameter data used for Fig. 9.2.

The many friends at ICI Paints who have helped us with all of the editions of this book. To those gratefully acknowledged in previous editions, we must add K. Boyle, P. Collins, J. Harrison and N. Martin.

Mr G. Light, whose illustrations have so self-evidently stood the test of time.

Carol and Nina Bentley for their comments and for the insight given by their various school and undergraduate texts into current use of names and units.

Finally, we acknowledge that all the errors in this book are ours.

J.B.
G.P.A.T.

Names and units

This introduction to the naming of chemicals and measurements in this book is to make readers aware that they are likely to meet different names for chemicals, and different units of measurement to those they may be used to.

With chemical names, schools and colleges now use 'systematic' names while industry often uses 'trivial' names. In the first part of this book, both names are generally given so that the reader will become familiar with both systems. In the second part of the book, generally, only the name most commonly used will be given. To provide a reference, the inside covers to this book cross-reference most of the names used in this book.

Some of the measurements used for paint and chemical purposes involve units which may be new, even to those familiar with the metric system of measurement. For example, there is the need to describe distances shorter than 1 millimetre. All the units of measurement to be found in the book are set out below. Again, just as for the chemical names, modern units known as SI units are used more and more, and the next page gives a list with some conversion factors.

Units of length, area and volume

1 metre (m) = 39.37 inches

$\quad\quad\quad$ = 1000 millimetres (mm) $\quad\quad\quad$ milli = 10^{-3}

$\quad\quad\quad$ = 1 000 000 (10^6) microns (μm) $\quad\quad$ micro = 10^{-6}

$\quad\quad\quad$ = 1 000 000 000 (10^9) nanometres (nm) $\quad$ nano = 10^{-9}

1 thousandth of an inch (thou or mil) = 25.4 μm = 25 400 nm

1 litre (l) = 1000 millilitres (ml) = 1.76 pints (Imp)

1 gallon (Imp) = 4546 ml = 1.2 gal (US)

Units of weight

1 kilogram (kg) = 2.2 pounds = 1000 grams (g)   kilo = 10^3

1 ounce = 28.4 grams

Units of temperature

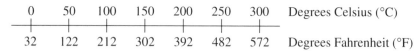

Translating from °F to °C:

$$x\,°F = (x - 32) \times \frac{5}{9}\,°C$$

Translating from °C to °F:

$$x\,°C = \frac{9x}{5} + 32\,°F$$

SI units

Those who already have some familiarity with scientific units of measurement will know some of the SI (Système Internationale) units, and also know that some of the older c.g.s. (centimetre, gram, second) units are still widely used. For those needing them, it will be helpful to have some of these listed, with their conversion.

	In SI units	In c.g.s. units	Conversion
Basic unit of length	metre (m)	centimetre (cm)	1 m = 100 cm
Basic unit of mass	kilogram (kg)	gram (g)	1 kg = 1000 g
Basic unit of time	second (s)	second (s)	–

Many other units may be derived from these three. Others of interest, including those in this book, are:

Unit of volume	cubic metre (m^3)	cubic centimetre (cm^3)	$1\,m^3 = 10^6\,cm^3$
Unit of pressure	pascal (Pa)	atmosphere (atm)	$101\,325\,Pa = 1\,atm$
			$1\,atm = 760\,mmHg$
Unit of force	newton (N)	dyne ($g\,cm\,s^{-2}$) (dyn)	$1\,N = 10^5\,dyn$
Unit of energy	joule (J)	calorie (cal)	$1\,J = 0.239\,cal$
Unit of viscosity	pascal-second (Pa s)	poise (P)	$1\,Pa\,s = 10\,P$
	millipascal-second (mPa s)	centipoise (cP)	$1\,mPa\,s = 1\,cP$
Unit of surface tension	newton per metre ($N\,m^{-1}$)	dyne per cm	$1\,N\,m^{-1} = 10^3\,dyn\,cm^{-1}$
Unit of density	kg per m^3	gram per cm^3	$1\,g\,cm^{-3} = 1000\,kg\,m^{-3}$
Unit of frequency	hertz (Hz)	cycles per second	$1\,Hz = 1\,c/s$

PART ONE

General science

One

Atoms to equations

the basis of chemical reactions

In this chapter we will describe the atomic nature of all substances and explain how this leads to understanding physical behaviour, molecular structure and chemical reactions.

Atoms and elements

If a portion of any pure chemical substance is divided into two and then one of the two pieces is again divided, and the process repeated so that the size of the piece is continually becoming smaller, eventually the substance cannot be sub-divided any further without being decomposed. At this stage, we have the smallest particle that can exist and still be identified as being of that substance.

For simple elementary substances the smallest particle is called an **atom**. An atom is very small indeed: a sheet of iron 1 μm (1000 nm) thick would be 4300 atoms thick. These simple substances, which have as their smallest particle a single atom, are called **elements**. There are 92 elements found in nature, including such commonplace materials as oxygen and carbon, sulphur and mercury, iron and lead. The names of elements, frequently their Latin names, are conveniently abbreviated to one or two letters. These abbreviations are internationally recognized as **symbols** for the elements, e.g. O for oxygen, C for carbon, S for sulphur, Hg for mercury (*hydrargyrum*), Fe for iron (*ferrum*) and Pb for lead (*plumbum*). The first 105 elements, including the 92 natural ones, are listed by their symbols in a table known as the Periodic Table (Fig. 1.1; Table 1.1). The elements listed in vertical columns behave similarly in the way they react as chemicals. The elements are given **atomic numbers** from 1 to 105.

The atoms of different elements are different in weight, size and behaviour. Atoms are built from a number of different electrically charged or neutral subatomic particles. An atom consists of a nucleus, which contains the heavier particles, some of which are positively charged (**protons**) and some neutral (**neutrons**), and this is surrounded by a number of much lighter negatively charged particles called **electrons**. The electrons are in constant

Group:

Period	I	II											III	IV	V	VI	VII	0
1	1 H																	2 He
2	3 Li	4 Be											5 B	6 C	7 N	8 O	9 F	10 Ne
3	11 Na	12 Mg											13 Al	14 Si	15 P	16 S	17 Cl	18 Ar
4	19 K	20 Ca	21 Sc	22 Ti	23 V	24 Cr	25 Mn	26 Fe	27 Co	28 Ni	29 Cu	30 Zn	31 Ga	32 Ge	33 As	34 Se	35 Br	36 Kr
5	37 Rb	38 Sr	39 Y	40 Zr	41 Nb	42 Mo	43 Tc	44 Ru	45 Rh	46 Pd	47 Ag	48 Cd	49 In	50 Sn	51 Sb	52 Te	53 I	54 Xe
6	55 Cs	56 Ba	57 La ■	72 Hf	73 Ta	74 W	75 Re	76 Os	77 Ir	78 Pt	79 Au	80 Hg	81 Tl	82 Pb	83 Bi	84 Po	85 At	86 Rn
7	87 Fr	88 Ra	89 Ac ■■	104 Rf	105 Ha													

■ Lanthanide series:

58 Ce	59 Pr	60 Nd	61 Pm	62 Sm	63 Eu	64 Gd	65 Tb	66 Dy	67 Ho	68 Er	69 Tm	70 Yb	71 Lu

■■ Actinide series:

90 Th	91 Pa	92 U	93 Np	94 Pu	95 Am	96 Cm	97 Bk	98 Cf	99 Es	100 Fm	101 Md	102 No	103 Lr

Fig. 1.1 The Periodic Table.

Table 1.1 The chemical elements

Atomic no.	Symbol	Element	Atomic no.	Symbol	Element
1	H	Hydrogen	54	Xe	Xenon
2	He	Helium	55	Cs	Caesium
3	Li	Lithium	56	Ba	Barium
4	Be	Beryllium	57	La	Lanthanum
5	B	Boron	58	Ce	Cerium
6	C	Carbon	59	Pr	Praesodymium
7	N	Nitrogen	60	Nd	Neodymium
8	O	Oxygen	61	Pm	Promethium
9	F	Fluorine	62	Sm	Samarium
10	Ne	Neon	63	Eu	Europium
11	Na	Sodium	64	Gd	Gadolinium
12	Mg	Magnesium	65	Th	Terbium
13	Al	Aluminium	66	Dy	Dysprosium
14	Si	Silicon	67	Ho	Holmium
15	P	Phosphorus	68	Er	Erbium
16	S	Sulphur	69	Tm	Thulium
17	Cl	Chlorine	70	Yb	Ytterbium
18	Ar	Argon	71	Lu	Lutetium
19	K	Potassium	72	Hf	Hafnium
20	Ca	Calcium	73	Ta	Tantalum
21	Sc	Scandium	74	W	Tungsten
22	Ti	Titanium	75	Re	Rhenium
23	V	Vanadium	76	Os	Osmium
24	Cr	Chromium	77	Ir	Iridium
25	Mn	Manganese	78	Pt	Platinum
26	Fe	Iron	79	Au	Gold
27	Co	Cobalt	80	Hg	Mercury
28	Ni	Nickel	81	Tl	Thallium
29	Cu	Copper	82	Pb	Lead
30	Zn	Zinc	83	Bi	Bismuth
31	Ga	Gallium	84	Po	Polonium
32	Ge	Germanium	85	At	Astatine
33	As	Arsenic	86	Rn	Radon
34	Se	Selenium	87	Fr	Francium
35	Br	Bromine	88	Ra	Radium
36	Kr	Krypton	89	Ac	Actinium
37	Rb	Rubidium	90	Th	Thorium
38	Sr	Strontium	91	Pa	Protoactinium
39	Yt	Yttrium	92	U	Uranium
40	Zr	Zirconium	93	Np	Neptunium
41	Nb	Niobium	94	Pu	Plutonium
42	Mo	Molybdenum	95	Am	Americium
43	Tc	Technetium	96	Cm	Curium
44	Ru	Ruthenium	97	Bk	Berkelium
45	Rh	Rhodium	98	Cf	Californium
46	Pd	Palladium	99	Es	Einsteinium
47	Ag	Silver	100	Fm	Fermium
48	Cd	Cadmium	101	Md	Mendelevium
49	In	Indium	102	No	Nobelium
50	Sn	Tin	103	Lr	Lawrencium
51	Sb	Antimony	104	Rf	Rutherfordium
52	Te	Tellurium	105	Ha	Hahnium
53	I	Iodine			

motion about the nucleus and the roughly spherical region in which they move defines the volume of the atom. The electrical charges on protons and electrons are equal, but opposite in sign. The number of electrons in an atom is always exactly the same as the number of protons, so that an atom is electrically neutral. Atoms of different elements differ in the number of protons and electrons which they contain; the atomic number of an element is the number of protons (or electrons) in one atom of that element. It is well known that it is possible to split certain atoms, but since the products contain fewer protons and neutrons that the original atom, they are therefore atoms of *different* (lighter) elements. Thus it is still true that the smallest particle of an element is an atom of that element; anything smaller does not possess the properties of that element.

Molecules and compounds

All substances have the elements as their basic constituents. However, some substances, known as **compounds**, have as their smallest particle, a **molecule**. A molecule consists of two or more atoms held together by chemical bonds (see below). A molecule of a compound must contain atoms of at least two different elements.

Compounds, like elements, can be described with the aid of the symbols used for atoms and these reveal the proportions of the atoms of different elements present in the molecule. Thus water shown as H_2O, contains two atoms of hydrogen and one of oxygen per molecule. This group of symbols is also described as the chemical **formula** of water.

It is customary, in most chemical formulae of inorganic compounds (Chapter 2), to write metallic atoms first (if there are any), e.g. NaCl, sodium chloride, or common salt. In the Periodic Table, the metallic elements extend from the left to almost three-quarters of the way across the table, i.e. all of groups I and II, all of the centre block of elements, and the lower parts of groups III–VI. In spite of this, the non-metallic elements have the larger role to play in chemistry, as we shall see (Chapters 3–16).

A *mixture* of elements, without chemical bonding between the different types of atoms, is not a compound. Mixtures of elements can often be separated by simple physical means, e.g. a magnet removes iron from a mixture of iron filings and sulphur. But the main distinction between mixtures and compounds is that the proportions of the elements in a mixture may vary considerably from sample to sample; in a given pure compound the proportions are always the same.

The states of matter

The knowledge that all matter is composed of atoms and molecules permits an explanation of the physical properties of matter. At any particular temperature any substance exists as either a solid, a liquid or a gas. If the temperature is varied, the same substance may pass through all states, provided that it remains stable and does not decompose or react. The changes are best followed on a molecular scale.

Identical molecules possess forces of attraction for one another at close range. At the same time, all molecules at temperatures above the absolute zero (approximately −273 °C) possess energy, received from their surroundings in the form of heat, light or radiation, or transferred from other molecules in the course of molecular collisions. The molecules are in a constant state of motion as a result of the energy they possess. These facts can explain the states of matter.

The solid state

In the solid state the forces of attraction are strong enough to virtually prevent the motion of the molecules, so that they are closely packed in an ordered arrangement. Nevertheless they are moving by vibrating backwards and forwards (Fig. 1.2). As the solid is heated, the vibration becomes more vigorous, until finally the molecules have sufficient energy to overcome the attractive forces. As they gain freedom of movement, the rigid pattern of atoms is broken and the structure collapses. The solid has melted. Melting occurs at a fixed temperature called the **melting point** and a fixed amount of heat, known as the **latent heat of fusion**, is required to melt each gram of the pure solid compound, changing it from a solid to a liquid, while it is at its melting point.

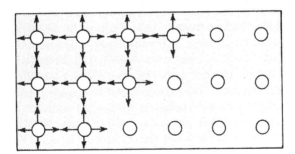

Fig. 1.2 The arrangement of molecules in a solid.

The liquid state

Although the molecules can now move about, they are still close to one another and the attractive forces prevent complete escape. The molecules at the surface of any volume of liquid are in a special position. They are not attracted by neighbours in all directions, since on one side their only

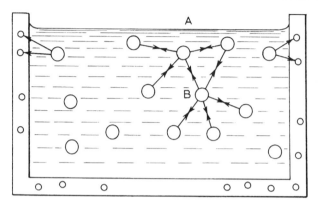

Fig. 1.3 The arrangement of molecules in a liquid; surface tension. (A) Surface molecules attracted to centre of liquid. (B) Interior molecules attracted in all directions.

neighbours are occasional gas molecules in the air. There is little pull in this direction: the attractive forces tend to draw surface molecules together and in towards the centre of the liquid. The liquid behaves as if it had a skin and the surface molecules are described as being under a **surface tension** (Fig. 1.3). Surface tension is a force expressed in dynes per centimetre (or milli-newtons per metres in SI units – the values are numerically identical). Surface tensions of simple liquids vary from 73 for water through 30 for xylene and 25 for methyl ethyl ketone to 18 for *n*-hexane (see Chapters 3, 4 and 9).

Viscosity is the resistance, drag or friction on any object moving through a liquid, or the resistance (due to the chemical attraction and physical shape of molecules) which hinders the flow of the liquid. Imagine pouring a collection of similarly shaped stones from a sack. They jostle against one another, there is drag and physical and mechanical resistance to flow. If we also imagine each stone as a small magnet, and further as having a built-in motor driving it along in a straight line, then we are nearer to the conditions operating inside a liquid. The attempt to impart movement to molecules in a given direction is hindered by (1) collisions with other molecules, (2) physical entanglement with the other molecules, due to irregularities in shape, and (3) attraction for other molecules in the immediate vicinity. The result of all this is viscosity.

The scientific measure of viscosity, the **coefficient of viscosity** of a liquid, η (pronounced 'eta'), is given by Newton's equation

$$\eta = \frac{\text{stress}}{\text{rate of shear}}$$

The **stress** is the frictional force operating per unit area, while the **rate of shear** is the variation in velocity of the moving liquid with depth of liquid. If we imagine a flowing river, at any instant the water moving fastest is at the surface, while the water in contact with the river bed is not moving at all. Considering the layers in between, the nearer we get to the top, the faster the water is flowing.

Since

$$\text{stress} = \frac{\text{force}}{\text{area}} = \frac{\text{mass} \times \text{acceleration}}{\text{length}^2}$$

$$= \frac{\text{mass} \times (\text{length}/\text{time}^2)}{\text{length}^2} = \frac{\text{mass}}{\text{length} \times \text{time}^2}$$

$$\left(\text{in SI units } \frac{\text{kg}}{\text{m} \times \text{s}^2}\right)$$

and

$$\text{rate of shear} = \frac{\text{velocity}}{\text{length}} = \frac{\text{length}/\text{time}}{\text{length}} = \frac{1}{\text{time}}$$

$$\left(\text{in SI units } \frac{1}{\text{s}}\right)$$

then η is measured in

$$\frac{\text{kg}}{\text{m} \times \text{s}^2} \times \text{s} = \frac{\text{kg}}{\text{m} \times \text{s}}$$

These units are known as **pascal-seconds** after the physicist Pascal. Still in regular use are units one-tenth of this size and these units are called **poises** after the physicist, Poiseuille. In scientific terms, we therefore speak of the coefficient of viscosity, or simply the viscosity, of a liquid as being so many pascal-seconds, or alternatively, so many poises.

The viscosity of a liquid varies appreciably with small temperature changes. A temperature rise provides the molecules with more energy, so that they move faster. They are thus more capable of overcoming the attractive forces and getting away from their neighbours. This means that the liquid will flow more readily and the viscosity will drop.

As the liquid is heated the energy is shared out unequally between the molecules so that, at the surface, some are eventually moving fast enough to escape the attractions of their neighbours and fly off into the air. This process is known as **evaporation**. Finally, if temperature rises sufficiently, all the surface molecules can escape readily and **boiling** takes place. The temperature of the liquid remains steady at the **boiling point**, until all the liquid has vaporized. A fixed amount of heat, known as the **latent heat of evaporation**, is required to boil each gram of pure liquid to vapour at the boiling point.

The gaseous state

The molecules in a gas are widely spaced and the forces of attraction come
into play only when chance encounters are made by moving molecules.
Between collisions they travel in straight lines at several hundred kilometres
an hour. The movements of the molecules are limited by the walls of the
container in which the gas is placed and the molecules exert *pressure* on
those walls by their repeated impacts on them. The higher the temperature,
the more energy the molecules possess, the faster they move and the greater
the impact. Hence gas pressure increases with temperature.

If the gas is cooled, the molecules get slower and slower until eventually
they cannot escape from one another after chance encounters. The attractive
forces prevent this. When sufficient molecules are in close proximity, they
start to form drops of liquid and the gas is said to have condensed. **Condensation** takes place at or below the boiling point. Further cooling leads to
further slowing down of the molecules, so that the attractive forces re-form
the rigid pattern of molecules of the solid state. Solidification takes place
at the *melting (or freezing) point.*

Solutions, suspensions and colloids

If a solid and liquid are mixed, one of three things can happen.

- A chemical reaction may occur.
- If there is a strong attraction between liquid and solid molecules, liquid
  molecules will penetrate the solid structure and surface solid molecules
  will break away and mingle with the liquid molecules. Gradually the
  solid structure is eroded away until it is no longer present, because the
  liquid and solid molecules are uniformly mixed. The solid had dissolved
  and a **solution** has formed.
- If the attraction between liquid and solid molecules is less than that
  between solid molecules, the liquid can do no more than separate solid
  particles from one another. The nature of the mixture then depends on
  the solid. This may have been crushed to a powder form, so that it is
  composed of particles, each containing from hundreds to millions of
  molecules depending on particle size. These will be dispersed throughout
  the liquid on shaking. The result is a **suspension** of solid particles, which
  will ultimately settle out on standing.

Solutions

There is a limit to the amount of solid that will dissolve in 100 g of liquid.
This quantity in grams is known as the **solubility** of the solid. Solubility
varies with temperature, usually increasing with a rise in temperature. A
solution containing the maximum amount of dissolved solid is a **saturated**

solution. The liquid in any solution is known as the **solvent** and the solid as the **solute**.

Colloidal dispersions

At its limit, a suspension may contain particles too small to be distinguished by the naked eye. These particles are significantly larger than large molecules, but can be seen with the aid of a very powerful optical microscope (magnification ×900), only if the particle diameter is 0.4 μm (400 nm) or more. Smaller particles are not directly visible, because 400 nm is the lower wavelength limit of the visible spectrum (Chapter 6), but they can be seen with the ultramicroscope or with the electron microscope. As the particles attempt to settle out, they are constantly bombarded by the liquid molecules, which are in continual motion. Such collisions do not affect large particles, but the small particles rebound slightly and the frequency of the collisions causes them to alter course and follow a zigzag path. This is known as **Brownian motion**. Such random movement delays settlement. The particle surfaces may also bear electric charges. When two particles come close together, the like charges repel one another. This also delays settlement. Such fine particles are said to be in the **colloidal** state and colloidal systems can have extreme stability and resist settling out for very long periods. These smaller size systems are generally called **colloidal dispersions** rather than suspensions, a name which reflects the generally longer term stability against settlement of these systems.

A colloidal dispersion of high solid content may be opaque (e.g. pigment in paint), but if the solid content is low it may appear transparent, like a solution. However, light entering the colloidal dispersion from the side will be scattered and the eye can detect the particles as scintillations of scattered light. This is called the **Tyndall effect**.

Another form of colloidal dispersion can be produced from two liquids that will not mix. Agitation will disperse one of them as droplets in the other. If coarse in size, these droplets will be referred to as a suspension as before, but if fine, the product is an **emulsion**. An emulsion can be very stable if the dispersed droplets are electrically charged or have absorbed a surfactant (Chapter 10) at their surfaces. Milk is an emulsion of fat in water.

Valency and chemical bonds

We have been discussing the behaviour of molecules during certain physical changes, but now we shall consider the internal structure of molecules in more detail. How are the atoms bonded together and what determines the proportions of the different kinds of atoms?

The proportions are determined by the valencies of the elements concerned. The **valency** of an element is the number of chemical bonds that one atom of an element can make. If two atoms come so close together that the spaces in which their electrons move overlap, then at that overlap the negative electrons are attracted by both positive atomic nuclei. Under these circumstances a pair of electrons become 'fixed' at an equilibrium position between the two nuclei. The atoms are not easily separated: a **chemical bond** has formed between them. In a **covalent bond** one electron is supplied by each atom. In a **coordinate bond** both electrons are provided by the same atom. There is also an ionic bond which will be described in the next chapter.

The electrons concerned are known as the **valency electrons** and there is a fixed number (fewer than eight) in each atom. The number varies from element to element. For some elements the valency is equal to the number of valency electrons, i.e. the number of electrons available for 'pairing up' in chemical bonds. This number is the Periodic Table group number of the element. For others, the valency is decided by the number of 'spaces' available for bonding electrons from other atoms (eight minus the group number). These elements have fixed valencies; some of the more common elements in these categories are given in Table 1.2.

Other elements (e.g. P, N and S) sometimes have the valency permitted by the number of valency electrons and, at other times, the valency permitted by the number of 'spaces'. A number of elements in groups in the middle of the table have atomic structures which permit an increase in the number of valency electrons under some circumstances. All of these elements also possess more than one valency. The two types are classified together as 'elements of variable valency' (Table 1.3). Chemistry text books now use the first form shown, but the older forms are still in very common use. Note that the latter, the *-ous* ending, is used to describe the lower valency and *-ic* the higher valency in each case.

Elements of variable valency usually exhibit their lowest valency, but will change valency under the right chemical conditions (see Oxidation, Chapter 2).

An element always uses its full valency to form chemical bonds. Should one of these bonds break in the course of chemical reaction (see below), the element becomes highly reactive and the full number of bonds is soon restored by a reaction involving bonding with another atom or molecule. In a pure sample of an element, the valencies are satisfied by bonding between atoms of the same kind. The only elements that exist in their normal states as

Table 1.2 Elements of fixed valency

Elements donating valency electrons	Valency	Elements accepting valency electrons
H, Na, K, Ag	1	H, F, Cl, Br, I
Mg, Ca, Ba, Zn	2	O
B, Al	3	
C, Si	4	C, Si

Table 1.3 Elements of variable valency

Valency	Elements	Names of metallic valency states
1 or 2	Cu	Copper(I) or cuprous, copper(II) or cupric
	Hg	Mercury(I) or mercurous, mercury(II) or mercuric
1 or 3	Au	Gold(I) or aurous, gold(III) or auric
2 or 3	Fe	Iron(II) or ferrous, iron(III) or ferric
	Co	Cobalt(II) or cobaltous, cobalt(III) or cobaltic
2 or 4	Sn	Tin(II) or stannous, tin(IV) or stannic
	Pb	Lead(II) or plumbous, lead(IV) or plumbic
	Ni	Nickel(II) or nickelous, nickel(IV) or nickelic
2 or 6	S	
3 or 5	N	
	P	
2, 3 or 6	Cr	Chromium(II) or chromous, chromium(III) or chromic
2, 4 or 7	Mn	Manganese(II) or manganous

single, unattached atoms are the noble gases (group 0, Periodic Table), which are elements of zero valency.

This electronic picture of valency is not necessary for visualizing the bonding in a compound of known formula and structure. The formula can be written as a diagram in which the atoms are circles, a line attached to one circle is a valency and a line attached at both ends is a chemical bond.

Thus water, H_2O, is composed of hydrogen, –Ⓗ, and oxygen, Ⓞ⟨. The formula can be drawn, so that there are no unsatisfied valencies, in the following manner:

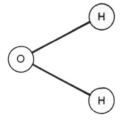

This picture tells us which atom is attached to which.

Slightly more complicates is rust, iron(II) or ferric oxide, Fe_2O_3. The iron, Ⓕₑ⟨, and oxygen, Ⓞ⟨ are combined in the following manner:

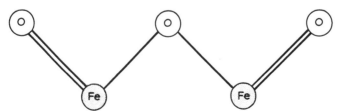

Once familiar with the system, we can dispense with the circles and abbreviate the picture. An example of variable valency in one compound is dinitrogen oxide or nitrous oxide, N_2O ('laughing gas'), with nitrogen valencies of three and five:

$$N \equiv N = O$$

These last two examples show that multiple bonds, both double and triple, are possible in compounds. It is in fact very common, particularly for oxygen, to form double bonds to other elements as shown above.

It is occasionally possible to work out a formula from scant information. Asked to find a formula for calcium chloride, we assume that it contains only calcium, Ca<, and chlorine, −Cl. Thus the formula must be $CaCl_2$ and cannot be CaCl or $CaCl_3$.

Electronegativity

We have seen that bonding is a consequence of the attraction that the positive nucleus of an atom has for the negative electrons of other atoms that come within its 'sphere of influence'. This attraction is stronger in some atoms than others. It depends upon the number of positive charges (protons) in the nucleus, the distance from the nucleus to the extremities of the atom and the number of non-valency electrons in between. These electrons provide a negative 'screen' around the nucleus, reducing its power of attraction for valency electrons. We say that those elements with a strong attraction for valency electrons are highly **electronegative**. The American chemist Pauling calculated numerical values for electronegativity and placed the elements in a series, in which those with the highest values are the most electro-negative.

Part of Pauling's scale of electronegativity

F	4.0	S	2.5	B	2.0	Ca	1.0
O	3.5	C	2.5	Si	1.8	Na	0.9
Cl	3.0	I	2.4	Sn	1.7	Ba	0.9
N	3.0	P	2.1	Al	1.5	K	0.8
Br	2.8	H	2.1	Mg	1.2		

We should note that the last seven elements are all metals and the first 12 all non-metals. Also, oxygen is particularly greedy for electrons and nitro-gen only a little less so. Carbon and hydrogen are moderately electro-negative, but do not differ much in their attractive power for electrons. Electronegativity determines much of the behaviour of elements and their compounds.

Chemical reactions and equations

Let us examine bonding in action and see how the process – chemical reaction – is written down. If the two elements iron and sulphur are heated together a dark powder is produced, ferrous sulphide. Sulphur has a valency of two; iron in this instance is in the lower [iron(II) or ferrous] valency state, also two.

This pictorial representation can be replaced by symbols only, to give the neater description:

$$Fe + S \longrightarrow FeS$$

This is a **chemical equation**. It states how many atoms (or molecules) of each substance are required and how many are produced. Since it is an equation, the two sides equal one another; they contain the same number of atoms of each element, but differently combined. This is because, in a chemical reaction, matter is never destroyed; it is simply converted into something else. Thus burnt carbon 'disappears' by conversion into the transparent gas, carbon dioxide:

$$C + O_2 \longrightarrow CO_2$$

In chemical equations, elementary gases are always written as molecules rather than atoms; hence O_2 and not $2O$ (two atoms).

Some reactions involve several molecules of each substance:

$$4Fe + 3O_2 + 2H_2O \longrightarrow 2Fe_2O_3 \cdot H_2O$$

This equation contains four atoms of iron, eight of oxygen and four of hydrogen on each side, i.e. it 'balances'. It denotes that two molecules of iron III oxide are produced from four atoms of iron and three molecules of oxygen and that each molecule of iron III oxide has one molecule of water associated with it in the solid crystal.

Reactions also occur between compounds. Common salt is made by mixing sodium hydroxide and hydrochloric acid.

$$Na{\vdots}OH \quad + \quad H{\vdots}Cl \quad \longrightarrow \quad NaCl + H_2O$$

sodium   hydrochloric

hydroxide   acid

Note that in this and many other reactions between compounds, the reaction simply involves exchanging halves of the molecules. This is known as **double**

Table 1.4 Groups of atoms frequently exchanged

Name	Formula	Valency
Carbonate	$=CO_3$	2
Hydrogencarbonate	$-HCO_3$	1
Sulphate	$=SO_4$	2
Phosphate	$\equiv PO_4$	3
Nitrate	$-NO_3$	1
Hydroxide	$-OH$	1

decomposition. Again

$$Ca{\vdots}CO_3 \quad + \quad 2H{\vdots}Cl \quad \longrightarrow \quad Ca{\vdots}Cl_2 + H_2{\vdots}CO_3$$
$$\text{calcium} \qquad\qquad\qquad\qquad\qquad \text{carbonic}$$
$$\text{carbonate} \qquad\qquad\qquad\qquad\qquad \text{acid}$$

Carbonic acid decomposes at normal temperatures, with carbon dioxide evolving:

$$H_2CO_3 \longrightarrow CO_2{\uparrow} + H_2O$$

Note that the half exchanged can consist of a group of atoms. Some groups that frequently occur are given in Table 1.4. The group valency is the sum of the valency remaining unlinked after the atoms in the group are bonded together as completely as possible, e.g.

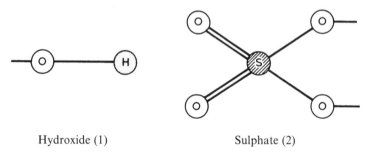

Hydroxide (1)   Sulphate (2)

Since these groups are found only in compounds, the valencies are never left unsatisfied.

Two

Inorganic chemistry
acids, bases and salts

Inorganic chemistry is the study of all the elements except carbon, though that element itself and a few simple carbon compounds are included under the heading. We have space here only to define some important ideas in inorganic chemistry and to have a brief look at the compounds of the four elements that we will meet most frequently in Part Two: carbon, hydrogen, oxygen and nitrogen.

Acids, bases and salts

All compounds are either acidic, basic or neutral. The nature of a substance in water is easily determined. *Pure* water is neutral and has no effect on litmus paper, be it red or blue. A solution of an acid, however, turns blue litmus red, while a basic solution turns red litmus blue. We recognize acids by their sharp taste, and by their ability to dissolve metals such as magnesium, at the same time giving off hydrogen. Bases generally do not do this; one characteristic is the soapy feel to their solution.

Another test is as follows. If we place two inert electrodes in the solution (which we now call the **electrolyte**) and allow a small direct current to flow through it, hydrogen bubbles will appear at the negative electrode (the **cathode**) if the solution is acid; if it is basic, then oxygen bubbles will appear at the positive electrode (the **anode**). This process is known as **electrolysis**.

It occurs because of the existence of electrically charged atoms or groups known as **ions**. If a chemical bond is formed between two atoms of elements at the *opposite extremes* of the electronegativity series (see Pauling's table in Chapter 1), then the following can occur. The more electronegative element exerts a very strong attraction on the valency electron from the weakly electronegative element and that element has little counter-attraction to retain it. The electron is therefore completely drawn into the cloud of electrons surrounding the electronegative atom and becomes part of it. The atom is now negatively charged, having more electrons than protons, and being charged, is now called an ion (an **anion**, because it is attracted to the anode). The atom which has lost an electron becomes a positively charged ion (a **cation**, because it is attracted to the cathode). Between two ions of opposite charge, when these exist in solids, there exists a strong electrostatic attraction and this bond between them is called an **ionic bond**. This bond is the third type, along with covalent and coordinate bonds described in Chapter 1.

An ionic bond may also be formed between two groups of atoms, or between an atom and a group. We will see in Chapter 5 that these are the forces found holding inorganic crystals together. It is not always easy to predict such an ionic bond from the molecular formula, since the electronegativity of the main atom in a group is influenced and modified by the other atoms covalently bonded to it.

If some of the solid ionic substance is put into water, then the high dielectric constant, or insulating property, of that liquid permits the ions to **dissociate** from one another and exist separately in the solution. Once the electric current is switched on, the ions are attracted by and move towards the electrode of opposite charge. They reach it and the charge is lost or discharged.

We can see why the electrolysis products are what they are when we know that all acidic solutions contain hydrogen ions, H^+, and all basic solutions contain hydroxide ions, OH^-. Thus for acids, at the cathode (negative), cations lose their charge by gaining electrons (written as e^-):

$$2H^+ + 2e^- \longrightarrow 2H \longrightarrow H_2\uparrow$$

In the hydroxide ion, the electronegative oxygen is responsible for the charge (see Chapter 1 to refamiliarize on this topic). Thus for bases, at the anode (positive), anions discharge by losing electrons:

$$4OH^- - 4e^- \longrightarrow 4OH \longrightarrow 2H_2O + 2O \longrightarrow O_2\uparrow$$

The up-arrows signify that the gases are evolved, bubbling from the surface of the electrode.

Hydrochloric acid is clearly capable of producing hydrogen ions by dissociation:

$$HCl \rightleftharpoons H^+ + Cl^-$$

The two-way double arrow is written to indicate that while ions are forming, some are recombining to form molecules. The bias in favour of ions rather than molecules is indicated by the longer arrow in that direction. Sulphur trioxide, SO_3, is an acidic gas, though it contains no hydrogen. On dissolution in water it is converted to sulphuric acid:

$$SO_3 + H_2O \longrightarrow H_2SO_4 \rightleftharpoons 2H^+ + SO_4^{2-}$$

The two negative charges on the sulphate ion are obtained via the two bonds between separate oxygen atoms and the two hydrogen atoms (page 16).

Again, sodium hydroxide (caustic soda), NaOH, gives hydroxide ions by

$$NaOH \rightleftharpoons Na^+ + OH^-$$

but ammonia gas, NH_3, is basic because in solution

$$NH_3 + H_2O \longrightarrow NH_4OH \rightleftharpoons NH_4^+ + OH^-$$
$$\text{ammonium}$$
$$\text{hydroxide}$$

The reverse direction of the long arrow indicates little dissociation.
    Water itself is very weakly dissociated:

$$H_2O \rightleftharpoons H^+ + OH^-$$

but is neutral because the numbers of hydrogen and hydroxide ions exactly balance.

Acids

The best known strong acids are the three mineral acids:

$$HCl \qquad HNO_3 \qquad H_2SO_4$$
$$\text{hydrochloric acid} \quad \text{nitric acid} \quad \text{sulphuric acid}$$

The first two are the strongest, strength being indicated by the degree of dissociation to H^+ ions in solution. They are highly corrosive in concentrated solution and sulphuric acid in particular has the ability to dilute itself, by extracting water from the structures of other chemicals. Orthophosphoric acid, H_3PO_4, is a moderately strong acid. Chromic acid, H_2CrO_4, exists only in aqueous solutions.
    Weak acids are usually organic chemicals, e.g. ethanoic or acetic acid (vinegar) and citric acid (fruit juices). Exceptions are nitrous (HNO_2) and sulphurous (H_2SO_3) acids, which are stable only in solution, where they are slightly dissociated. Solutions of carbon dioxide behave as weak acids, because they have extremely low carbonic acid contents:

$$CO_2 + H_2O \rightleftharpoons H_2CO_3$$

Bases

The strong bases are the **alkalis**, such as sodium and potassium hydroxides (NaOH and KOH). Strength, in this case, is the degree of dissociation to OH^- ions in solution. In concentrated solutions these are also corrosive. Weak bases include ammonia (above) and its derivatives in organic chemistry (Chapter 4).

Salts

When an acid and a base are mixed, chemical reaction occurs and a salt and water are formed, e.g.

$$2KOH + H_2SO_4 \longrightarrow K_2SO_4 + H_2O$$
$$\text{potassium}$$
$$\text{sulphate}$$

Salts take their names from the parent acids, as in the examples in Table 2.1. Under the older naming scheme, a salt with 'bi' in the name contains an atom of hydrogen from the original acid, e.g. sodium bisulphite, $NaHSO_3$; the name now used is sodium hydrogensulphite. Phosphoric acid with three hydrogen atoms forms primary (NaH_2PO_4), secondary (Na_2HPO_4) and tertiary (Na_3PO_4) phosphates.

Salts dissociate in water to produce ions. Potassium sulphate is neutral:

$$K_2SO_4 \longrightarrow 2K^+ + SO_4^{2-}$$

Thus if there is no excess of H_2SO_4 or KOH molecules after the formation of K_2SO_4, the acid and alkali **neutralize** one another. The formation of the salt can be considered as the mutual neutralization of the H^+ and OH^- ions to form water:

$$H^+ \quad + \quad OH^- \rightleftharpoons H_2O$$
$$\text{(from } H_2SO_4) \quad \text{(from KOH)}$$

Other salts are not necessarily neutral. Ammonium chloride, NH_4Cl, has an acid solution.

$$NH_4Cl \rightleftharpoons NH_4^+ + Cl^- \quad \text{and} \quad H_2O \rightleftharpoons H^+ + OH^-$$

but OH^- ions from water are removed by combination with NH_4^+ to form the weakly dissociated ammonium hydroxide. This leaves H^+ and Cl^- ions in solution, i.e. a dilute solution of fully dissociated hydrochloric acid. The salt is formed from a weak base and a strong acid: in solution it is the H^+

Table 2.1 Acids and their salts

Acid	Salt
Nitric acid	Nitrate
Nitrous acid	Nitrite
Sulphuric acid	Sulphate or hydrogensulphate
Sulphurous acid	Sulphite or hydrogensulphite
Hydrochloric acid	Chloride
Carbonic acid	Carbonate or hydrogencarbonate
Hydrogen sulphide	Sulphide
Orthophosphoric acid	Orthophosphate
Chromic acid	Chromate

ions from the latter which are in excess over OH^- and give the salt its acid character.

Similarly, sodium hydrogensulphite, $NaHSO_3$, although it contains hydrogen (which might be thought to be a source of H^+ ions), has a basic solution:

$$NaHSO_3 \rightleftharpoons Na^+ + HSO_3^- \quad \text{and} \quad H_2O \rightleftharpoons H^+ + OH^-$$

but H^+ ions from water are removed by combination with hydrogensulphite ions to form the weakly dissociated sulphurous acid. This leaves Na^+ and OH^- ions in solution, i.e. a dilute solution of fully dissociated sodium hydroxide. Sodium hydrogensulphite is the salt of a strong base and a weak acid.

These examples emphasize that acidity is the ability to produce H^+ ions in solution and depends on no other property. The more H^+ ions produced, the stronger the acid. Basicity is associated with OH^- ions, and when OH^- ion concentration is high, H^+ ion concentration is low. Acidic or basic strength is measured on a scale of 0 to 14 called the **pH scale**. A pH of 7 is exactly neutral, 1 is strongly acid and 13 strongly alkaline.

Titration

By the use of **indicators** – compounds which change colour as their environment changes from an acidic to a basic one (or vice versa) – the precise moment of neutralization can be determined. For example, a measured quantity of acid solution of unknown concentration is placed in a flask with an indicator (e.g. litmus). A solution of base of known concentration is run into the flask until, with the addition of *one excess drop* of base, the colour changes. The volume of base run in has been measured and it is now possible to calculate the concentration of the acid solution. The method of calculation can be found in any elementary chemistry book and depends on knowledge of the **equivalent weights** of acids and bases. Titrations are used in paint chemistry to determine the acid value of a resin (Chapter 12) and it will be seen from the definition of acid value that no knowledge of equivalent weights is required for this determination; it is sufficient to know the concentration of the alkali solution.

Other acidic and basic compounds

A large number of materials – particularly oxides – are not obviously acidic or basic from their formulae, but yet behave in an acidic or basic manner. The oxides of non-metals are invariably neutral or *acidic*. The acidic oxides form acids in water and neutralize alkalis, e.g.

$$2NaOH + B_2O_3 \longrightarrow 2NaBO_2 + H_2O$$
$$\quad \text{boric} \quad\quad \text{sodium}$$
$$\quad \text{oxide} \quad\quad \text{borate}$$

Most metallic oxides are *basic* and neutralize acids, e.g.

$$Fe_2O_3 + 6HCl \longrightarrow 2FeCl_3 + 3H_2O$$
$$\text{iron III (ferric)}$$
$$\text{chloride}$$

but a number of metallic oxides can neutralize both acids and bases and, because of this double character, are called **amphoteric oxides**, e.g.

$$Al_2O_3 + 6HCl \longrightarrow 2AlCl_3 + 3H_2O$$

aluminium   aluminium
oxide   chloride

$$2NaOH + Al_2O_3 \longrightarrow 2NaAlO_2 + H_2O$$

sodium
aluminate

Polarity and the hydrogen bond

If a bond exists between two atoms of similar electronegativity, e.g. C and H, the electrons in the bond are fairly evenly shared and reasonably centrally situated between the two atomic nuclei. Such a bond is said to be **non-polar**. If, however, there is a marked difference in electronegativity, e.g. C and O or H and O, the electrons are displaced towards the more electro-negative atom. Ionization may not occur if the displacement is insufficient, but the bond will not be electrically symmetrical, e.g. $\overset{+}{C}:\overset{-}{O}$, where : are the electrons in the bond. Such an arrangement of atoms is said to be **polar** and gives the group a slight electrical charge as shown.

In a polar $-\overset{-}{O}-\overset{+}{H}$ bond, the hydrogen atom, which consists solely of one proton and one electron, has practically become a hydrogen ion, i.e. a proton. Should another electronegative atom in a polar group come near this proton, there is an attraction between the negative and positive charges sufficient to form a weak chemical bond between the two groups. In water, for example:

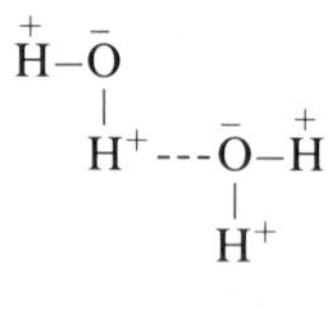

$$\overset{+}{H}-\overset{-}{O}$$
$$|$$
$$\overset{+}{H}\text{---}\overset{-}{O}-\overset{+}{H}$$
$$|$$
$$\overset{+}{H}$$

The **hydrogen bond**, as it is called, is shown by a broken line. The effect of this weak bonding between molecules is to raise the boiling point, since heat energy is required to break the bond before the molecules can be separated. Thus H_2O boils at 100 °C, while H_2S (sulphur is immediately below oxygen in the Periodic Table), though a heavier molecule, boils at −61 °C. Electro-negativities are O 3.5, S 2.5, and H 2.1. Hydrogen bonding affects solvency in paints (Chapter 9).

Oxygen and hydrogen

Oxidation and reduction

The chemistry of these elements is closely associated with a very important type of chemical reaction: **oxidation and reduction**. An atom which has been oxidized is one that, in the course of a chemical reaction, loses control – totally or partially – over one or more of its valency electrons, in favour of a more electronegative atom. Reduction *occurs simultaneously*, since the reduced atom gains control over an electron from a less electronegative atom.

This definition may become clearer after some examples, but in the process, simpler practical definitions can be made. Thus, *oxidation is the addition of oxygen to an element or compound*, e.g.

$$2Mg + O_2 \xrightarrow{\text{burns}} 2MgO$$
$$\text{magnesium}$$
$$\text{oxide}$$

and

$$2SO_2 + O_2 \xrightarrow[\text{Pt catalyst}]{\text{heat}} 2SO_3$$

Reduction is the opposite process: *the removal of oxygen from a compound*. Hydrogen is often used to do this, e.g.

$$CuO + H_2 \xrightarrow{\text{heat}} Cu + H_2O$$
$$\text{copper}$$
$$\text{oxide}$$

In these reactions, oxygen is the **oxidizing agent** and hydrogen the **reducing agent**. In the last example, while the CuO is reduced, the hydrogen is oxidized to water and in the other examples, oxygen is reduced to an oxide. The processes of oxidation and reduction always occur together.

The oxidizing agent need not be oxygen, e.g.

$$C + 2H_2SO_4 \longrightarrow CO_2 + 2H_2O + 2SO_2$$

Carbon is oxidized to CO_2 and H_2SO_4 is reduced to SO_2. Similarly, hydrogen need not be the reducing agent. Carbon is in the above example and again in

$$PbO + C \xrightarrow{\text{heat}} Pb + CO\uparrow$$
$$\text{lead} \qquad\qquad \text{carbon}$$
$$\text{monoxide} \qquad \text{monoxide}$$
$$\text{(litharge)}$$

While oxidation is the addition of oxygen, it is also the *removal of hydrogen*, e.g.

$$Cl_2 + H_2S \longrightarrow 2HCl + S$$

H_2S is oxidized to sulphur by chlorine (the oxidizing agent), which is itself *reduced* to HCl by the *addition of hydrogen*.

Finally, if an element of variable valency is caused to increase its valency, it has been oxidized, but if its valency is reduced, the element has been reduced, e.g.

$$2FeCl_2 + Cl_2 \longrightarrow 2FeCl_3 \quad \text{oxidation}$$
$$Fe = 2 \qquad\qquad\quad Fe = 3$$
$$\text{iron II (ferrous) chloride} \quad \text{iron III (ferric) chloride}$$

$$2FeCl_3 + H_2 \longrightarrow 2FeCl_2 + 2HCl \quad \text{reduction}$$
$$Fe = 3 \qquad\quad Fe = 2$$

This shows why **oxidation state** is a term used about an element where its valency is variable.

The original definition is most easily understood if applied to the last example. Consider the ionizations

$$FeCl_2 \rightleftharpoons Fe^{2+} + 2Cl^-$$
$$FeCl_3 \rightleftharpoons Fe^{3+} + 3Cl^-$$

Oxidation changes Fe^{2+} to Fe^{3+}: loss of one electron. Reduction returns that electron. The other examples all fit the definition, though where ionization does not occur, electronegativities must be used to see which element has lost and which has gained some control over a valency electron.

Oxygen is made in the laboratory by heating potassium chlorate.

$$2KClO_3 \longrightarrow 2KCl + 3O_2\uparrow$$

About 10% of manganese dioxide, MnO_2, is added to speed up the reaction. It is unchanged by the heating and can be recovered afterwards. This ability to accelerate the reaction without being used up in the process is called **catalysis** and the substance with this ability is a **catalyst**. Industrially, oxygen is separated from nitrogen and other atmospheric gases by cooling air to the liquid state. Its boiling point, $-182.5\,°C$, differs from those of the other gases.

Hydrogen is made by the reaction between acids and metals, e.g.

$$Zn + H_2SO_4 \longrightarrow ZnSO_4 + H_2\uparrow$$
$$\text{(dilute)} \qquad \text{zinc}$$
$$\text{sulphate}$$

Hydrogen peroxide

Oxygen and hydrogen are found in countless chemical compounds. One of the most interesting is their joint compound, hydrogen peroxide, H_2O_2. It is made by several methods, one of which is

$$BaO_2 + H_2SO_4 \xrightarrow{\ 0\,°C\ } BaSO_4{\downarrow} + H_2O_2$$

barium   (dilute)   barium
peroxide   sulphate

The barium sulphate is insoluble in water and settles out, as indicated by the down arrow ↓. The valencies indicate that the atoms must be connected thus:

$$H-O-O-H$$

The compound is not very stable and is decomposed by heat and light. When a compound decomposes, chemical bonds are broken. A covalent bond may break in two ways: either the two electrons in the bond will be shared between the atoms joined by the bond (homolytic dissociation), or they will be retained by the more electronegative of those atoms (heterocyclic dissociation).

$$\text{homolytic } A : B \longrightarrow A\cdot + \cdot B$$
$$\text{heterolytic } A : B \longrightarrow A^+ + :B^-$$

Heterolytic dissociation produces ions. Homolytic dissociation – which is the more likely method if the two fragments are equally electronegative – produces neutral atoms or groups, each with an unsatisfied valency or unpaired valency electron. This electron is denoted by a dot. Such groups are known as **free radicals**. Most free radicals are highly reactive, satisfying their valencies by forming new chemical bonds at the earliest opportunity.

$H-O-O-H$ is a very symmetrical formula, so that the usual method of decomposition is homolytic:

$$H_2O_2 \longrightarrow 2HO\cdot$$

The final decomposition products are

$$2H_2O_2 \longrightarrow 2H_2O + O_2{\uparrow}$$

unless some other chemical is present with which the free radicals may react. Since the oxygen in the free radical requires another electron to form a bond, and oxygen is highly electronegative, H_2O_2 is a powerful oxidizing agent, e.g.

$$PbS + 4H_2O_2 \longrightarrow PbSO_4 + 4H_2O$$

lead   lead
sulphide   sulphate
(black)   (white)

This reaction is used for 'cleaning' paintings, where lead pigments have discoloured.

Carbon

Although the element is found in many impure forms, such as soot, lamp-black, charcoal, coal and coke, the pure substance exists in two very different forms: diamond and graphite (the substance used with clay to form pencil 'leads').

The outstanding property of the element carbon is its ability to bond *with itself* to form chains of atoms. Other elements joined in this way give unstable compounds after a chain of three or four atoms, but carbon has no practical limit. The wide variety of atoms and groups that may be attached to the valencies not forming the chain leads to an infinite variety of carbon compounds, whose study is the subject 'organic chemistry'.

The inorganic compounds of carbon include the oxides CO and CO_2, carbonic acid and its salts and the carbides.

In a restricted supply of air

$$2C + O_2 \xrightarrow{\text{800 °C}} 2CO$$

Carbon monoxide is a neutral gas, highly dangerous because it is poisonous and has no smell. It is used as a fuel in coal gas.

$$2CO + O_2 \longrightarrow 2CO_2$$

Carbon dioxide is more conveniently made from dilute acid and a carbonate:

$$CaCO_3 + 2HCl \longrightarrow CaCl_2 + CO_2\uparrow + H_2O$$

The gas is easily liquefied by compression and becomes solid at $-79\,°C$. The 'dry ice' formed is used for cooling. The compressed gas is used in fire extinguishers, since nothing will burn in it. Water dissolves almost its own volume of CO_2 at normal temperatures and pressure (carbonic acid). With alkali:

$$2NaOH + CO_2 \longrightarrow Na_2CO_3 + H_2O$$
$$\text{sodium}$$
$$\text{carbonate}$$
$$\text{(washing soda)}$$

and

$$Na_2CO_3 + H_2O + CO_2 \longrightarrow 2NaHCO_3$$
$$\text{sodium}$$
$$\text{hydrogencarbonate}$$
$$\text{(baking powder)}$$

These salts dissolve in water, together with potassium and ammonium carbonates. Carbonates of metals outside Group I in the Periodic Table are insoluble:

$$Pb(NO_3)_2 + Na_2CO_3 \longrightarrow PbCO_3\downarrow + 2NaNO_3$$
$$\text{lead} \qquad\qquad\qquad \text{lead} \quad\; \text{sodium}$$
$$\text{nitrate} \qquad\qquad\; \text{carbonate} \; \text{nitrate}$$

All the other salts in the equation are soluble. Alkali metal carbonates are stable, but others decompose on heating, e.g.

$$PbCO_3 \longrightarrow PbO + CO_2\uparrow$$

Sodium and potassium hydrogencarbonates are less soluble than the carbonates. Magnesium, calcium, strontium and barium hydrogencarbonates exist only in solution. Attempts to remove the water destroy the compounds:

$$Ca(HCO_3)_2 \xrightarrow{\text{100 °C}} CaCO_3\downarrow + H_2O + CO_2\uparrow$$

calcium hydrogencarbonate   ('fur' in
(temporary hardness in water)   a kettle)

The alkali hydrogencarbonates are also decomposed by heat.
   Some metal oxides form carbides when heated with carbon (coke)

$$CaO + 3C \xrightarrow{\text{2000 °C}} CaC_2 + CO\uparrow$$

calcium   calcium
oxide   carbide
(lime)

This compound produces acetylene gas by reaction with water:

$$CaC_2 + 2H_2O \longrightarrow Ca(OH)_2 \quad + \quad C_2H_2\uparrow$$

calcium   ethyne
hydroxide   (acetylene)
(slaked lime)

Nitrogen

This is an element essential to plant growth. Air is four-fifths nitrogen, and nitrogen can be obtained from the air, either by removing the oxygen by burning it to form a solid oxide, or by liquefying the air (see Oxygen). In the laboratory it is made thus:

$$NH_4NO_2 \xrightarrow{\text{heat}} N_2\uparrow + 2H_2O$$

ammonium
nitrite

It is a colourless, odourless, neutral gas. It is very unreactive, but forms nitrides when strongly heated with some metals, e.g.

$$3Mg + N_2 \longrightarrow Mg_3N_2$$

magnesium
nitride

and can be made to react with hydrogen to form the basic gas, ammonia.

$$N_2 + 3H_2 \xrightarrow[\text{very high pressure}]{500\,°C} 2NH_3$$

Salts of ammonium hydroxide, NH_4OH, all contain the ammonium ion, NH_4^+. On heating with a base, they give ammonia, e.g.

$$2NH_4Cl + Ca(OH)_2 \longrightarrow 2NH_3\uparrow + CaCl_2 + 2H_2O$$

The chief oxides of nitrogen are made as follows:

$$NH_4NO_3 \xrightarrow{\text{heat}} N_2O\uparrow \ + \ 2H_2O \qquad \text{a neutral oxide}$$

ammonium	dinitrogen (or
nitrate	nitrous) oxide
	(laughing gas)

$$3Cu + 8HNO_3 \longrightarrow 3Cu(NO_3)_2 + 2NO\uparrow + 4H_2O$$

(50% with   nitrogen
water)   monoxide
or nitric oxide

Nitric oxide is a neutral, colourless gas, but mere contact with oxygen oxidizes it to the acidic, brown gas, nitrogen dioxide or peroxide. At normal temperatures this is a mixture of 'single' and 'double' molecules:

$$2NO + O_2 \longrightarrow N_2O_4 \underset{-10\,°C}{\overset{140\,°C}{\rightleftharpoons}} 2NO_2$$

$$2NO_2 \ (\text{or } N_2O_4) + H_2O \longrightarrow HNO_3 + HNO_2$$

Nitrous acid is stable only at lower temperatures and decomposes thus:

$$3HNO_2 \longrightarrow HNO_3 + H_2O + 2NO$$

$$\Big\downarrow {\scriptstyle +O_2}$$

$$2NO_2 \text{ etc.}$$

The salts (nitrites) are more stable than the acid.
   Nitric acid is made industrially via NO:

$$4NH_3 + 5O_2 \xrightarrow{\text{red hot Pt}} 4NO + 6H_2O \quad (\text{Pt is a catalyst})$$

or

$$N_2 + O_2 \xrightarrow[\text{high temp.}]{\text{electric arc}} 2NO$$

Dilute nitric acid shows normal acidic properties, but the concentrated acid is also an oxidizing agent. Note the action on copper (above) when hydrogen is *not* produced (see normal acid–metal reaction under Hydrogen).

Nitrates differ in the way they decompose on heating. NH_4NO_3 is shown above. Alkali nitrates give oxygen:

$$2KNO_3 \longrightarrow 2KNO_2 + O_2\uparrow$$

potassium	potassium
nitrate	nitrite
(saltpetre)	

Other nitrates lose oxygen *and* nitrogen dioxide:

$$2Pb(NO_3)_2 \longrightarrow 2PbO + 4NO_2 + O_2\uparrow$$

All nitrates are soluble in water.

Three

Organic chemistry

alkanes to oils

Organic chemistry is the study of carbon compounds. The principal element associated with carbon is these compounds is hydrogen. All the compounds of interest here contain these elements, while oxygen, nitrogen and the halogens (principally chlorine) are present in some compounds. This chapter is concerned solely with **aliphatic** compounds – those deriving from fatty materials. The simplest of these compounds contain carbon and hydrogen only.

Aliphatic hydrocarbons

Alkanes

The valency of carbon is four, that of hydrogen, one. The simplest possible compound is, therefore

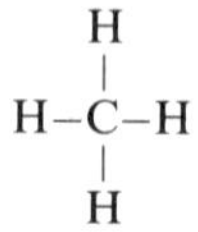

$$
\begin{array}{c}
\text{H} \\
| \\
\text{H}-\text{C}-\text{H} \\
| \\
\text{H}
\end{array}
$$

methane

With a chain of two carbon atoms we have

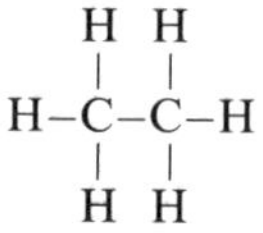

$$
\begin{array}{c}
\text{H} \quad \text{H} \\
| \quad | \\
\text{H}-\text{C}-\text{C}-\text{H} \\
| \quad | \\
\text{H} \quad \text{H}
\end{array}
$$

ethane

with three,

$$
\begin{array}{c}
\text{H} \quad \text{H} \quad \text{H} \\
| \quad | \quad | \\
\text{H}-\text{C}-\text{C}-\text{C}-\text{H} \\
| \quad | \quad | \\
\text{H} \quad \text{H} \quad \text{H}
\end{array}
$$

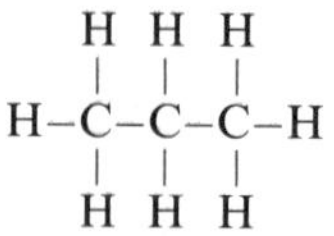

propane

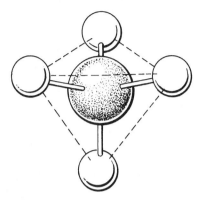

Fig. 3.1 A methane molecule.

and so on. Note that in each compound the number of H atoms is *twice* the number of C atoms *plus two*, so the general formula for the alkanes is C_nH_{2n+2} (where n is any whole number). These compounds are also referred to as paraffins.

It is important at once to get a true picture of what these molecules look like in three dimensions. We must realize that in these compounds the carbon valencies have fixed directions and the bonds are equidistant in space, pointing from the centre to the corners of a tetrahedron (Fig. 3.1).

Thus in propane (Fig. 3.2), the carbon atoms are not connected together in a straight line as in the two-dimensional formula.

Furthermore, the other attached groups or atoms can rotate about any single bond (Fig. 3.3). A model of a molecule can be made from wooden balls (atoms) containing sockets, into which wooden or metal rods (chemical bonds) may be fitted. If the rods fit loosely, rotation is easily demonstrated.

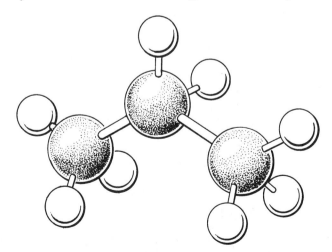

Fig. 3.2 A propane molecule.

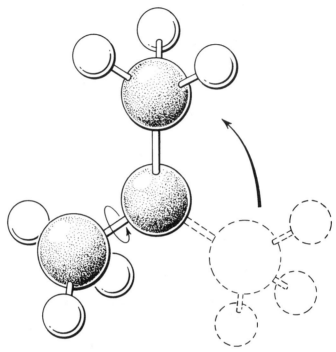

Fig. 3.3 Rotation about a single bond. The hydrogen atoms on the central carbon atom are omitted to simplify the figure. The arrows show the movement with rotation about the central carbon.

These models are a simplification of reality. Other models are preferred for complex structures, but this type is ideal for demonstrating rotation. In Fig. 3.3, the hydrogen atoms on the central carbon atom are omitted to simplify the figure. The arrows show the movement with rotation about the central carbon. Thus a long hydrocarbon chain is able to coil by rotation around the single bonds (Fig. 3.4).

Coming back to the series of alkanes, a fourth carbon atom could be attached to propane at an end carbon or alternatively at the centre carbon. Thus two C_4 alkanes are possible:

$$
\begin{array}{cccc}
H & H & H & H \\
| & | & | & | \\
H-C & -C & -C & -C-H \\
| & | & | & | \\
H & H & H & H
\end{array}
\quad \text{and} \quad
\begin{array}{ccc}
H & H & H \\
| & | & | \\
H-C & -C & -C-H \\
| & | & | \\
H & & H \\
  & H-C-H & \\
  & | & \\
  & H &
\end{array}
$$

n- or normal butane

isobutane or
2-methyl propane

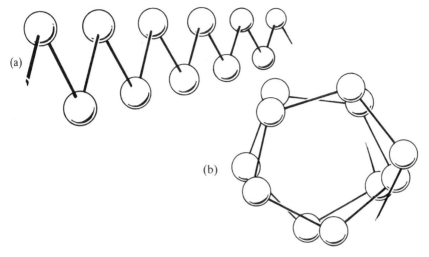

Fig. 3.4 The chain of carbon atoms: (a) extended, (b) coiled.

n-Butane is a **straight chain** compound, while isobutane, now more properly referred to as 2-methyl propane, contains a **branched chain**. Both have the overall formula C_4H_{10}, but the atoms are arranged differently. This is enough to give the compounds some noticeable differences in properties, e.g. boiling points: butane $-0.5\,°C$, 2-methyl propane $-11.7\,°C$. Compounds with the same overall formula, but a different structure, are called **isomers**. The straight chain alkane isomer is the **normal** form.

In organic chemistry there are so many isomers that overall formulae are useless and are not used. Structural formulae are used but, for ease of printing, every bond is not shown. The butanes, for example, are abbreviated to:

$$CH_3 \cdot CH_2 \cdot CH_2 \cdot CH_3$$

or

$$CH_2 \cdot (CH_2)_2 \cdot CH_3$$
n-butane

$$\begin{array}{c} CH_3 \cdot CH \cdot CH_3 \\ | \\ CH_3 \end{array}$$

or

$$CH_3 \cdot CH(CH_3) \cdot CH_3$$
isobutane or
2-methyl propane

With five carbons in the molecule there are three isomers:

$$CH_3 \cdot CH_2 \cdot CH_2 \cdot CH_2 \cdot CH_3$$
n-pentane

$$\begin{array}{c} CH_3 \\ | \\ CH_3 \cdot CH \cdot CH_2 \cdot CH_3 \end{array}$$
isopentane or
2-methyl butane

$$\begin{array}{c} CH_3 \\ | \\ H_3C \cdot C \cdot CH_3 \\ | \\ CH_3 \end{array}$$
neopentane or
2,2-dimethyl propane

and formation of a stable ring is possible:

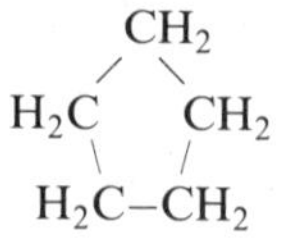

$$
\begin{array}{c}
CH_2 \\
\diagup \quad \diagdown \\
H_2C \qquad CH_2 \\
\diagdown \qquad \diagup \\
H_2C-CH_2
\end{array}
$$

cyclopentane

There is no known limit to the number of carbon atoms that can be linked together in this way. In naming all of these structures, it is usual for them to take their names from the longest straight chain of carbon atoms, which is numbered, and the side chains are described as being attached to such-and-such number of carbon atoms, e.g.

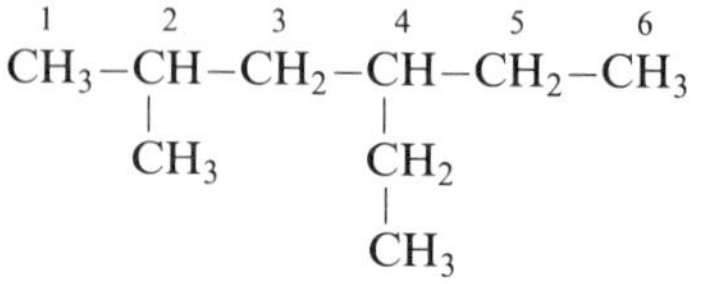

$$
\overset{1}{CH_3}-\overset{2}{CH}-\overset{3}{CH_2}-\overset{4}{CH}-\overset{5}{CH_2}-\overset{6}{CH_3}
$$

$$
\begin{array}{ccc}
& | & \quad | \\
& CH_3 & \quad CH_2 \\
& & \quad | \\
& & \quad CH_3
\end{array}
$$

is 2-methyl 4-ethyl hexane, even though, with nine carbon atoms, it is a complex nonane. Note that it is 2,4-substituted, not 3,5-, since the order of numbering is chosen to give the lowest numbers. Also note the -yl ending for the attached groups: methyl from methane, and ethyl from ethane. Longer normal alkane side groups are called propyl, butyl, hexyl, etc. These groups are **alkyl** groups, since they are derived from alkanes.

The physical state of the alkane alters as the number of carbon atoms increases. Heavier molecules require more energy to escape from their neighbours and hence their melting points and boiling points increase, as shown in Table 3.1.

It will have been seen already that many compounds have been given two names, and the reason is that compounds, having been given individual 'trivial' names in the past, now all have 'systematic' names. From now on,

Table 3.1 Melting and boiling points of alkanes

Alkane	Formula	Boiling point (°C)	Melting point (°C)
Methane	CH_4	−161	
Butane	$CH_3 \cdot (CH_2)_2 \cdot CH_3$	−0.5	
Pentane	$CH_3 \cdot (CH_2)_3 \cdot CH_3$	36	
2-Methyl butane (isopentane)	$CH_3 \cdot CH(CH_3) \cdot CH_2 \cdot CH_3$	28	
2,2-Dimethyl propane (neopentane)	$C \cdot (CH_3)_4$	9	
Pentadecane	$CH_3 \cdot (CH_2)_{13} \cdot CH_3$	273	10
Octadecane	$CH_3 \cdot (CH_2)_{16} \cdot CH_3$	318	28

in Part One of this book, in most instances both names are given with the systematic name first, and both are likely to be encountered in use in many practical situations. In Part Two, the older names will be used in instances where these are more customary. The names more customarily used are often those which are shortest and, more importantly, least ambiguous to the user. Tables are given inside the covers to help the reader to translate from one system to the other.

The chief natural sources of alkanes are natural gas (mainly methane) and petroleum. The mixture in petroleum is so complex that the alkanes are not completely separated from one another, but are collected in groups according to **boiling range**, by fractional distillation. Table 3.2 lists the 'fractions' collected.

Chemical properties

Alkanes react with chlorine, under the influence of light, heat or catalysts, and to a lesser extent with bromine:

$$C_nH_{2n+2} + Cl_2 \longrightarrow C_nH_{2n+1}Cl + HCl$$

In any alkane the carbon atoms will have different numbers of hydrogen atoms attached. There are four types of carbon atom, with names corresponding to the number of bonds made *with other carbon atoms*:

$$
\begin{array}{cccc}
\text{H} & \text{H} & \text{H} & \\
| & | & | & | \\
-\text{C}-\text{H} & -\text{C}- & -\text{C}- & -\text{C}- \\
| & | & | & | \\
\text{H} & \text{H} & & \\
\end{array}
$$

primary   secondary   tertiary   quaternary carbon

Chlorine removes hydrogen most easily from a tertiary carbon and least readily from a primary carbon. Where the first three types are present, a

Table 3.2 Fractions of alkanes

Boiling range (°C)	Fraction	C content	Use
Below 20	Refinery gas	C_1–C_4	Heating, cooking fuel
20–60	Light petroleum	C_5–C_6	Solvent (Chapter 9)
60–120	Light naphtha	C_6–C_8	Solvent, dry cleaning
70–200	Petrol, gasoline	C_6–C_{11}	Motor fuel
200–300	Paraffin oil (kerosene)	C_{12}–C_{16}	Lighting, jet fuel
Above 300	Gas oil (heavy oil)	C_{13}–C_{18}	Diesel, fuel oil
	Lubricating oil, greases	C_{18}–C_{34}	Lubricant, grease
	Vaseline, paraffin wax	C_{25}–C_{40}	Petroleum jelly, polishing waxes
	Residue (asphaltic bitumen)	$>C_{40}$	Road surfacing

Table 3.3 Well known chloro-alkanes

		Use	Boiling point (°C)
CH_2Cl_2	Dichloromethane (methylene chloride)	Paint strippers	40
$CHCl_3$	Trichloromethane (chloroform)	Former anaesthetic; Chemical intermediate	61
CCl_4	Tetrachloromethane (carbon tetrachloride)	Dry cleaner	77

mixture will be produced:

$$3CH_3 \cdot CH_2 \cdot CH(CH_3)_2 + 3Cl_2 \longrightarrow CH_3 \cdot CH_2 \cdot CCl(CH_3)_2$$
$$CH_3 \cdot CHCl \cdot CH(CH_3)_2 + 3HCl$$
$$CH_2Cl \cdot CH_2CH(CH_3)_2$$

with further attack on other atoms possible. Well-known **chloro-alkanes** are given in Table 3.3.

Between 150 and 475 °C alkanes react as vapour with nitric acid, e.g.

$$2CH_3 \cdot CH_2 \cdot CH_3 + 2HNO_3 \xrightarrow{400\,°C}$$
$$CH_3 \cdot CH_2 \cdot CH_2 \cdot NO_2 + CH_3 \cdot CH(NO_2) \cdot CH_3 + 2H_2O$$

($+ CH_3 \cdot CH_2 \cdot NO_2 + CH_3 \cdot NO_2$ due to 'splitting' of the alkane). The products are called **nitroalkanes**, some of which have found some use as paint solvents (see Table 9.1).

Alkanes burn, forming CO_2 and H_2O, hence their use as fuels.

Alkenes

In order to produce more petrol, the refiners pass some of the higher boiling fractions of petroleum through a heating process with either high pressure, or a catalyst. Large molecules are 'split' into smaller ones and the process is called **cracking**. As well as alkanes, compounds called **alkenes** are produced. These contain the group $>C=C<$, i.e. carbon atoms jointed by two bonds. The group is referred to as a 'double bond' or 'unsaturation'. These compounds are also referred to as olefins.

A whole family of alkenes exists, e.g.

$CH_2=CH_2$	$CH_2=CH-CH_3$	$CH_2=CH \cdot CH_2 \cdot CH_3$	$CH_3 \cdot CH=CH \cdot CH_3$
ethene	propene	but-l-ene	but-2-ene
(ethylene)	(propylene)	(butylene)	(isobutylene)

The prefix number indicates the *lower* numbered carbon atom involved in the double bond, numbering from one end of the molecule as with alkanes. Two or more double bonds in one molecule are possible, e.g.

$$CH_2=CH \cdot CH=CH_2 \quad \text{buta-1,3-diene} \quad \text{(1,3-butadiene)}$$

Note the -ene ending, denoting an alkene, and the -di-, indicating the number of double bonds. Alkenes with one double bond have the general formula C_nH_{2n}.

Rotation is possible about a single bond, but not about a double bond. If the wooden balls and metal rods are used again to make the model, a double bond consists of two wooden balls joined by *two* metal rods. Even with the rods fitting loosely in their sockets, it is not possible to twist the two balls in opposite directions if the rods are rigid. Thus

$$\begin{array}{cc} H \quad\quad H \\ \diagdown\,\diagup \\ C{=}C \\ \diagup\,\diagdown \\ Cl \quad\quad Cl \end{array} \quad \text{and} \quad \begin{array}{cc} H \quad\quad Cl \\ \diagdown\,\diagup \\ C{=}C \\ \diagup\,\diagdown \\ Cl \quad\quad H \end{array}$$

cis-1,2-dichloroethene   trans-1,2-dichloroethene
(*cis*-dichloroethylene)   (*trans*-dichloroethylene)

are different compounds and are known as **geometrical isomers** of 1,2-dichloroethene.

The double bond is very reactive, the second bond breaking readily and the free valencies attaching themselves to other reactive substances, e.g.

$$CH_2{=}CH_2 \begin{cases} + H_2 & \longrightarrow CH_3{\cdot}CH_3 \quad\quad \text{ethane} \\ + Cl_2 & \longrightarrow CH_2Cl{\cdot}CH_2Cl \quad \begin{array}{l}\text{1,2-dichloroethane}\\\text{(ethylene dichloride)}\end{array} \\ + H_2O + [O] & \longrightarrow CH_2OH{\cdot}CH_2OH \quad \begin{array}{l}\text{ethane-1,2-diol}\\\text{(ethylene glycol)}\end{array} \\ \text{v. dilute } KMnO_4 \end{cases}$$

Alkynes

A third group of aliphatic hydrocarbons exists, called the **alkynes**, also known as the acetylenes. Like the simplest compound in the group, ethyne itself ($HC{\equiv}CH$), all these compounds contain at least one **triple bond**. Ethyne is better recognized by its more common name acetylene. These are also unsaturated compounds and even more reactive than the alkenes, but they are not found in the later chapters of this book.

Terpenes

A special family of unsaturated compounds exists with the general formula $(C_5H_8)_n$, where $n = 2$ or more. These compounds are called **terpenes** and are composed of two or more molecules of 2-methyl buta-1,3-diene (isoprene) bonded together.

$$\begin{array}{c} CH_3 \\ | \\ \text{head} \quad H_2C{=}C{-}CH{=}CH_2 \quad \text{tail} \end{array}$$

2-methyl buta-1,3-diene
(isoprene)

The isoprene molecules are usually (though not always) joined head-to-tail. The simplest terpenes (monoterpenes) are the chief constituents of the essential volatile oils obtained from the sap and tissues of certain plants and trees. These have been used in the manufacture of perfumes from earliest times. In particular, the monoterpenes extracted from pine trees, or obtained as by-products in paper pulp manufacture, have been used for many years as paint solvents (Chapter 9) and anti-oxidants (Chapter 10).

The solvents are mixed terpenes, but some of the main ingredients are

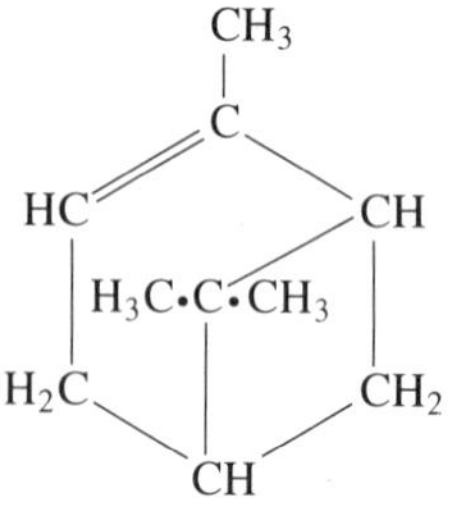

α-pinene
(turpentine)

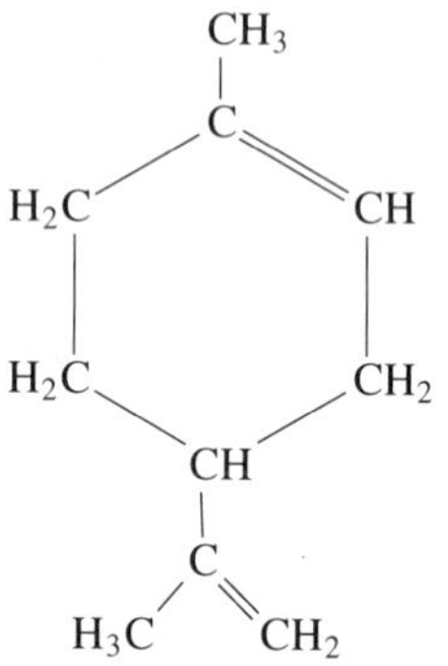

limonene
(*p*-metha-1,8-diene
or dipentene)

Pine oil itself is a very complex mixture of alcohols (see below) derived from terpenes, e.g.

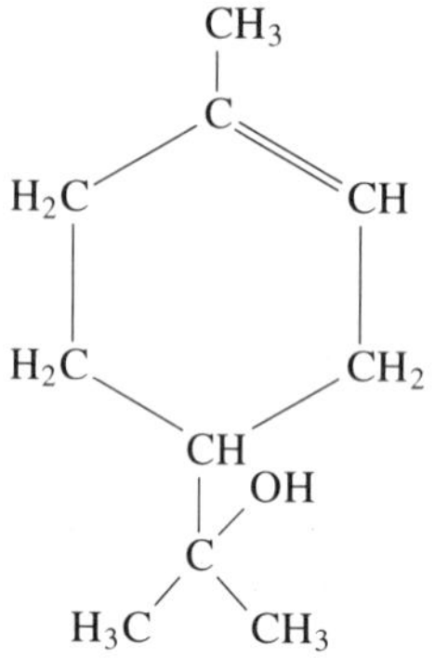

α-terpineol (note
similarity to limonene)

Alcohols

If we imagine an aliphatic hydrocarbon, in which one hydrogen atom is removed and an −O−H group put in its place, we have an alcohol. The oxygen is bonded to carbon by a covalent bond and so, as distinct from a hydroxide ion, the group is called a **hydroxyl** group. A number of isomeric alcohols can be obtained, e.g. from the butanes, as shown in Table 3.4.

There are thus three types of alcohol: primary ($-CH_2OH$), secondary ($>CHOH$) and tertiary ($≥COH$), the name deriving from the type of carbon atom on which the theoretical substitution has been made.

Table 3.4 Alcohols obtained from butanes

Alkane isomer	Type of C atom chosen for substitution	Alcohol produced
(1) $CH_3 \cdot CH_2 \cdot CH_2 \cdot CH_3$ butane	Primary	$CH_3CH_2CH_2CH_2OH$ butan-1-ol (*n*-butyl alcohol or *n*-butanol)
(2) $CH_3 \cdot CH_2 \cdot CH_2 \cdot CH_3$ butane	Secondary	$CH_3 \cdot CH_3 \cdot CHOH \cdot CH_3$ butan-2-ol (secondary or *sec*-butyl alcohol, *sec*-butanol)
(3) $CH_3 \cdot CH(CH_3) \cdot CH_3$ 2-methyl propane (isobutane)	Primary	$CH_3 \cdot CH(CH_3) \cdot CH_2OH$ 2-methylpropan-1-ol (isobutyl alcohol or isobutanol)
(4) 2-methyl propane (isobutane)	Tertiary	$(CH_3)_3 \cdot C \cdot OH$ 2-methylpropan-2-ol (tertiary or *tert*-butyl alcohol, *tert*-butanol)

The lower alcohols have fewer isomers:

CH_3OH	$CH_3 \cdot CH_2OH$	$CH_3 \cdot CH_2 \cdot CH_2OH$	$CH_3 \cdot CHOH{-}CH_3$
methanol (methyl alcohol)	ethanol (ethyl alcohol)	propan-1-ol (*n*-propyl alcohol or *n*-propanol)	propan-2-ol (isopropyl alcohol or isopropanol)

There may, of course, be more than one hydroxyl group in the molecule:

- two in a **glycol** (or dihydric alcohol), e.g. $CH_2OH \cdot CH_2OH$
  ethane-1,2-diol (ethylene glycol)
  – used in antifreeze

- three in a trihydric alcohol, e.g. $CH_2OH \cdot CHOH \cdot CH_2OH$
  1,2,3-propane-triol
  (glycerine or glycerol)

- four in a tetrahydric alcohol, e.g.
$$CH_2OH$$
$$HOCH_2{-}\overset{|}{\underset{|}{C}}{-}CH_2OH$$
$$CH_2OH$$

  2,2-bis(hydroxymethyl)-propane-1,3-diol (pentaerythritol)

and so on. In the last of the list above, the reader will see an example of a systematic name which is far longer than the trivial name, and is an example where the latter is used almost exclusively. We will come across these polyhydric alcohols (polyols) again in Part Two.

Organic chemistry: alkanes to oils

Methanol is prepared industrially by the reaction

$$CO + 2H_2 \xrightarrow[\text{400 °C, v. high pressure}]{\text{zinc chromite}} CH_3OH$$

Ethanol is what is often referred to just as 'alcohol'. It is prepared by the fermentation of sugar using yeast. Alternatively, ethene (ethylene) may be absorbed in concentrated sulphuric acid under pressure and at 75–80 °C:

$$CH_2{=}CH_2 + HO{\cdot}SO_2{\cdot}OH \rightarrow CH_3{\cdot}CH_2{\cdot}O{\cdot}SO_2{\cdot}OH$$
$$\text{ethyl sulphuric acid}$$

Ethyl sulphuric acid is an ester (see below) of ethanol and the alcohol can be recovered when the ester is hydrolysed (see below) by dilution with water and the mixture is distilled.

Carboxylic acids

Oxidation of primary alcohols produces a type of acid peculiar to organic chemistry, the carboxylic acid, e.g.

$$CH_3{\cdot}CH_2OH \xrightarrow{[O]} CH_3{\cdot}C\overset{\diagup H}{\underset{\diagdown\!\!\!\!\text{O}}{}} \xrightarrow{[O]} CH_3{\cdot}C\overset{\diagup O{-}H}{\underset{\diagdown\!\!\!\!\text{O}}{}} \quad (\text{or } CH_3{\cdot}COOH)$$

ethanal   ethanoic acid
(acetaldehyde)   (acetic acid)
$+H_2O$

An oxidizing agent, such as potassium permanganate, $KMnO_4$, is used rather than actual oxygen, hence the notation [O].

As with all types of organic compound, a large family of carboxylic acids, with different numbers and arrangements of carbon atoms, exists. All contain the carboxyl group $-CO{\cdot}OH$.

They are acids in all meanings of the word, e.g.

$$CH_3{\cdot}COOH \rightleftharpoons CH_3{\cdot}COO^- + H^+$$

They are usually weak acids, though some, e.g. $CCl_3{\cdot}COOH$ (trichloro-ethanoic acid), are very strong.

Carboxylic acids may incorporate the features of organic compounds discussed previously in this chapter, e.g.

• An alkyl group, in $CH_3{\cdot}CH_2{\cdot}COOH$, propanoic (propionic) acid.
• A double bond, in $CH_2{=}CH{\cdot}COOH$, propenoic (acrylic) acid.
• A chlorine atom, in $CH_3{\cdot}CHCl{\cdot}COOH$, 2-chloropropanoic
                β α (α-chloropropionic) acid.

Note: The carbon atoms which in the older system were labelled by Greek letters in alphabetical order, beginning next to the carboxyl group, are now numbered from the carboxyl group, but with that numbered as 1, so that substitution begins on the next or '2' carbon.

- A hydroxyl group, in $CH_3 \cdot CHOH \cdot COOH$, 2-hydroxy-propanoic (lactic) acid, found in sour milk.

Just as alcohols can have more than one hydroxyl group, so acids can have more than one carboxyl group in the molecule, i.e. they may be **polybasic**, e.g. dibasic acids:

$CH_2 \cdot COOH$	$CH \cdot COOH$	$HOOC \cdot CH$
$\|$	$\|\|$	$\|\|$
$CH_2 \cdot COOH$	$CH \cdot COOH$	$CH \cdot COOH$
butanedioic	*cis*-butenedioic	*trans*-butenedioic
(succinic) acid	(maleic) acid	(fumaric) acid

The reader will recall from earlier in this chapter how free rotation is normally possible around a normal single $C-C$ bond, but that free rotation is not possible about the $>C=C<$ bond, and hence the two butenedioic acids are distinctly different in properties, and are distinguished by the *cis-* and *trans-* prefixes which relate here to the carboxyl groups being respectively adjacent and opposite.

The reader will notice that the organic molecules are becoming more complicated. Be reassured; on the whole, the reactions of a group (e.g. $>C=C<$, $-OH$, $-COOH$) are not affected by the complexity of the molecule. The main thing is to learn the important reactions of each group and apply them where possible.

The carboxylic acids with alkane chains as shown, are known as the **fatty acids**, because the larger molecules of the series occur in oils and fats. The simpler members of the series are:

$H \cdot COOH$	$CH_3 \cdot COOH$	$CH_3 \cdot CH_2 \cdot COOH$	$CH_3 \cdot CH_2CH_2 \cdot COOH$
methanoic	ethanoic	propanoic	butanoic (butyric)
(formic) acid	(acetic) acid	(propionic) acid	acid – smell of
– nettle stings	– vinegar		rancid butter

Typical reactions of the carboxyl group are:

- Salt formation with alkalis, e.g.

$$CH_3 \cdot COOH + NaOH \longrightarrow CH_3COONa + H_2O$$
sodium ethanoate
(acetate)

If a higher fatty acid is used, a **soap** is formed, e.g. $C_{17}H_{35} \cdot COONa$, sodium octadecanoate (sodium stearate). Soaps are the oldest and simplest detergents and surfactants (Chapter 10).

- Forms acid chlorides, e.g.

$$CH_3 \cdot C \underset{O}{\overset{OH}{\diagup}} + PCl_5 \longrightarrow CH_3 \cdot C \underset{O}{\overset{Cl}{\diagup}} + POCl_3$$

phosphorus	ethanoyl	phosphorus
pentachloride	(acetyl) chloride	oxychloride

This type of acid (or **acyl**) chloride is frequently used in reactions instead of the acid. It is more reactive than the acid and with water reforms the acid:

$$CH_3 \cdot COCl + H_2O \longrightarrow CH_3 \cdot COOH + HCl$$

- Salts and acid chlorides together give **acid anhydrides** (acids without water), e.g.

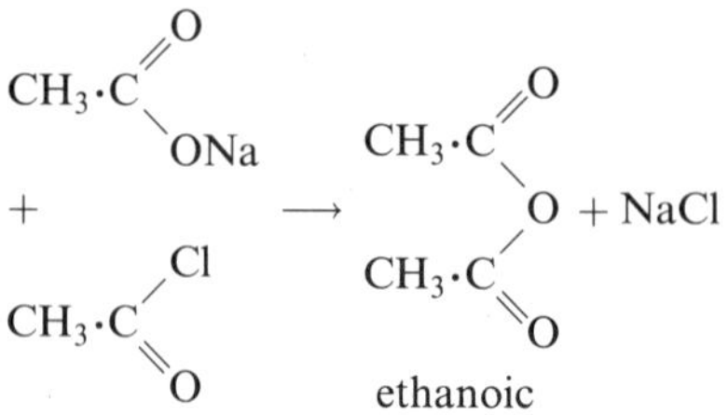

ethanoic
(acetic) anhydride

Anhydrides are also more reactive than acids and again are converted to acids by water:

$$(CH_3 \cdot CO)_2O + H_2O \longrightarrow 2CH_3 \cdot COOH$$

- Acids and alcohols react in a manner reminiscent of salt formation. The products are called **esters**, e.g.

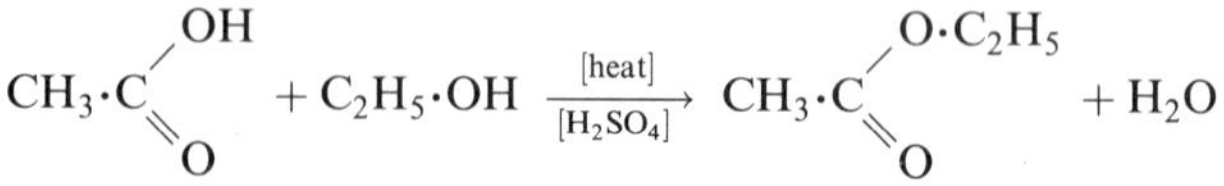

ethyl ethanoate (acetate)

This can also be made by

$$CH_3 \cdot COCl + C_2H_5 \cdot OH \longrightarrow CH_3 \cdot CO \cdot OC_2H_5 + HCl$$
$$(CH_3 \cdot CO)_2O + 2C_2H_5OH \longrightarrow 2CH_3 \cdot CO \cdot OC_2H_5 + H_2O$$

Esters

The whole range of acids and alcohols may be reacted to produce an enormous number of esters. They are found in a large number of natural and synthetic scents and perfumes. Even the simplest compounds, which are used as lacquer solvents (Chapters 9 and 11), have characteristic and generally pleasant smells. For example:

- ethyl ethanoate, $CH_3 \cdot CO \cdot C_2H_5$ (ethyl acetate)   fruity
- butyl ethanoate, $CH_3 \cdot CO \cdot O \cdot C_4H_9$ (butyl acetate)   pear drops
- 1-pentyl ethanoate, $CH_3 \cdot CO \cdot O \cdot C_5H_{11}$ (amyl acetate)   pear drops
- 3-methyl-1-butyl ethanoate, $CH_3 \cdot CO \cdot O(CH_2)_2 \cdot CH(CH_3)_2$ banana
  (isoamyl acetate)

- 1-methylpropyl propanoate,   rum
  $C_2H_5 \cdot CO \cdot O \cdot CH(CH_3) \cdot CH_2 \cdot CH_3$ (isobutyl propionate)
- butyl butanoate, $C_3H_7 \cdot CO \cdot O \cdot C_4H_9$ (butyl butyrate)   pineapple

Note that the ester names are derived first from the alcohol, then the acid portion and end in -ate. All contain the ester linkage,

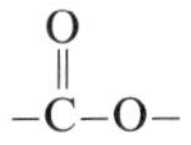

$$\overset{\displaystyle O}{\underset{\displaystyle }{\overset{\displaystyle \|}{-C-O-}}}$$

Esters can be converted back into the original acids and alcohols by **saponification** (soap formation) with alkali:

$$CH_3 \cdot CO \cdot O \cdot C_2H_5 + NaOH \longrightarrow CH_3 \cdot CO \cdot ONa + C_2H_5OH$$

$$\downarrow + HCl$$

$$CH_3 \cdot CO \cdot OH + NaCl$$

or often with some difficulty, by **hydrolysis** in boiling water:

$$CH_3 \cdot COOC_2H_5 + H_2O \longrightarrow CH_3 \cdot COOH + C_2H_5OH$$

Because these reactions proceed so readily, the ester linkage between acids and alcohols is a relatively weak one and paint films containing ester linkages are attacked by bases generally (Chapters 12 and 16).

Alcohols also form esters with inorganic acids, e.g.

$$C_2H_5OH + H_2SO_4 \longrightarrow C_2H_5O \cdot SO_2 \cdot OH + H_2O$$
$$\text{ethyl sulphuric acid}$$
$$\text{(ethyl hydrogensulphate)}$$

and it is the reverse of this reaction (hydrolysis) that has already been quoted as a step in the manufacture of ethanol.

With hydrochloric acid the reaction is

$$C_2H_5OH + HCl \longrightarrow C_2H_5 \cdot Cl + H_2O$$
$$\text{chloroethane}$$
$$\text{(ethyl chloride)}$$

and with nitric acid

$$C_2H_5OH + HNO_3 \longrightarrow C_2H_5 \cdot O \cdot NO_2 + H_2O$$
$$\text{ethyl nitrate}$$

Note the difference between this compound and the nitroalkane, nitroethane:

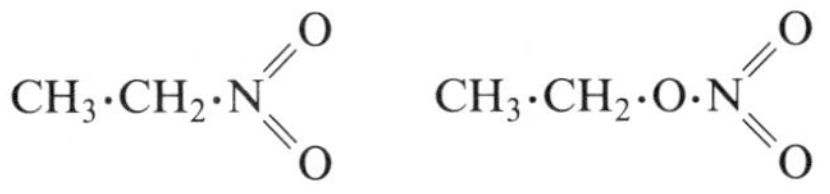

$$CH_3 \cdot CH_2 \cdot N\!\!\overset{\displaystyle O}{\underset{\displaystyle O}{\diagup\!\!\!\diagdown}} \qquad CH_3 \cdot CH_2 \cdot O \cdot N\!\!\overset{\displaystyle O}{\underset{\displaystyle O}{\diagup\!\!\!\diagdown}}$$

$$\text{nitroethane} \qquad\qquad \text{ethyl nitrate}$$

Nitrate esters find wide use of explosives, e.g. 1,2,3-propane-triyl trinitrate (nitroglycerin),

$$CH_2 \cdot O \cdot NO_2$$
$$|$$
$$CH \cdot O \cdot NO_2$$
$$|$$
$$CH_2 \cdot O \cdot NO_2$$

Note that, although commonly called *nitro*glycerin, this compound as implied by its systematic name is a nitrate, and its trivial name would more correctly be glycerin *nitrate*. An equally misleading name is applied to another explosive that we will encounter in Chapter 11: nitrocellulose.

Oils

The oils that are of particular interest to the paint chemist are the 'fatty' oils, which are largely vegetable oils pressed or extracted from the seeds or fruit of many types of vegetable matter. These are esters of 1,2,3-propane triol (glycerol), in which all three hydroxyl groups are reacted with fatty acids. The majority of these acids contain 18 carbon atoms. The triple esters are normally referred to as **triglycerides**.

The usefulness of an oil is determined by the nature of the fatty acids present. Oils are classified according to their ability to dry to a solid film, when spread thinly and exposed to the air. **Drying oils** form a film at normal temperatures, **semi-drying oils** require heat, while non-drying oils will not form a film. More is said about the mechanism of the drying process in Chapter 12. Drying oils are found to contain a substantial proportion of fatty acids with three double bonds, semi-drying oils contain principally fatty acids with two double bonds, while non-drying oils contain only minor proportions of fatty acids with more than one double bond. Table 3.5 gives the fatty acid distributions of average samples of vegetable oils that are or have been used in paint formulation. Formulae of the unsaturated acids (all contain 18 carbon atoms) are as follows:

- Octadeca-9-enoic (oleic)

$$CH_3 \cdot (CH_2)_7 \cdot CH = CH \cdot (CH_2)_7 \cdot COOH$$

- Octadeca-12-hydroxy-9-enoic (ricinoleic)

$$CH_3 \cdot (CH_2)_4 \cdot CH_2 \cdot CHOH \cdot CH_2 \cdot CH = CH \cdot (CH_2)_7 \cdot COOH$$

- Octadeca-9,12-dienoic (linoleic)

$$CH_3 \cdot (CH_2)_4 \cdot CH = CH \cdot CH_2 \cdot CH = CH \cdot (CH_2)_7 \cdot COOH$$

- Octadeca-9,11-dienoic (conjugated linoleic)

$$CH_3 \cdot (CH_2)_5 \cdot CH = CH \cdot CH = CH \cdot (CH_2)_7 \cdot COOH$$

- Octadeca-9,12,15-trienoic (linolenic)

$$CH_3 \cdot CH_2 \cdot CH{=}CH \cdot CH_2 \cdot CH{=}CH \cdot CH_2 \cdot CH{=}CH \cdot (CH_2)_7 \cdot COOH$$

- Octadeca-5,9,12-trienoic (pinolenic)

$$CH_3 \cdot (CH_2)_4 \cdot CH{=}CH \cdot CH_2 \cdot CH{=}CH \cdot (CH_2)_2 \cdot CH{=}CH \cdot (CH_2)_3 \cdot COOH$$

- Octadeca-9,11,13-trienoic (α-eleostearic)

$$CH_3 \cdot (CH_2)_3 \cdot CH{=}CH \cdot CH{=}CH \cdot CH{=}CH \cdot (CH_2)_7 \cdot COOH$$

Note that the double bonds may be **conjugated** ($-CH{=}CH-CH{=}CH-$), as in α-eleostearic, or **non-conjugated** ($-CH{=}CH-CH_2-CH{=}CH-$), as in linoleic and linolenic acids. Conjugated double bonds are directly joined by a single bond. Conjugated fatty acids give oils that dry faster and are less prone to yellowing (see p. 179). Fatty acids derived from oils (e.g. linseed oil fatty acids) can be obtained separately, if so desired. Catalytically treated versions of these fatty acids can also be obtained, in which the proportions of conjugated double bonds have been increased deliberately by an isomerization process.

Table 3.5 Composition of various oils

Oil	Fatty acids (%)					
	Saturated acids	Oleic	Ricinoleic	Linoleic	Linolenic	α-eleostearic
	No. of double bonds/acid molecule					
	0	1	1	2	3	3
Drying						
Linseed	10	22	–	17	51	–
Tung	5	9	–	3	3	80
Semi-drying						
Soya bean	13	28	–	54	5	–
Tall	3	32	–	54	11	–
Safflower	10	14	–	76	–	–
Sunflower	7	34	–	59	–	–
Cottonseed	27	24	–	49	–	–
Dehydrated						
castor	2	8	–	87	3	–
Non-drying						
Castor	2	8	87	3	–	–
Coconut	92	6	–	2	–	–

Since each oil contains several acids in esterified form, one glycerol molecule is seldom esterified by three similar fatty acid molecules. It is not certain whether the distribution of fatty acids among the glycerol molecules is completely random, or influenced by some chemical factor to give a preponderance of particular arrangements. The exact distribution affects the drying considerably, since it affects the functionality of the molecules (Chapters 5 and 12).

Tall oil fatty acids are a by-product of paper making, and there is no naturally occurring oil. Tall oil of Scandinavian origin contains the trienoic isomer pinolenic acid rather than linolenic acid as shown in Table 3.5.

Castor oil is non-drying, but if water is removed from the ricinoleic acid by heating above 250 °C with sulphuric acid, a second double bond is introduced:

$$
\underset{\substack{|\\ \text{OH}}}{-\overset{12}{CH_2}-CH-\overset{11}{CH_2}-\overset{10}{CH}=\overset{9}{CH}-} \xrightarrow{-H_2O}
\begin{array}{c}
\overset{12}{-CH}=\overset{11}{CH}-\overset{10}{CH_2}-\overset{9}{CH}=CH- \\[2pt]
\text{non-conjugated } (67-75\%) \\[6pt]
\overset{12}{-CH_2}-\overset{11}{CH}=\overset{10}{CH}-\overset{9}{CH}=CH- \\[2pt]
\text{conjugated } (25-33\%)
\end{array}
$$

The resulting **dehydrated castor oil** (DCO), although it contains no fatty acids with three double bonds, has such a high proportion of acids with two double bonds (linoleic acid), including some conjugated unsaturation, that it is a drying oil. It does, however, dry to a film with a noticeable surface tack.

Castor oil can also be modified by reduction with hydrogen to produce **hydrogenated castor oil** (HCO)

$$
\underset{\substack{|\\ \text{OH}}}{-CH_2-CH-CH_2-CH=CH-} \xrightarrow{H_2} \underset{\substack{|\\ \text{OH}}}{-CH_2-CH-CH_2-CH_2-CH_2-}
$$

This has a special use as a paint additive (Chapter 10).

More will be said about oils and drying in Chapter 12.

Four

Organic chemistry

ethers to isocyanates

This chapter continues the study of carbon compounds, and here we describe the basic chemistry of organic compounds, introducing further structural and reactive groups containing oxygen, and some important ones containing nitrogen. We are also going to introduce the chemistry of **aromatic** compounds, those containing the benzene ring structure.

Ethers

These compounds contain the characteristic ether linkage $C-O-C$, and can be thought of either as alcohols in which the hydrogen atom of the hydroxyl group is replaced by an alkyl group, or as 'alcohol anhydrides'. Ethoxyethane and some other ethers are made by a dehydration reaction:

$$2C_2H_5OH \xrightarrow[140\,°C]{[H_2SO_4]} C_2H_5OC_2H_5 + H_2O$$

ethoxyethane
(diethyl ether)

This proceeds via the esterification of ethanol by sulphuric acid. Alternatively the dehydration can be carried out over aluminium oxide at 240–260 °C under high pressure.

In spite of this resemblance to anhydrides, ethers are very resistant to hydrolysis, and heating under pressure with dilute sulphuric acid is required to bring about this breakdown:

$$R-O-R + H_2O \longrightarrow 2ROH$$

The ether linkage is more readily broken by heating with halogen acids, particularly hydrogen iodide:

$$CH_3 \cdot O \cdot C_3H_7 \quad + \quad HI \quad \longrightarrow \quad CH_3I \quad + \quad C_3H_7OH$$

| methoxypropane | hydrogen | methyl | propan-1-ol |
| (methyl propyl ether) | iodide | iodide | (propyl alcohol) |

A variety of ethers with similar and dissimilar alkyl groups attached to the oxygen atom are possible and the ether linkage is found in many more complex compounds. Ethers are fairly unreactive compounds, though the hydrogen atoms of the methylene group next to the oxygen are more easily substituted by chlorine than the hydrogen atoms in paraffin.

Epoxides

The oxirane or epoxide ring

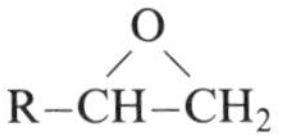

$$R-\overset{\displaystyle O}{\overset{\displaystyle \diagup \diagdown}{CH-CH_2}}$$

which at first sight resembles a cyclic ether, is formed by direct reaction of ethene (ethylene) and oxygen,

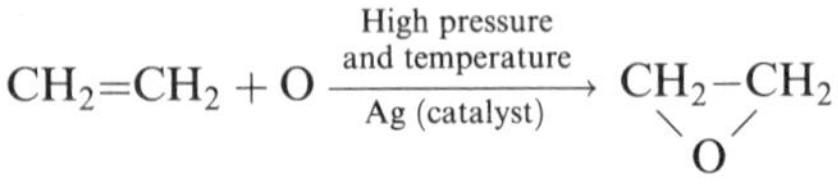

$$CH_2{=}CH_2 + O \xrightarrow[\text{Ag (catalyst)}]{\substack{\text{High pressure}\\ \text{and temperature}}} CH_2{-}CH_2 \underset{O}{\diagdown \diagup}$$

epoxyethane
(ethylene oxide)

or via a **chlorohydrin**, formed between an alkene and chlorine in alkaline solution

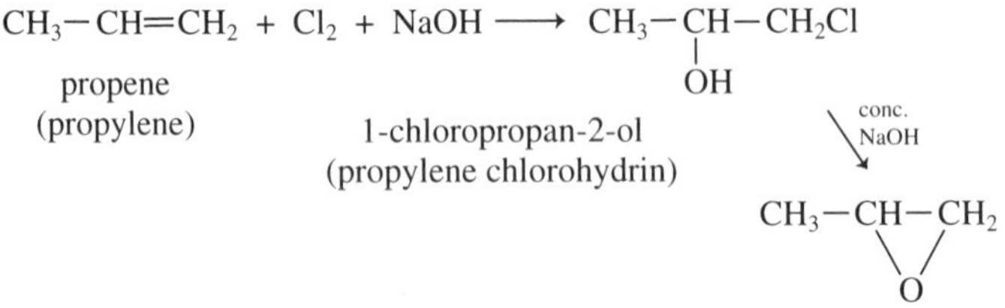

$$CH_3{-}CH{=}CH_2 + Cl_2 + NaOH \longrightarrow CH_3{-}\underset{\underset{OH}{|}}{CH}{-}CH_2Cl$$

propene
(propylene)

1-chloropropan-2-ol
(propylene chlorohydrin)

$$\xrightarrow[\text{NaOH}]{\text{conc.}} CH_3{-}CH{-}CH_2 \underset{O}{\diagdown \diagup}$$

1,2-epoxypropane
(propylene oxide)

The three-membered ring is as reactive as a double bond and is readily opened by compounds containing an 'active' hydrogen atom (i.e. one attached to a strongly electronegative atom, such as N or O). In the ring opening reaction, the hydrogen attaches itself to the ring's oxygen atom, e.g.

(1) 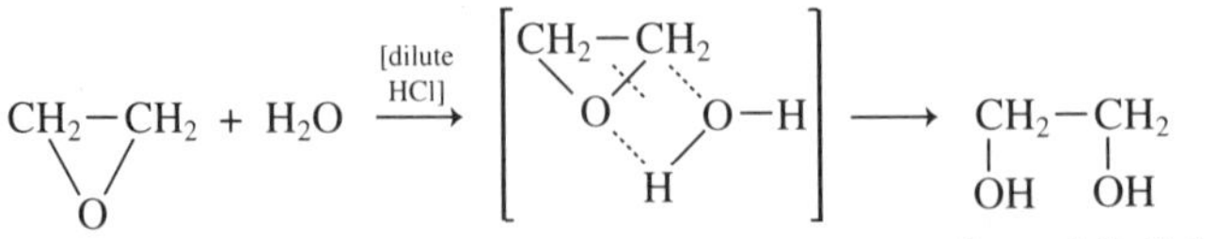
$$CH_2{-}CH_2 + H_2O \xrightarrow{\substack{[\text{dilute}\\ \text{HCl}]}} \left[\begin{array}{c} CH_2{-}CH_2 \\ O \diagdown \diagup O{-}H \\ H \end{array} \right] \longrightarrow \underset{\underset{OH}{|}}{CH_2}{-}\underset{\underset{OH}{|}}{CH_2}$$

epoxyethane
(ethylene oxide)

ethane-1,2-diol
(ethylene glycol)

In this reaction, we see direct hydrolysis, just as above for a normal ether, but occurring much more readily

(2)

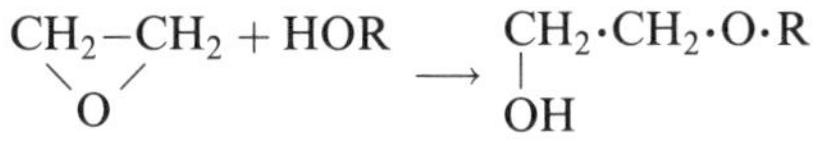

$$CH_2{-}CH_2 + HOR \underset{O}{\diagdown\diagup} \longrightarrow \underset{OH}{\overset{CH_2 \cdot CH_2 \cdot O \cdot R}{|}}$$

an ether of ethane-1,2-diol or glycol ether
– an ether-alcohol, see Chapter 9

where ROH is an alcohol.

(3)

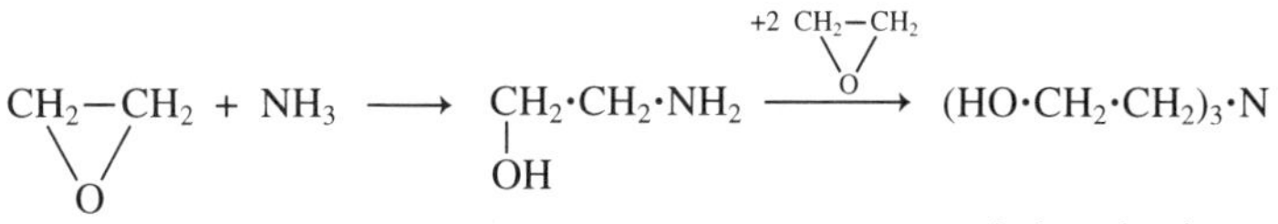

$$CH_2{-}CH_2 \underset{O}{\diagdown\diagup} + NH_3 \longrightarrow \underset{OH}{\overset{CH_2 \cdot CH_2 \cdot NH_2}{|}} \xrightarrow{+2\ CH_2{-}CH_2 \diagdown\diagup O} (HO \cdot CH_2 \cdot CH_2)_3 \cdot N$$

1-aminoethan-2-ol   triethanolamine
(ethanolamine)   – an alcohol-amine

(4)

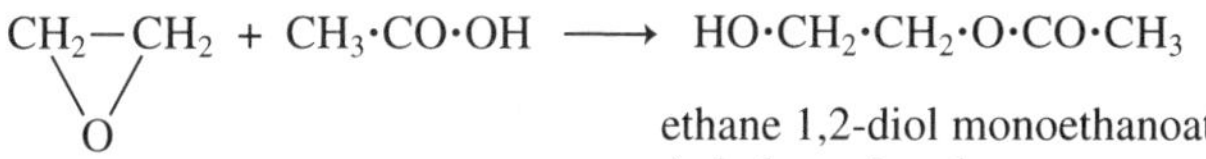

$$CH_2{-}CH_2 \underset{O}{\diagdown\diagup} + CH_3 \cdot CO \cdot OH \longrightarrow HO \cdot CH_2 \cdot CH_2 \cdot O \cdot CO \cdot CH_3$$

ethane 1,2-diol monoethanoate
(ethylene glycol monoacetate)
– an alcohol-ester

The oxirane or epoxide ring is the reactive group attached to epoxide resins which form the basis of the 'epoxy' paints.

Aldehydes and ketones

We have already encountered the keto- or carbonyl group

$$\underset{O}{\overset{-C-}{\underset{\|}{}}}$$

as part of the carboxyl group.

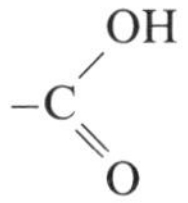

$$-C\overset{\nearrow OH}{\underset{\searrow O}{}}$$

It is also the basis of two other families of compounds, the aldehydes and ketones. These have the general formula

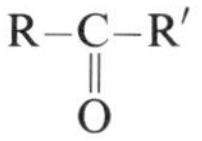

$$\underset{O}{\overset{R{-}C{-}R'}{\underset{\|}{}}}$$

R is an alkyl or aryl group. If R′ is a hydrogen atom, the compound is an aldehyde, but if it is also an alkyl or aryl group, then the compound is a ketone. There is also the special case of methanal (formaldehyde) where R and R′ are both hydrogen atoms. (The meaning of aryl will become clear near the end of this chapter.)

Thus the simplest aldehydes are

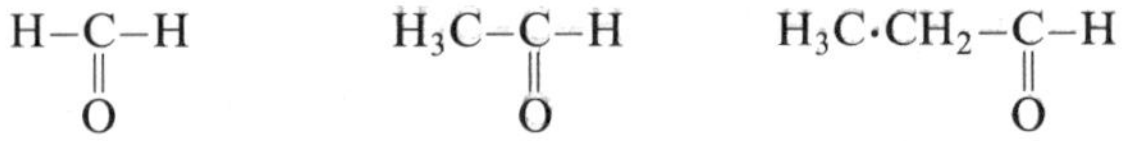

$$
\underset{\substack{\| \\ O}}{H-C-H} \qquad \underset{\substack{\| \\ O}}{H_3C-C-H} \qquad \underset{\substack{\| \\ O}}{H_3C \cdot CH_2-C-H}
$$

methanal	ethanal	propanal
(formaldehyde)	(acetaldehyde)	(propionaldehyde)

Some better known members of the ketone family are

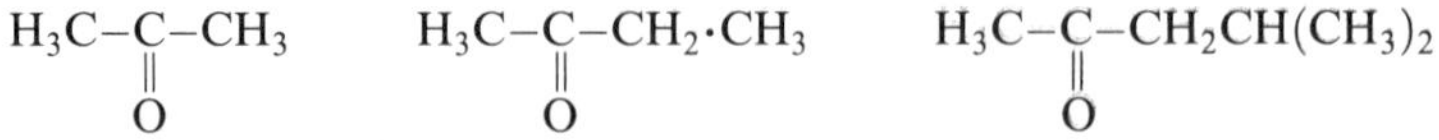

$$
\underset{\substack{\| \\ O}}{H_3C-C-CH_3} \qquad \underset{\substack{\| \\ O}}{H_3C-C-CH_2 \cdot CH_3} \qquad \underset{\substack{\| \\ O}}{H_3C-C-CH_2CH(CH_3)_2}
$$

propanone	butanone	2-methylpentan-4-one
(acetone)	(methyl ethyl ketone)	(methyl isobutyl ketone)

These ketones are all paint solvents and our interest in ketones is limited, in this book, to their performance as such. Aldehydes are not solvents, but the reactions of methanal, in particular, give rise to several families of new resins (Chapter 13).

Aldehydes are intermediate products in the oxidation of primary alcohols (Chapter 3), whereas ketones are produced by oxidation of secondary alcohols.

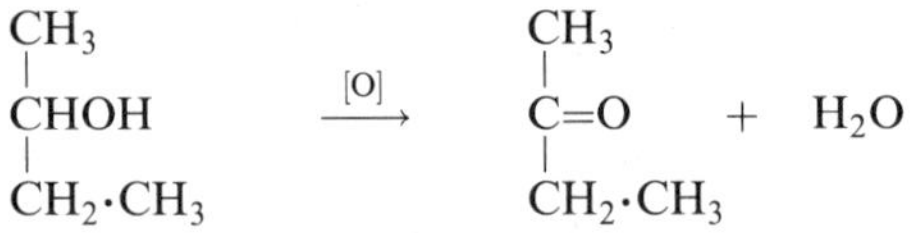

$$
\begin{array}{c} CH_3 \\ | \\ CHOH \\ | \\ CH_2 \cdot CH_3 \end{array} \quad \xrightarrow{[O]} \quad \begin{array}{c} CH_3 \\ | \\ C{=}O \\ | \\ CH_2 \cdot CH_3 \end{array} \quad + \quad H_2O
$$

butan-2-ol	butanone
(*sec*-butyl alcohol)	(methyl ethyl ketone)

Although the reactions of the two families do differ at some points, they are largely the reactions of the carbonyl group, which readily accepts hydrogen from other compounds, such as another aldehyde molecule, an alcohol or even an inorganic compound:

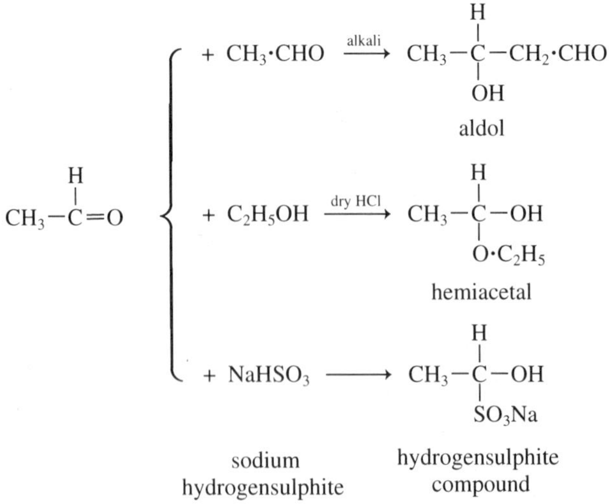

$$
\underset{\substack{| \\ CH_3-C=O}}{\overset{H}{}} \left\{ \begin{array}{l}
+ \; CH_3 \cdot CHO \; \xrightarrow{\text{alkali}} \; \underset{\substack{| \\ OH}}{\overset{H}{CH_3-C-CH_2 \cdot CHO}} \\[4ex]
\hspace{6em} \text{aldol} \\[2ex]
+ \; C_2H_5OH \; \xrightarrow{\text{dry HCl}} \; \underset{\substack{| \\ O \cdot C_2H_5}}{\overset{H}{CH_3-C-OH}} \\[4ex]
\hspace{6em} \text{hemiacetal} \\[2ex]
+ \; NaHSO_3 \; \longrightarrow \; \underset{\substack{| \\ SO_3Na}}{\overset{H}{CH_3-C-OH}}
\end{array} \right.
$$

sodium	hydrogensulphite
hydrogensulphite	compound

Sometimes, although the first step in the reaction may be addition of hydrogen to carbonyl, the overall reaction involves elimination of the carbonyl oxygen as water:

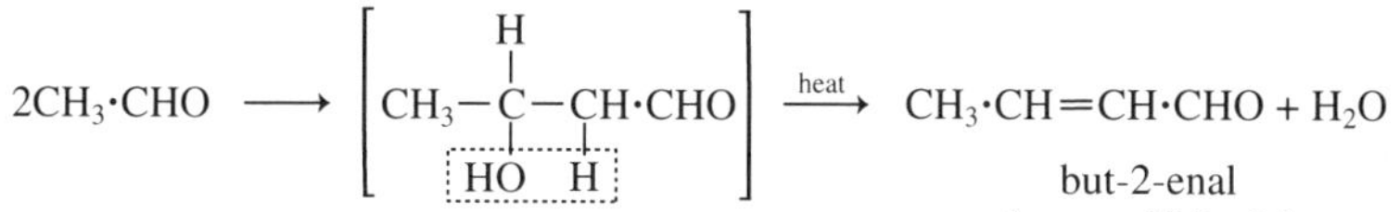

$$2CH_3 \cdot CHO \longrightarrow \left[CH_3 - \underset{\underset{HO\ \ H}{|}}{\overset{\overset{H}{|}}{C}} - CH \cdot CHO \right] \xrightarrow{\text{heat}} CH_3 \cdot CH = CH \cdot CHO + H_2O$$

but-2-enal
(crotonaldehyde)

Amines

Compounds containing carbon, hydrogen and nitrogen are introduced by a family of chemicals called amines. Amines are derived from ammonia, and primary, secondary and tertiary forms are possible:

NH_3	$CH_3 \cdot NH_2$	$(CH_3)_2 \cdot NH$	$(CH_3)_3 \cdot N$
ammonia	methylamine	dimethylamine	trimethylamine
	(primary)	(secondary)	(tertiary)

Like ammonia, these compounds are basic and therefore form salts with acids

$$\left. \begin{array}{l} NH_3 + HCl \longrightarrow NH_4Cl \\ CH_3 \cdot NH_2 + HCl \longrightarrow CH_3 \cdot NH_3Cl \end{array} \right\} \begin{array}{l} \text{nitrogen} \\ \text{expanding its} \\ \text{valency to 5} \end{array}$$

methylamine
hydrochloride

The strength of the base increases with the number of alkyl groups attached to nitrogen. Amines can be prepared from ammonia and an alkyl halide:

$$C_2H_5Br + NH_3 \longrightarrow C_2H_5 \cdot NH_3Br \xrightarrow{+ NaOH} C_2H_5NH_2 + NaBr + H_2O$$

ethylamine

But the C_2H_5Br can react with the other hydrogen atoms in ethylamine, forming the secondary and the tertiary amine and finally a **quaternary ammonium salt**:

$$(C_2H_5)_3N + C_2H_5Br \longrightarrow (C_2H_5)_4\overset{+}{N}\overset{-}{Br}$$

quaternary ethyl
ammonium bromide

Quaternary ammonium hydroxides are strong bases.

Amines react with oxiranes (epoxides) in the manner of ammonia,

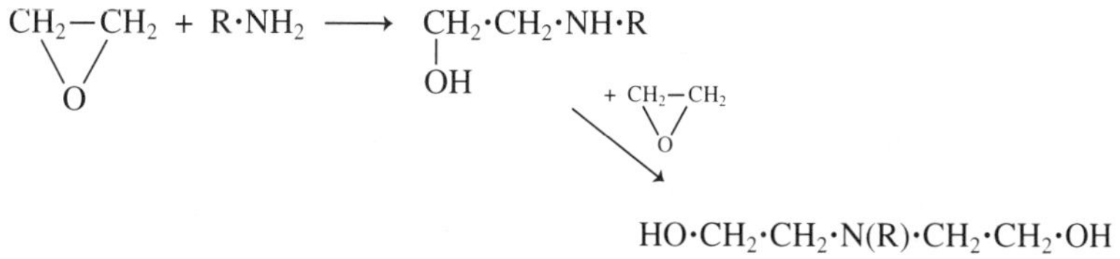

$$\underset{\underset{O}{\diagdown\diagup}}{CH_2 - CH_2} + R \cdot NH_2 \longrightarrow \underset{\underset{OH}{|}}{CH_2 \cdot CH_2 \cdot NH \cdot R}$$

$$\xrightarrow{+ \ CH_2 - CH_2 \atop \diagdown\diagup \atop O}$$

$$HO \cdot CH_2 \cdot CH_2 \cdot N(R) \cdot CH_2 \cdot CH_2 \cdot OH$$

and with ketones, eliminating water,

$$
\begin{array}{c}
CH_3 \\
| \\
C=O + H_2N \cdot CH_3 \\
| \\
CH_3
\end{array}
\longrightarrow
\begin{array}{c}
CH_3 \\
| \\
C=N \cdot CH_3 + H_2O \\
| \\
CH_3
\end{array}
$$

propanone   a ketoxime
(acetone)

Amines can react with esters, yielding the amide of the ester and eliminating the ether or alcohol (and also ammonia):

$$
\underset{\text{methyl ester}}{R-\overset{\overset{\displaystyle O}{\|}}{C}-OCH_3} \xrightarrow{\ R'NH_2\ } \underset{\text{amide}}{R-\overset{\overset{\displaystyle O}{\|}}{C}-NH_2} + \underset{\text{methyl ether}}{CH_3OR'}
$$

Amides

Amides are neutral or mildly basic compounds containing the group $-CO \cdot NH_2$. They are produced by reaction between acids, acid chlorides, acid anhydrides or esters and ammonia:

$$
CH_3 \cdot C \overset{\displaystyle O}{\underset{\displaystyle OH}{\Big\langle}} + NH_3 \longrightarrow CH_3 \cdot C \overset{\displaystyle O}{\underset{\displaystyle ONH_4}{\Big\langle}} \xrightarrow{\text{[heat]}} CH_3 \cdot C \overset{\displaystyle O}{\underset{\displaystyle NH_2}{\Big\langle}} + H_2O
$$

ethanoic acid   ammonium ethanoate   ethanamide
(acetic acid)   (ammonium acetate)   (acetamide)

$$
CH_3 \cdot CO \cdot Cl + NH_3 \longrightarrow CH_3 \cdot CO \cdot NH_2 + HCl
$$

ethanoyl chloride
(acetyl chloride)

$$
\begin{array}{c}
CH_3 \cdot CO \\
\diagdown \\
O + NH_3 \longrightarrow \\
\diagup \\
CH_3 \cdot CO
\end{array}
\quad
\begin{array}{c}
CH_3 \cdot CO \cdot NH_2 \\
+ \\
CH_3 \cdot CO \cdot OH
\end{array}
$$

ethanoic anhydride
(acetic anhydride)

$$
CH_3 \cdot CO \cdot O \cdot C_2H_5 + NH_3 \longrightarrow CH_3 \cdot CO \cdot NH_2 + C_2H_5OH
$$

ethyl ethanoate
(ethyl acetate)

Substituted amides are formed by reaction of amines with acid chlorides or anhydrides:

$$R \cdot NH_2 + (CH_3 \cdot CO)_2O \longrightarrow R \cdot NH \cdot CO \cdot CH_3 + CH_3 \cdot CO \cdot OH$$

N-substituted ethanamide
R = alkyl

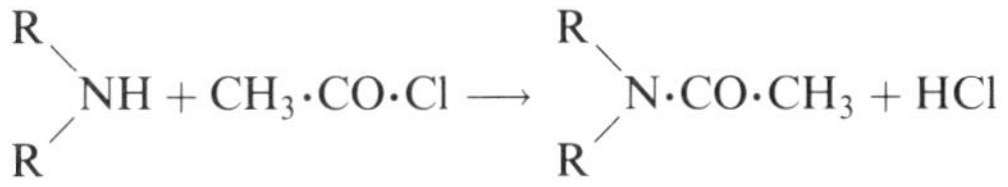

$$\begin{matrix} R \\ \diagdown \\ \quad NH + CH_3 \cdot CO \cdot Cl \longrightarrow \\ \diagup \\ R \end{matrix} \quad \begin{matrix} R \\ \diagdown \\ \quad N \cdot CO \cdot CH_3 + HCl \\ \diagup \\ R \end{matrix}$$

di-N-substituted ethanamide

The $(-NH-CO-)$ amide linkage is found in polyamide resins.

Amides are resistant to hydrolysis, but this can be brought about by boiling with ethanoic and hydrochloric acids:

$$R \cdot CO \cdot NH_2 + H_2O \longrightarrow R \cdot CO \cdot OH + NH_3$$

Carbamide (urea)

Carbamide is the diamide of carbonic acid,

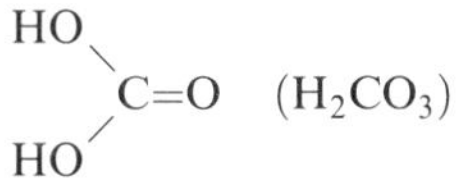

$$\begin{matrix} HO \\ \diagdown \\ \quad C{=}O \quad (H_2CO_3) \\ \diagup \\ HO \end{matrix}$$

Carbonyl chloride may be considered to be the acid dichloride and it reacts with ammonia to give carbamide:

$$COCl_2 + NH_3 \longrightarrow H_2N \cdot CO \cdot NH_2 + 2HCl$$

<table>
<tr><td>carbonyl chloride
(phosgene)</td><td>carbamide
(urea)</td></tr>
</table>

or alternatively it may be made directly from carbon dioxide:

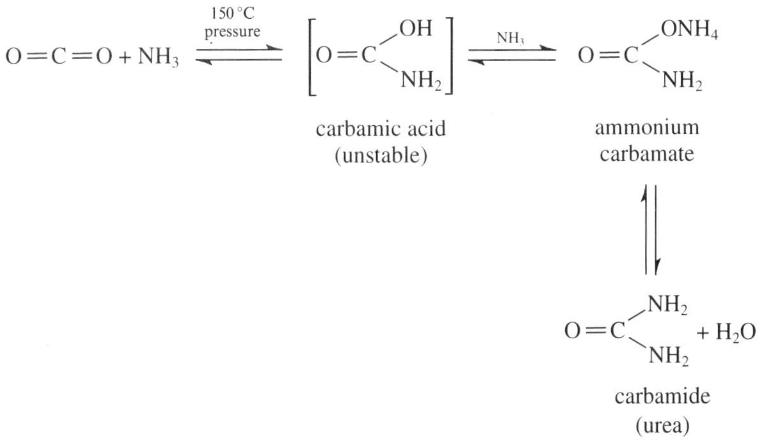

$$O{=}C{=}O + NH_3 \underset{\text{pressure}}{\overset{150\,°C}{\rightleftharpoons}} \left[O{=}C\begin{matrix}\diagup OH \\ \diagdown NH_2\end{matrix} \right] \overset{NH_3}{\rightleftharpoons} O{=}C\begin{matrix}\diagup ONH_4 \\ \diagdown NH_2\end{matrix}$$

carbamic acid   ammonium
(unstable)   carbamate

$$O{=}C\begin{matrix}\diagup NH_2 \\ \diagdown NH_2\end{matrix} + H_2O$$

carbamide
(urea)

Carbamide is the basic ingredient of urea formaldehyde resins.

Arenes or aromatic compounds

These are an important series of compounds derived from benzene. The earliest known examples were marked by a spicy aroma, though the majority are not. Benzene is a liquid boiling at 80 °C. It is obtained by distillation of coal tar, one of the four products of the destructive distillation of coal.

The structure of benzene is the key to the distinctive properties of these compounds. The overall formula is C_6H_6, and the carbon atoms form a ring. However, the formula which correctly uses all the valencies does not fit the reactions of the compound:

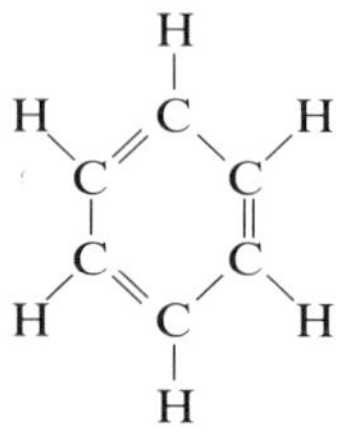

This formula suggests that benzene is an olefin and therefore various reagents should readily add themselves to the double bonds in the ring. In fact, only with difficulty can chlorine and hydrogen be made to react in this way to produce respectively $C_6H_6Cl_6$ (one form of which is the insecticide 'Gammexane') and C_6H_{12} (cyclohexane), but it is relatively easy to *substitute* other atoms or groups from the hydrogen atoms, while the six-membered carbon ring stays intact. Compare the reactions of sulphuric acid with ethene and benzene:

$$CH_2{=}CH_2 + H_2SO_4 \longrightarrow CH_3{\cdot}CH_2{\cdot}HSO_4$$

ethyl hydrogensulphate

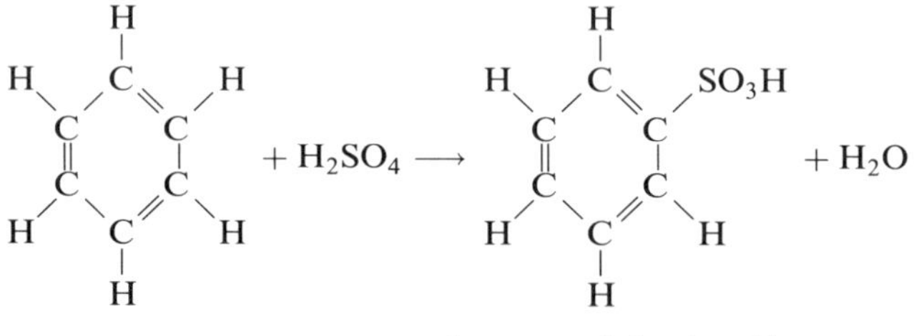

benzene sulphonic acid

Benzene, therefore, is not an ordinary alkene, yet is does have some alkene reactions. The above formula is not entirely satisfactory. Two more reasons can be given why this is so:

- If we substitute any two groups 'X' and 'Y' for the hydrogen atoms on two adjacent carbon atoms, we should obtain two isomers:

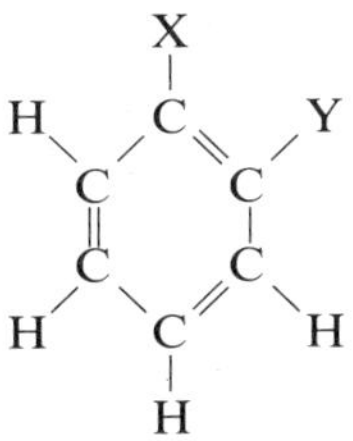

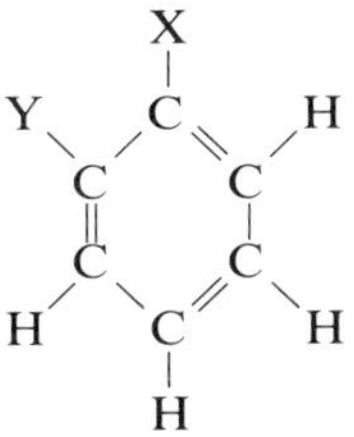

But these are never found.

- By the technique of X-ray crystallography, the distances between atoms can be measured. The distance between carbons (C–C) in ethane is 0.154 nm, and in ethene (C=C) it is 0.133 nm. Yet in benzene all adjacent carbon atoms are *equally* spaced 0.139 nm apart, a distance *between* the normal single and double bond lengths.

The full explanation of these facts is based on wave mechanics and the modern theories of electron distribution in chemical bonds. A simplified explanation is that the carbon atoms in benzene are held together by $1\frac{1}{2}$ valencies (three electrons). Two valency electrons are firmly located in a bond directly between each pair of carbon atoms, but the remaining electrons are 'pooled' (or delocalized) between all six atoms and are free to move to particular atoms when the molecule is influenced by some outside reagent. This arrangement has much higher stability because of this electron sharing:

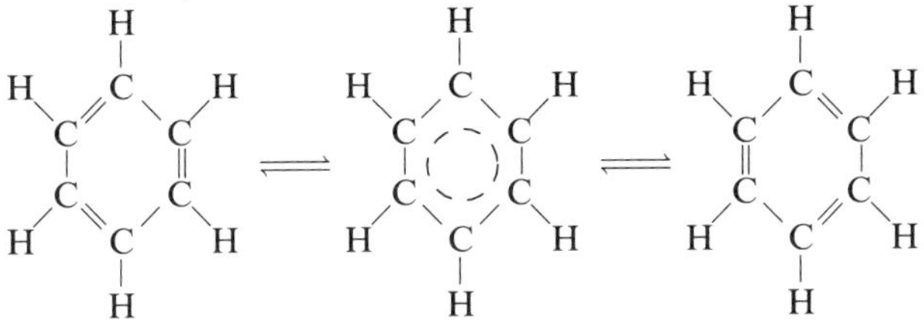

In practice, unless we wish to draw attention to a particular double bond or hydrogen atom, we depict the formula as a hexagon,  to which substituent groups may be added, e.g. benzene sulphonic acid,

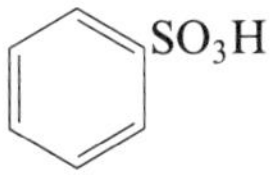

Benzene is a toxic chemical, causing anaemia and cancer if inhaled in prolonged doses. For this reason, it is not used as a paint solvent.

Benzene substitution products

Since some or all of the six hydrogen atoms of benzene may be replaced by other atoms or groups, a large variety of aromatic compounds is possible and many of these will be isomers. Isomers are distinguished from one another by numbering the carbon atoms from 1 to 6, so that the atoms on which substitution has occurred can be clearly defined

Mono-substitution

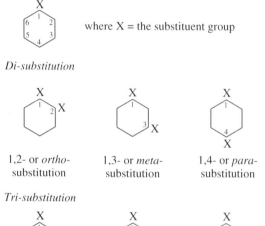

where X = the substituent group

Di-substitution

1,2- or *ortho-* substitution	1,3- or *meta-* substitution	1,4- or *para-* substitution

Tri-substitution

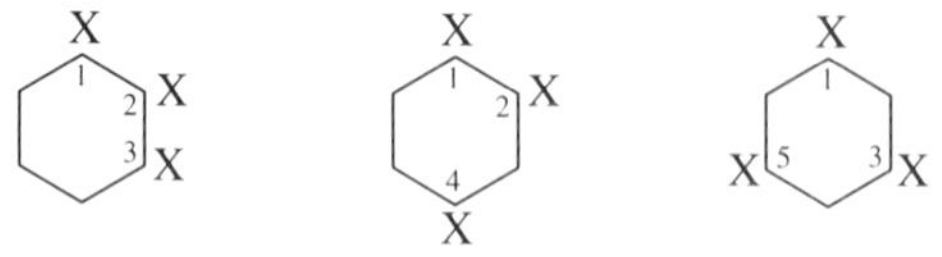

1,2,3-substitution   1,2,4-substitution   1,3,5-substitution

and so on.

Note that the first group is always given position 1 and the others are given the lowest numbers possible.

Every type of organic compound covered in the aliphatic series is also possible in the aromatic series: hydrocarbons, alcohols, acids, esters, ethers, aldehydes, ketones, amines, amides and others. On the whole, groups behave exactly as they do in aliphatic compounds, but sometimes they have their behaviour slightly modified by the direct attachment to the benzene ring. Aromatic alcohols, for example, are weakly acid in character:

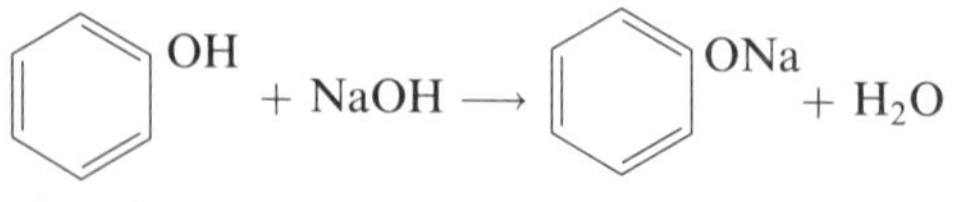

phenol   sodium phenoxide

Note that the C_6H_5- group is called **phenyl**.

Aromatic alcohols do not give esters with carboxylic acids, but will do so with acid chlorides. They differ markedly from alcohols in the ease with which substitution takes place (in the 2 and 4 positions), e.g.

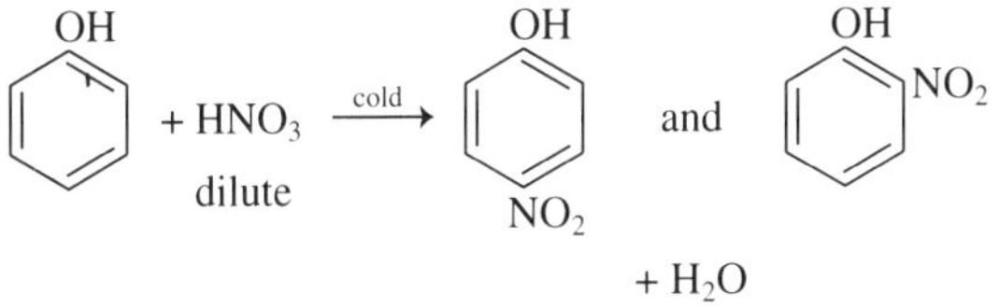

$$+ H_2O$$

1,4 or p-(*para*)   1,2 or o-(*ortho*)
nitrophenol   nitrophenol

Nitrates are not formed. Because of these differences, the family of compounds is called **phenols**, not alcohols.

The following are aromatic compounds that will be encountered later in the book.

Hydrocarbons

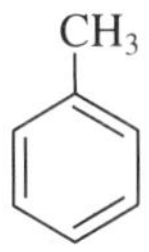

methylbenzene
(toluene)

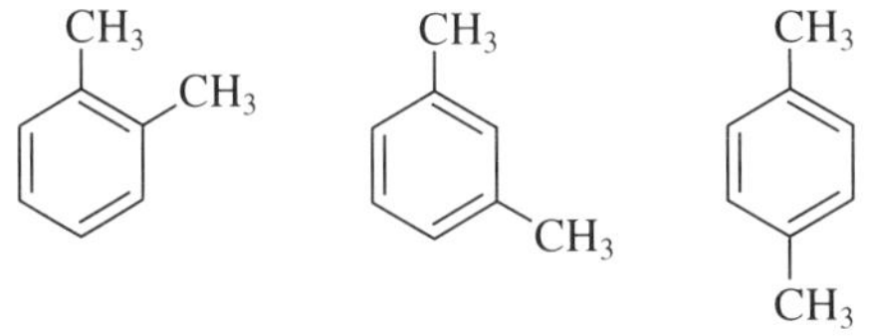

1,2   1,3   1,4

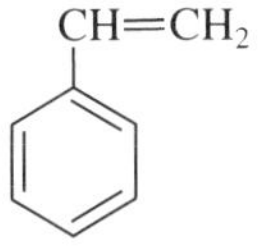

phenylethene
(styrene or vinyl
benzene)

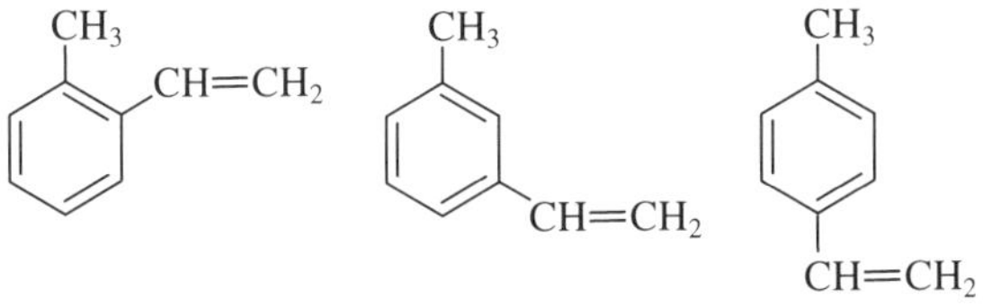

dimethyl benzene
(*o*-, *m*- or *p*-xylene)

1,2   1,3   1,4

methyl phenylethene
(*o*-, *m*- or *p*-vinyl toluene)

The aliphatic hydrocarbon groups attached behave in a typical manner. Note that just as methyl or ethyl groups are known as alkyl groups, so phenyl (or C_6H_5-) and substituted phenyl groups are known as aryl groups.

Acids

COOH

benzene carboxylic
acid
(benzoic acid)

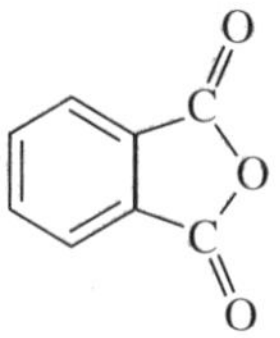

benzene 1,2-dicarboxylic
anhydride
(*o*-phthalic anhydride)

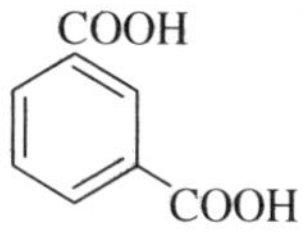

COOH

COOH

benzene 1,2-dicarboxylic
acid
(*m*- or *iso*-phthalic acid)

Alcohol

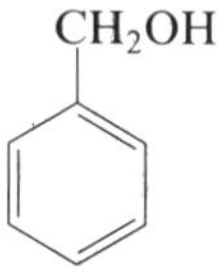

CH$_2$OH

phenylmethanol
(benzyl alcohol)

This is an aromatic-aliphatic compound. As the hydroxyl group is attached to the aliphatic portion, this is a true alcohol, not a phenol.

Ester

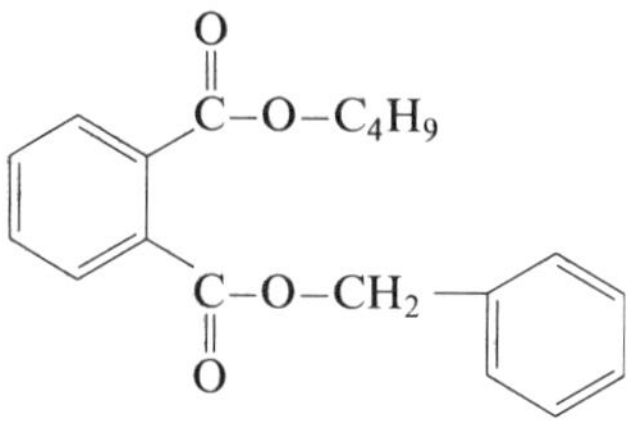

O
‖
C−O−C$_4$H$_9$

C−O−CH$_2$
‖
O

butyl benzyl phthalate

Amine

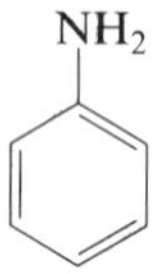

NH$_2$

phenylamine
(aniline)

Isocyanates

The isocyanates are a family of compounds existing in aliphatic and aromatic forms, and both are now used. All isocyanates contain the isocyanate group: −N=C=O, e.g.

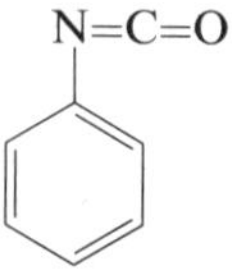

N=C=O

phenyl isocyanate

Isocyanates are prepared by the following route:

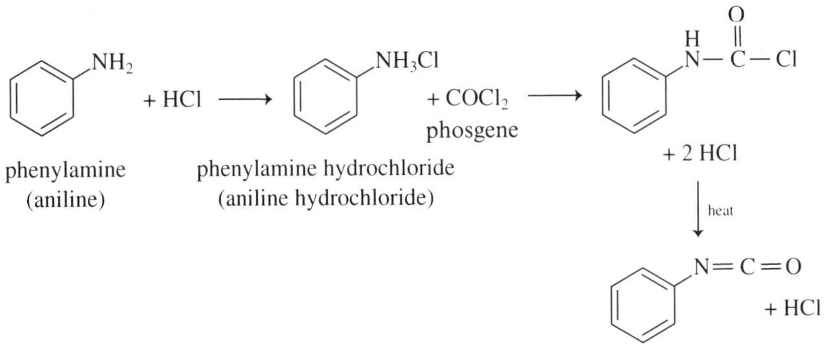

phenylamine   phenylamine hydrochloride
(aniline)   (aniline hydrochloride)

phenyl isocyanate

Two or more isocyanate groups may be included in a molecule. Some well-known compounds are:

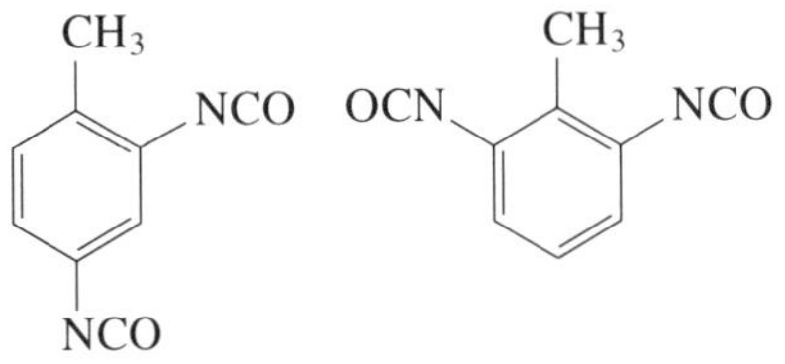

$OCN \cdot (CH_2)_6 \cdot NCO$

hexane 1,6-diisocyanate
(hexamethylene diisocyanate)

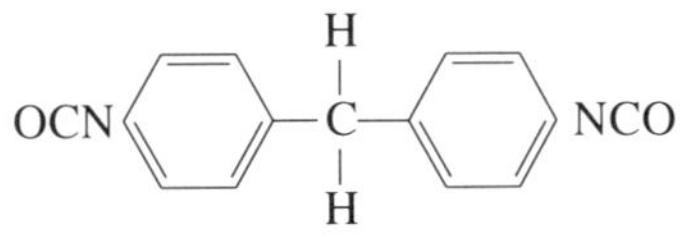

1-methylbenezene-2,4- and
2,6-diisocyanates
(toluene diisocyanates)

diphenyl methane -4,4′diisocyanate
(4′ implies substitution in the second ring)

The reactions of the isocyanate group occur with compounds containing an 'active' hydrogen atom. In each example, the hydrogen atom attaches itself to nitrogen in the $-NCO$ group, opening the $N=C$ double bond. The remainder of the reacting molecule then attaches itself to the spare valency which has been made available on the $-NCO$ carbon. Thus

$$R-N=C=O + H-X \longrightarrow \begin{bmatrix} R-N-C=O \\ \quad | \quad | \\ \quad H \\ \quad + \\ \quad -X \end{bmatrix} \longrightarrow R-N-C-X \\ \qquad\qquad | \quad || \\ \qquad\qquad H \quad O$$

In the following examples R is an alkyl or aromatic (aryl) group and R′ is some other alkyl or aryl group.

• $R-N=C=O + R' \cdot NH_2 \longrightarrow R-N-C-N-R'$
           amine   H O H

a substituted urea

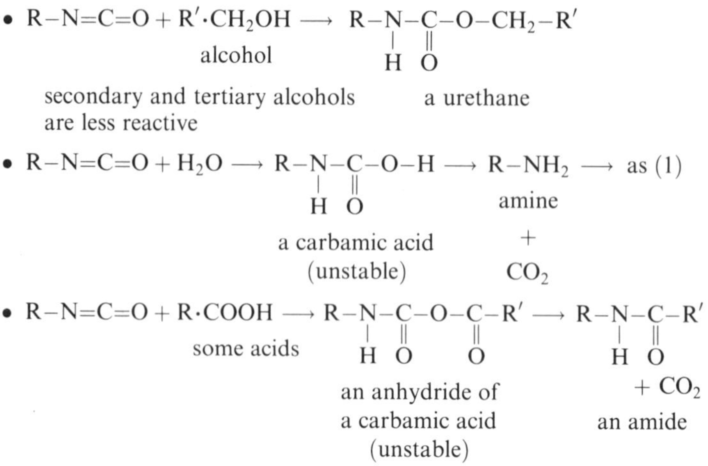

- $R-N=C=O + R' \cdot CH_2OH \longrightarrow$ R-N-C-O-CH$_2$-R'

  alcohol   H O

  secondary and tertiary alcohols   a urethane
  are less reactive

- $R-N=C=O + H_2O \longrightarrow$ R-N-C-O-H $\longrightarrow$ R-NH$_2$ $\longrightarrow$ as (1)

                       H O   amine

           a carbamic acid   +
            (unstable)   CO$_2$

- $R-N=C=O + R \cdot COOH \longrightarrow$ R-N-C-O-C-R' $\longrightarrow$ R-N-C-R'

        some acids   H O   O   H O

           an anhydride of   + CO$_2$
           a carbamic acid   an amide
            (unstable)

The first reaction proceeds the most readily and at the lowest temperature. Each subsequent reaction occurs a little less readily and at a slightly higher temperature, but all proceed at normal atmospheric temperatures. Further reactions can proceed with the 'active' hydrogens in the urea, urethane and amide products, but these are less important, except at paint stoving temperatures (100 °C and above). Catalysts, which speed reactions when added in small amounts, are frequently used for isocyanate reactions.

Five

Solid forms

polymers and solid structures

In this chapter we now come to the larger chemical and physical structures we shall encounter in paint. It will introduce the inorganic solids that occur in pigment and, most importantly, the large molecules, **polymers**, that are the most prominent and vital components of paint. This chapter is thus concerned with building large molecules, with the *arrangement* of groups in large molecules, and the arrangement of these large molecules in solids. We will include the polymerization processes and the characteristics of some basic polymers.

Crystalline solids

Pure, simple substances separate from solution on cooling as **crystals**. These crystals have a sharp melting point, which marks precisely the boundary between the liquid and solid states. They also have a definite and consistent shape. This shape is due to the pattern in which the atoms, ions or molecules are arranged. Thus crystals of common salt are always cubic. The arrangement of the Na^+ and Cl^- ion is in a cubic pattern (Fig. 5.1), which extends in all directions to fill and form the shape of a single crystal.

Other inorganic compounds have a variety of crystal shapes, in all of which the ions, atoms or molecules are arranged in a definite pattern, held together by ionic or covalent bonds, or by intermolecular attractions, which arise from the polarity (or electrical asymmetry) of the molecules. In diamond, for example, each carbon atom is covalently bonded to four neighbours (Fig. 5.2) so that the crystal is one 'giant molecule'. This in part accounts for its great strength and hardness.

Organic crystals contain more complex molecules arranged in patterns decided by the molecular shape and the intermolecular attractions. Thus urea has the tetragonal crystal shape (Fig. 5.3), with molecules arranged as shown in Fig. 5.4, each amino group being directed towards a carbonyl oxygen atom, so that $N-H\cdot\cdot O$ hydrogen bonding takes place.

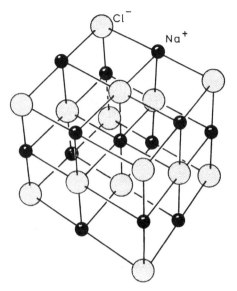

Fig. 5.1 Arrangement of ions in a crystal of common salt.

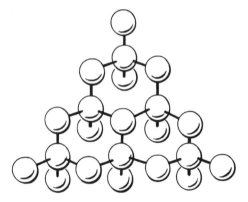

Fig. 5.2 The structure of a diamond crystal.

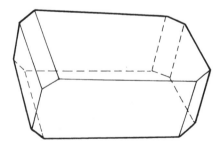

Fig. 5.3 Tetragonal shape of a urea crystal.

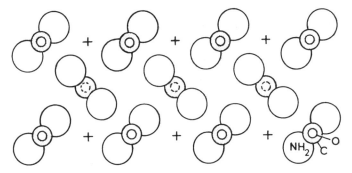

Fig. 5.4 Arrangement of molecules in a urea crystal.

Amorphous solids

In many solids the molecules are not arranged in any pattern at all. Instead they contain the random arrangement of molecules typical of a liquid, which appears to have become 'fixed' as a solid by cooling. Since there is no pattern of molecules, there is no typical shape to the solid, which is known as an **amorphous** (shapeless) **solid**.

The act of crystallization is a sudden process, which occurs at the precise temperature at which the energy of movement yields to the forces of attraction. In an amorphous solid, however, the molecules gradually become slower and more closely attracted to one another, until they are so closely packed that they completely impede one another and the substance will not flow. It is impossible to say *exactly* when this occurs, or even whether the substance has changed from flowing very slowly indeed to not flowing at all. There is no melting point and the boundary between the solid and the liquid state is blurred.

As the amorphous solid is warmed, the energy it receives is, at first, only sufficient to allow some movement of the less impeded portions of the large molecules that usually form these solids. Freedom of movement spreads to other parts of the molecules as heating continues until whole molecules can move. At this stage the substance is obviously in the liquid state.

The temperature at which the molecules begin to obtain partial freedom of movement is known as the **glass** (or second-order transition) **temperature**, normally represented as T_g. This can be determined by plotting a graph of, for example, the density of the substance against temperature. There is a sharp change in the *slope* of the plot at the glass temperature.

Various **softening temperatures** are often quoted for these substances. These correspond approximately to the state of obvious fluidity, but vary appreciably with the apparatus and method used for measurement, and especially the rate of temperature change used.

Examples of amorphous solids are glass, tar and the naturally occurring gums and resins. Such resins formed the basis of many early paints, because the distinctive property of an amorphous solid is that, when a melt is poured into a tray, or when a solution of the solid is poured upon a surface and the solvent is allowed to evaporate, the resultant solid forms a *continuous* film. Because it has no natural shape of its own, it takes the shape into which it has been cast. A crystalline material would give a discontinuous film, consisting of many hundreds of tiny individual crystals. It is the amorphous continuity that makes the resin film extremely suitable for protecting surfaces.

Polymers

Natural resins are complex mixtures of different substances with fairly large molecules. Nowadays many synthetic resins are used in paints. Synthetic resins have many properties that their natural predecessors did not, but they also consist of mixtures of different molecules, all of which are large.

Synthetic resins are **polymers**. 'Poly-mer' means 'many parts'. A polymer molecule is composed of many smaller parts, contributed by similar or dissimilar molecules, which are joined together until there are hundreds or thousands of atoms in the polymer molecule.

Chain growth polymers

Ethylene (ethene), $CH_2=CH_2$, is an example of a simple molecule that can be reacted to form a polymer or become **polymerized**. If ethylene gas is highly compressed (1500 atm), heated to 200 °C and allowed to come into contact with a carefully controlled trace of oxygen, the following occurs:

- Oxygen attacks a double bond, opening it. Temporarily a free radical has been created:

$$O_2 + CH_2=CH_2 \longrightarrow \underset{\substack{| \\ O \\ | \\ O.}}{CH_2-\dot{C}H_2} \quad (\text{or } R-CH_2\cdot)$$

- The highly reactive free radical attacks another double bond:

$$
\begin{array}{c}
CH_2=CH_2 \\
+ \\
\dot{C}H_2 \\
| \\
R
\end{array}
\quad \longrightarrow \quad
\begin{array}{c}
CH_2-\dot{C}H_2 \\
| \\
CH_2 \\
| \\
R
\end{array}
\quad \xrightarrow{+\,CH_2=CH_2} \quad
\begin{array}{c}
CH_2-\dot{C}H_2 \\
| \\
CH_2-CH_2 \\
| \\
CH_2 \\
| \\
R
\end{array}
\quad \text{and so on}
$$

A **chain reaction** has been set in motion and the $-CH_2-CH_2-$ unit will repeat itself a thousand times or more in forming a huge molecule of the polymer, polyethylene or polyethene ('polythene').

Polyethylene (polyethene) is not used in paints, because of its marked crystallinity (see below). However, many compounds containing the vinyl ($CH_2=CH-$) or vinylidene ($CH_2=C<$) group can be polymerized by a free radical route and many of the products are useful in paints, as we shall see, e.g.

$CH_2=CHCl$ PVC (polyvinyl chloride)

vinyl chloride (chloroethene)

$CH_2=CH-O-CO \cdot CH_3$ PVAc (polyvinyl acetate)

vinyl acetate (ethanoyloxy-ethene)

$$CH_2=\underset{\underset{\displaystyle }{|}}{\overset{\overset{\displaystyle CH_3}{|}}{C}}-CO-O-CH_3$$

PMMA (polymethyl methacrylate, 'Perspex')

methyl methacrylate (methyl 2-methyl-propenoate)

$CH_2=CH-CO-O-(CH_2)_3-CH_3$ PBA (polybutyl acrylate)

butyl acrylate (butyl propenoate)

Usually the **monomer** (as the starting compound is called) is liquid and can be polymerized in bulk or in solution. The **initiator** of polymerization is not oxygen, but some compound which, on heating, decomposes into free radicals, e.g. an organic peroxide (Table 5.1). The first reaction of free radical with a monomer molecule is terms the **initiation step**.

The most frequently encountered peroxides are compounds which decompose at temperatures between 80 and 180 °C. They all possess the link $-O-O-$ found in the simple inorganic compound hydrogen peroxide, but with one or both hydrogen atoms substituted by other groups as shown on page 66. Those peroxides with only one hydrogen substituted are called **hydroperoxides**, and they are particular important in the oxidative drying of oils (Chapter 12).

Other compounds used are **azo compounds** with a $-N=N-$ bond; azo-diisobutyronitrile (ADIB) is a common example. These compounds decompose as follows:

ADIB

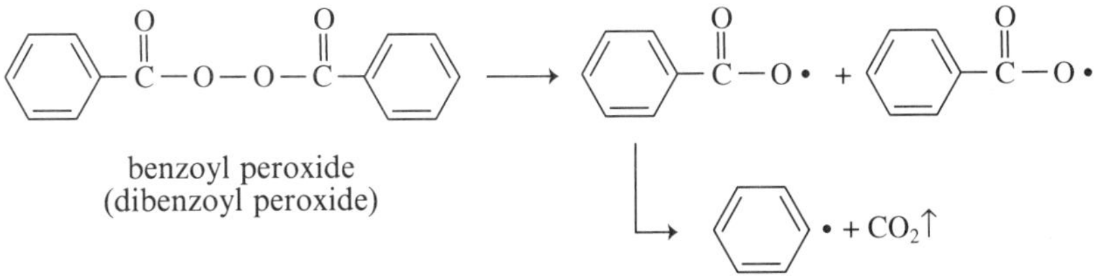

benzoyl peroxide
(dibenzoyl peroxide)

Table 5.1 Organic peroxides

Attached group(s)	Peroxide type	Comments	Example
1. 2R–C– acyl groups ($R-\overset{\displaystyle O}{\overset{\|}{C}}-$)	$R-\overset{O}{\overset{\|}{C}}-O-O-\overset{O}{\overset{\|}{C}}-R$ diacyl peroxide	R may be aliphatic or aromatic	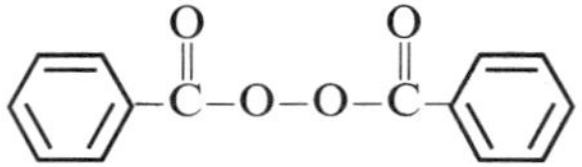 benzoyl peroxide (dibenzoyl peroxide)
2. $R-\overset{O}{\overset{\|}{C}}-$ and $-R'$ alkyl	$R-\overset{O}{\overset{\|}{C}}-O-O-R'$ peracid ester	ester of the peracid, $R-\overset{O}{\overset{\|}{C}}-O-O-H$ or aldehyde peroxide if R′ is H	tert-butyl perbenzoate $\overset{O}{\overset{\|}{C}}-O-O-C(CH_3)_3$
3. $\overset{R}{\underset{R'}{\diagdown}}C\diagup$ 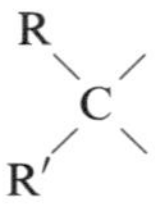	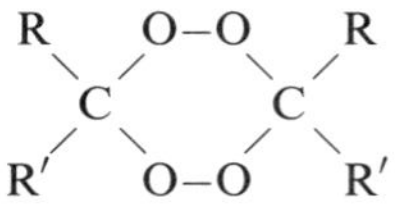 $\overset{R}{\underset{R'}{}}$ O–O $\overset{R}{}$ C C O–O $\overset{R'}{}$ ketone peroxide		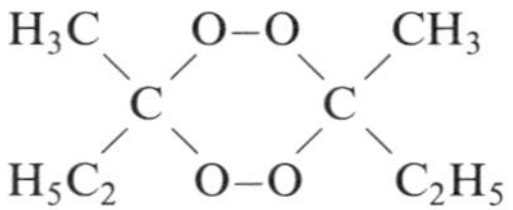 methyl ethyl ketone peroxide H_3C O–O CH_3 C C H_5C_2 O–O C_2H_5
4. R– and –H	R–O–O–H hydroperoxide		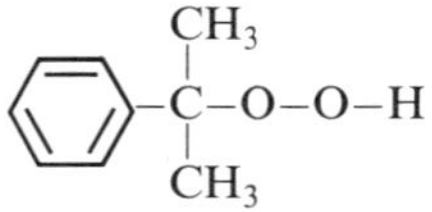 cumene hydroperoxide CH_3 $\overset{\|}{C}-O-O-H$ CH_3
5. 2R–	R–O–O–R dialkyl peroxide		$(CH_3)_3 \cdot C-O-O-C \cdot (CH_3)_3$ di tert-butyl peroxide

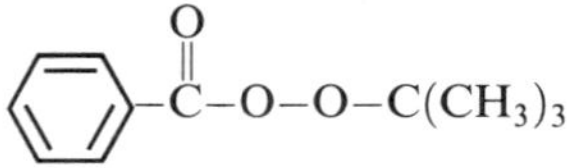

The sequence of growth steps, normally termed **propagation**, of the polymer chain is terminated by one of the following occurrences:

- **Combination**. An encounter between two polymer free radicals which satisfy one another's free valencies:

$$R \cdot + \cdot R' \longrightarrow R\!-\!R'$$

- **Disproportionation**. A similar encounter, in which one free radical removes a hydrogen atom from the other to become a saturated compound. The double radical rearranges to an unsaturated compound; neither free radical survives:

$$
\begin{array}{c}
R'\!-\!CH\!-\!CH_2\cdot \\
| \\
H \\
+ \dot{R}
\end{array}
\longrightarrow
\begin{array}{c}
R'\!-\!\dot{C}H\!-\!CH_2\cdot \\
+ \\
RH
\end{array}
\longrightarrow
\begin{array}{c}
R'\!-\!CH\!=\!CH_2 \\
+ \\
RH
\end{array}
$$

- **Chain transfer**. The polymer free radical is satisfied by the removal of a monovalent atom (usually hydrogen) from another molecule:

$$R \cdot + R'H \longrightarrow RH + R' \cdot$$

One polymer chain is ended, but another begins, as a result of the creation of a new free radical. Hence the chain reaction is 'transferred' to another molecule. The hydrogen atom may be removed from:

- another monomer molecule;
- a solvent molecule;
- another polymer molecule, in which case further polymer will grow from the transfer site resulting in a branched rather than a linear structure;
- a compound called a **chain transfer agent**, specially included in the preparation for this purpose.

A reactive material which may also be present but which hinders or prevents polymerization taking place is known as an **inhibitor**. Hydroquinone (benzene-1,4-diol),

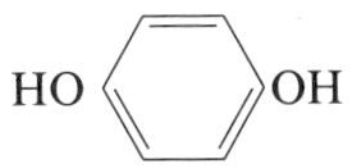

and other phenolic materials are effective in reacting with free radicals to destroy their free radical nature, e.g.

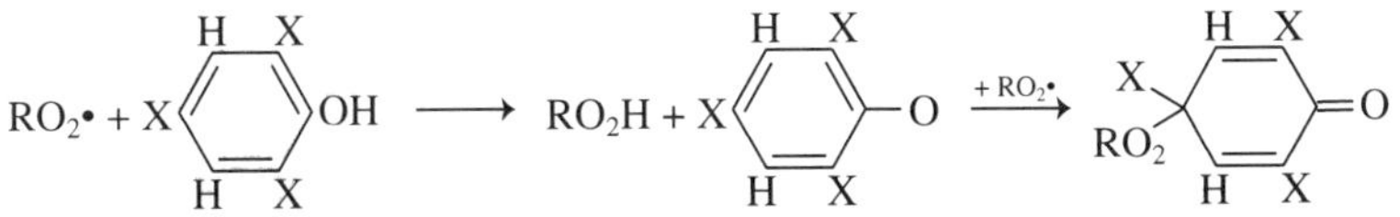

where X is an alkyl group and the inhibitor a trialkyl phenol.

While not useful in the polymerization stage, inhibitors have the important function of allowing safe storage of monomer when present in amounts of about 0.05%. They are important also in controlling the stability of unsaturated polyester compositions (Chapter 16), and most inhibitors function as anti-oxidants (Chapter 12) since oxidation processes are largely free radical processes.

It is already obvious that the molecules produced are large. They will also be different, even if only one monomer is used, because the chemist cannot control the size of each individual molecule. He or she must leave this to chance encounters between molecules in the reaction mixture. Any sample of polymer will contain a range of molecular sizes, some chains being terminated early and others later in their growth. The *average* size can be controlled by the amounts of initiator and chain transfer agent used (larger amounts of both giving smaller molecules) and by variation in the practical conditions under which the polymerization is performed.

The polymer molecular size is described by its **molecular weight**. If a hydro-gen atom is given a weight of one, carbon, which is approximately 12 times heavier, is 12, nitrogen 14, oxygen 16, and so on. The molecular weight is found by adding together the relative weights (called **atomic weights**) of all the atoms in the molecule. A polymer molecular weight might vary from about 1000 to over a million. By contrast, the ethylene (ethene) monomer, from which polyethylene is made, has a molecular weight of only 28.

Each of the polymers listed above is made from one monomer only. These polymers are called **homopolymers**. Any number of co-reactive vinyl or vinylidene monomers however can be polymerized together, or **copoly-merized**, and the product is called a **random copolymer**. In a copolymer molecule, the different monomer units are combined together in a semi-random way. As the polymer free radical grows, the next unit to be added is decided in part by chance: it can depend upon which monomer molecule encounters the right part of the chain first:

A + B ⟶ A.A.A.B.B.A.B.B.A.A.B.A.B.B.B.B. etc.

Individual molecules will differ in the proportions of the comonomer units, but the overall proportions will depend on the ratio of the monomers charged into the reaction vessel, if all the monomer is used up. Copolymers have properties somewhere between those of the homopolymers which could be made from the separate components. Thus polymethyl methacrylate is hard, relatively brittle and insoluble in petrol. Polybutyl methacrylate is softer, more flexible and soluble in petrol. A copolymer of both, yet predominantly methyl methacrylate, could remain insoluble in petrol (though softened by it) and yet be more flexible than polymethyl methacrylate.

Sometimes it is possible, by carrying out the polymerization in stages, to arrange the monomer units in runs or blocks, e.g.

...A.A.A.A.A.A.A.B.B.B.B.B.B... etc.

These more orderly copolymers are known as **block copolymers**. This is not possible with straightforward free radical-initiated polymerization, but requires special techniques which are beyond the scope of this book. An A–B block has molecules with a chain of type A monomer units joined at one end to a chain of type B units. A–B–A blocks have only type A chains at the ends of the molecules. A–B–A block copolymers can be used as associative thickeners (Chapter 10).

Step growth polymers

The polymerizations described so far have all been of one type: **chain growth polymerization**, where polymerization occurs through a chain reaction. The polymer molecule grows by the addition of another monomer unit without the loss of any atoms in the process. This type of polymerization has also been referred to as **addition polymerization**.

The alternative type of polymerization is known as **step growth polymerization**. Here each growth reaction is often, but not necessarily, accompanied by the production of a small molecule (usually water) which is excluded from the polymer chain as each link is formed, e.g.

$$HO \cdot OC \cdot R \cdot CO \cdot OH + HO \cdot R' \cdot OH \longrightarrow HO \cdot OC \cdot R \cdot CO \cdot O \cdot R' \cdot OH + H_2O$$

dibasic acid   dihydric alcohol

The esterification of one group from each molecule produces water and the product on the right, containing a carboxyl group at one end and a hydroxyl group at the other, is the repeating unit (or monomer) for the polymerization. This proceeds as follows:

$$HO \cdot R' \cdot OH + HO \cdot OC \cdot R \cdot CO \cdot O \cdot R' \cdot OH + HO \cdot OC \cdot R \cdot CO \cdot OH \longrightarrow$$

$$HO \cdot R' \cdot O \cdot OC \cdot R \cdot CO \cdot O \cdot R' \cdot O \cdot OC \cdot R \cdot CO \cdot OH + 2H_2O, \text{ and so on}$$

Each reaction produces another ester group and the resultant polymer is a polyester resin. This reaction with elimination of a small molecule is termed a condensation reaction, hence this type of polymerization has also been termed **condensation polymerization**.

Esterification need not be the reaction concerned; polyamides, epoxy resins, amino resins and others are produced by step growth polymerization. It is worth mentioning that, in these polymerizations, the by-product is almost always unwanted and must be removed before the resin can be used. Water, for example, is immiscible with polyester resins and can be removed by distillation during the polymerization. Removal of the by-product accelerates the polymerization, and is essential to moving the reaction along.

Chain and step growth polymerizations take very different courses. In chain polymerizations the highly reactive nature of a free radical means that, once one is formed, it will grow by chain reaction to termination in a matter of seconds. the polymerization proceeds by the addition of single small monomer molecules to a large growing chain. Completed polymer chains and unreacted monomer exist together until all monomer is used up and polymerization is 100%. The total time for polymerization, which may be a matter of hours, depends upon the rate of formation of new free radicals and the yield of polymer required, though the lifetime of radical chain ends may only be a few seconds. Under the right conditions, very high molecular weights (greater than 100 000) are easily obtained.

Step growth polymerizations, on the other hand, depend on more ordinary reactions between reactive organic groups. There is no chain reaction. All the original monomer molecules soon double their size and further growth depends upon the joining together of larger and still larger units. The concentration of reactive groups falls with each reaction and the viscosity rises so that, towards the end of the polymerization, the rate of reaction is very slow indeed, since the right groups have difficulty in finding one another. Consequently, although the yield of low molecular weight polymer is high after an hour or two and the monomers have disappeared, the attainment of molecular weights as high as 20 000 is very difficult and, even if it can be done, it involves very long reaction times. The total time for a step growth polymerization therefore depends upon the molecular weight required and not upon the yield.

Functionality

It should be noted that a molecule with the ability to react with only one other molecule cannot produce a polymer. Let us assume that X and Y are reactive groups and that X will react with Y, but two Xs and two Ys will not combine. Then

$$R-X + Y-R' \longrightarrow R-XY-R'$$

and no further reaction is possible since X and Y are no longer reactive. Again, if only one reactant has more than one reactive group,

$$R-X + Y-R'-Y + X-R \longrightarrow R-XY-R'-YX-R$$

reaction stops without polymer being formed.

The useful monomer must have the ability to react with a minimum of two other molecules. In step growth polymerization, this means two reactive groups per molecule; in chain polymerization, one double bond has the ability to produce two reactive free valencies:

$$CH_2=CH_2 \longrightarrow -CH_2-CH_2-$$

The number of molecules with which a monomer molecule may combine in the polymerization reaction concerned is known as its **functionality**. If the monomers present have a functionality of two, a linear polymer is

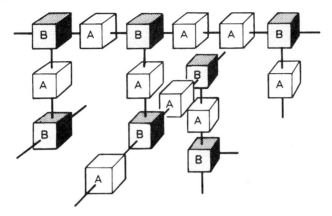

Fig. 5.5 A cross-linked polymer network.

produced:

$$-A-A-A-A-A\ldots, \text{etc.} \quad (-A- \text{ has a functionality of two})$$

If a small amount of a monomer with a functionality of *three or more* is included, branching can occur:

$$-A-A-A-A-A-B\overset{\diagup A-A-A-}{\underset{\diagdown A-A-}{}} \quad (-B\overset{\diagup}{\diagdown} \text{ has a functionality of three})$$

If a large amount of trifunctional monomer is present, a network in three dimensions is formed (Fig. 5.5). The polymer is then said to be **cross-linked**.

The effect of functionality is summarized in Table 5.2. The importance of this lies in the different properties that may be expected from the types of polymer produced. Let us consider a piece of solid polymer again.

If the polymer is linear, the long chains will be randomly packed, some coiled, some extended and several entangled with one another. Remember that, since rotation can occur about any single bond, a polymer chain is

Table 5.2 The effect of functionality of monomers

Functionality of monomer	Type of product
One	Simple compound
Two	Linear polymer
Three (*minor ingredient*, when principal monomer is difunctional)	Branched polymer
Three (*major ingredient*)	Cross-linked polymer

generally extremely flexible. The atoms will not all lie in a straight line and the chain may be fully extended (Fig. 3.4a), partially coiled, fully coiled (Fig. 3.4b) or even, possibly, knotted! However, heat sets the molecules moving and small liquid molecules may penetrate between the chains. Both help to sort out the tangle, i.e. the polymer will melt or dissolve. The same is true for lightly branched polymers of low molecular weight.

In a cross-linked polymer, every chain is probably connected to at least three others, so that the solid lump is one large molecule. To separate the chains we have to *break* a number of covalent bonds. Thus a cross-linked polymer is at first softened by heat, as the segments between cross-links get a limited freedom of movement, but the only further stage possible is not melting: it is decomposition.

Again, while liquid molecules can penetrate into the holes in the cage-like structure, causing it to swell, they cannot dissolve the polymer. Since dissolution implies separating the molecules from one another, this is clearly impossible. There is only one molecule.

Linear polymers, because of their ability to melt and solidify as many times as may be required, are spoken of as **thermoplastic** polymers. Linear polymers which subsequently cross-link on heating, or as a result of further reaction which forms cross-linked networks, are described as **thermosetting** polymers.

Crystallinity in polymers

Up to this point, no firm statement has been made about the nature of linear synthetic polymers as solids. Their resemblance to the amorphous natural resins has been mentioned and, in the previous section, reference to random packing has inferred an amorphous nature. At the same time, the first synthetic linear polymer described, polyethylene or polyethene, was said to be 'markedly crystalline'.

Since crystallinity requires the arrangement of atoms and molecules in a three-dimensional ordered pattern, and since polymers contain large, complex, flexible molecules, it may be difficult at first to see how polymers can be crystalline. Crystallinity in a polymer involves the parallel arrangement of lengthy segments of uncoiled uniform polymer chains on a three-dimensional scale, but it does not necessarily involve complete chain lengths. A crystalline area in a polymer can be likened to a bundle of rods. The following conditions are required for such an alignment to occur in a polymer.

• The lengths of chain concerned should consist of regular repeating units arranged head-to-tail and should contain no bulky side groups, which could prevent the close parallel packing of the extended chains.

- There should be sufficient intermolecular forces of attraction – resulting from polarity – to hold the chains together in the crystalline pattern, when the chains align as closely as their side groups will let them.

These conditions are quite exacting, so that it is not surprising that even polymers whose molecular structures appear to be ideal for crystallinity should have both crystalline and non-crystalline areas in the solid form. These latter areas are due to the presence of coiled and tangled molecules, which could not uncoil in response to intermolecular forces during the solidification process. Equally, less suitable polymers, forming largely amorphous solids, can show areas of crystallinity where, during the solidi-fication process, chance has decreed that chain segments of the right type should come together. Thus no polymer is wholly crystalline and few are completely amorphous.

Marked crystallinity introduces a number of objectionable features into a polymer, from a paint point of view:

- lack of transparency, due to refractive index differences between region of the polymer solid (Chapter 6);
- sharp, and often high, softening temperatures. Gradual softening over a range of temperature is advantageous in stoving paints. It will permit minor irregularities to flow out during the stoving process, without causing 'sags'. The glass temperature of a polymer used in an emulsion paint should be lower than the paint's drying temperature (Chapter 11);
- insolubility in all but the most polar liquids, which have sufficient attrac-tion for the polymer molecules to separate them. In some cases, not even these are effective.

Thus for paint purposes, polymers which are largely amorphous are required. The tendency for a polymer to crystallize can be reduced and the softening temperature lowered by:

- making the arrangement of repeating units an irregular one, e.g. by copolymerization;
- introducing bulky side groups;
- spacing out the polar groups that provide the strong intermolecular forces;
- introducing a plasticizing liquid of high boiling point, which reduces intermolecular attractions (being itself attracted by these forces) and increases the spacing between polymer chains.

Some polymer structures will now be considered for their effects on crystal-linity and general properties; their usefulness in coatings is also indicated.

Polyacrylonitrile

Polyacrylonitrile or polypropene nitrile has the ideal structure for crystallinity: uniform chain; relatively small, highly polar side groups. The intermolecular attractions are reinforced by hydrogen bonding:

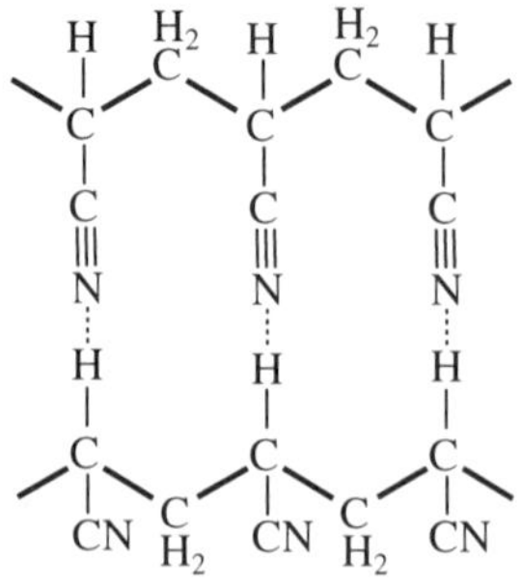

This polymer has high softening temperatures and is insoluble in all but a handful of highly polar liquids with marked hydrogen bonding capacity [e.g. $H \cdot CO \cdot N(CH_3)_2$, dimethyl formamide]. Both properties are the direct result of the extreme difficulty of separating the chains and breaking the crystal structure. The polymer is unsuitable for paints, but has ideal properties for a synthetic fibre ('Acrilan'). As a comonomer, however, acrylonitrile imparts improved toughness and solvent resistance.

Polyethylene

Polyethylene or polyethene is a non-polar polymer with weak intermolecular forces, but the simplicity of its structure

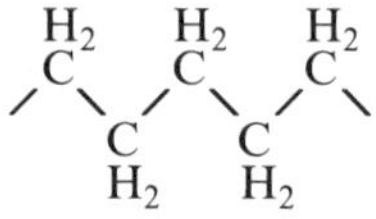

allows very close packing of the chains. The polymer is therefore markedly crystalline but, since relatively little energy is required to separate the molecules, it has a low softening temperature (105–115 °C). It is completely insoluble at room temperature, though swollen by hydrocarbons. As a homopolymer, it is therefore unsuitable for paints and fibres, but is a very useful plastic. It is only found in coatings as a comonomer in emulsion polymers with vinyl acetate, where it is the plasticizing component.

Polyvinyl chloride

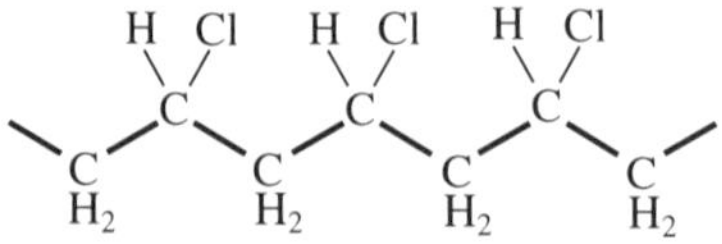

This polymer is similar to acrylonitrile, in that it contains small polar side groups on a regular chain. It is therefore markedly crystalline, with a high softening temperature and solubility in few solvents. It has, however, found limited use in paint in two ways.

In the first, the crystalline polymer is used in particulate form and the particles are dispersed in a liquid – solvent or plasticizer – capable of penetrating the particles at higher temperatures ($c. 170 °C$) and integrating them into a film. Crystallinity in the film is reduced and flexibility increased by the plasticizer, or by plasticizing resin dissolved in the solvent. Such mixtures are called **organosols** or **plastisols** (Chapter 11).

The second method is to copolymerize vinyl chloride with about 10–20% vinyl acetate. The copolymers are soluble in ketones and esters. A characteristic in coating formulations is the toughness vinyl chloride imparts.

Polyvinyl acetate

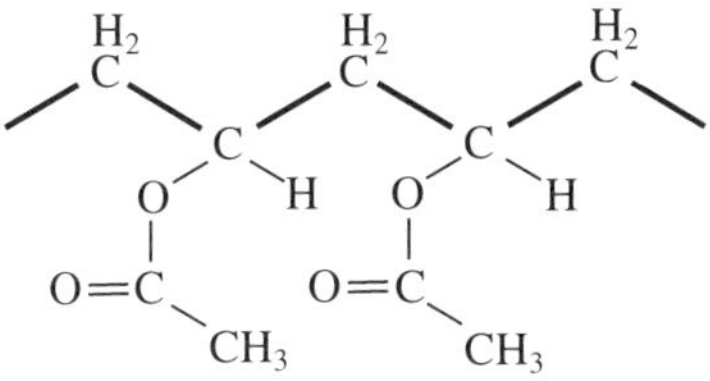

Although the side group of this polymer has some polarity, the bulk and flexibility of the group lead to irregularity and wide spacing of chains. The polymer is amorphous, readily dissolved and highly suitable for paints. The softening temperature varies from about $70 °C$, for low molecular weight samples, to $125 °C$ for higher molecular weight materials. This variation with molecular weight in ease of softening (and ease of solution) is common to all linear amorphous polymers. Higher molecular weights mean more polar groups per molecule, a greater number of points of attraction between molecules and an increased opportunity for tangling. Thus more needs to be done to separate the molecules. Vinyl acetate copolymers are encountered mostly in polymer emulsions for emulsion paint. A chemical weakness is the tendency for the side chain acetate groups to hydrolyse, and thus these copolymers are less durable, and less suitable where moisture is frequently encountered.

Polystyrene

This polymer also has a bulky side group, and one of relatively low polarity:

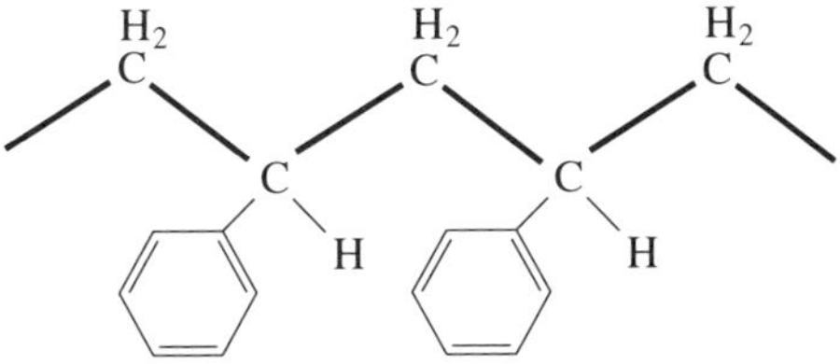

Table 5.3 Glass transition temperature and softening range of polymers

Polymer	Glass transition temperature (°C)	Softening range (°C)
Polyethylene (polyethene)	−125	105–115
Polybutyl acrylate	−56	35–50
Polyvinyl acetate	30	70–125
Polycaprolactam (Nylon 6)	40	215–235
Polyethylene terephthalate (Terylene)	70	260–280
Polyvinyl chloride	81	110–210
Polyacrylonitrile (polypropenenitrile)	97	220–250
Polystyrene	100	180–225
Polymethyl methacrylate	105	150–180

The polymer is amorphous and soluble. In this polymer, however, the side groups are rigid and they restrict considerably the bending and coiling of the polymer chains. Except at high molecular weights (unsuitable for paints), polystyrene is therefore weak and brittle, and styrene copolymers (sometimes styrene is copolymerized with drying oils and alkyds) are more suitable as paint resins.

This illustrates a general principle: rigid groups – in the polymer backbone, or attached to it – make a polymer inflexible.

Step growth polymers are governed by the same rules. In many of these polymers, chains are less regular, but more polar. Two polymer structures which lead to such high crystallinity that the polymers have been used for fibres are the polyamide structure of 'Nylon', and the polyester structure of 'Terylene'. The latter is a polymer derived from *ter*ephthalic acid and eth*ylene* glycol from which the name derives. Bulk nylon is a waxy solid of such toughness that a major use is in gear wheels in small machinery.

In Table 5.3 the glass transition temperatures and the softening temperatures of all the homopolymers referred to in this chapter are given. The glass transition temperature is a good guide to the hardness or softness of the polymer. It has particular importance to the coalescence of emulsion polymers, as we shall see in Chapter 11.

Six

Colour

physics and chemistry

In this chapter we describe the physical and chemical background to colour. The colour of paint depends on the physical and chemical nature of paint and on the properties of light. Both aspects will be covered in sufficient detail to aid understanding of later chapters.

Light

Light is a form of energy. Our chief source of light is the sun, a vast ball of matter, kept at a fierce heat by the energy produced in a continuous series of chemical and nuclear reactions proceeding in it. The energy released in these reactions is also felt by us as heat. Heat and light are produced in the same process; they are different forms of energy.

In fact heat and light are part of a large series of radiations, all different forms of energy. Energy is radiated in waves or pulses, alternating between maximum and zero intensity – the crests and troughs of the waves. These radiations all move at the same high speed: 300 000 km or 186 000 miles per second, when in air or vacuum. They differ, however, in **frequency**. The frequency is the *number* of waves that move past a fixed point in 1 second, and is measured in hertz.

Since all the radiations move at the same speed, the more waves that move past the point per second, the shorter they must be, i.e. the **wavelengths** must vary. The wavelength is the distance from crest to crest or trough to trough. These facts are summarized in Figure 6.1.

The radiations referred to above are known as **electromagnetic radiations**. Their wavelengths vary as follows:

Less than 0.01 nm	Gamma rays
0.001–10 nm	X-rays
10–400 nm	Ultraviolet light
400–750 nm	Visible light
750 nm–1 mm	Infrared ray (radiant heat)
Above 1 mm	Microwaves, radio waves

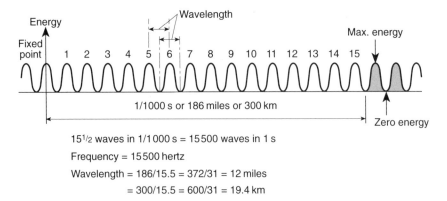

Fig. 6.1 Electromagnetic radiation: energy, frequency and wavelength.

It is difficult to imagine all these different forms of radiation as being of the same type and differing only in wavelength and frequency, because they are recognized in completely different ways. Visible light seems different and unique to us, because it is the only part of the series which affects the eyes in such a way that they transmit signals to the brain, which the brain in turn interprets as a picture. (Note that only visible light occupies a precise range in the spectrum. All of the other forms of radiation are defined by source or usage and their defined regions may vary or overlap.)

Reflection

That light travels in straight lines is evident from the fact that a beam can be produced if light is passed through a slit or hole, and from the well known explanation of eclipses of the sun and moon. If a beam of light shines onto a surface, it will be wholly or partly reflected. Light meeting a smooth surface at a certain angle will be reflected at a similar angle. This is usually expressed by saying that the **angle of incidence** is equal to the **angle of reflection** (Fig. 6.2).

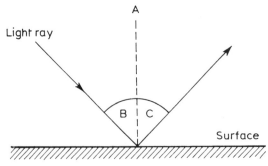

Fig. 6.2 Reflection: angles of incidence and reflection. A, the normal (at right angles to the surface); B, angle of incidence; C, angle of reflection.

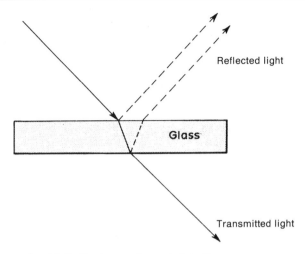

Fig. 6.3 Reflection at glass–air interfaces.

By 'surface', we mean a boundary between two different types of substance. For example, light will be reflected from the bottom face, as well as the top face, of a plate of glass, since both are glass–air boundaries (Fig. 6.3).

One of the results of this law of reflection is that it explains the 'gloss' on a paint film. A perfectly smooth paint film is like the surface in Fig. 6.2: light from an object is reflected uniformly to the eye of the observer. Since the eyes receive a complete picture of the object, the observer sees it in sharp outline and the film is hence said to be glossy. If the surface is not smooth (Fig. 6.4), parts of the object are seen clearly by reflection, but parts are 'lost' since the light is reflected elsewhere and not to the eye. The overall result is a blurred picture and the observer decides that the gloss is low.

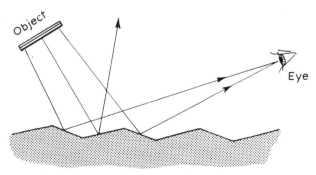

Fig. 6.4 Reflection at a surface of low gloss.

Refraction

When light meets a surface, some will be reflected, but the remainder will pass below the surface where either it is absorbed, and the object is said to be opaque, or it passes through the object, which is said to be transparent or translucent. Now the speed of light varies with the medium through which it is passing and its speed through a transparent object will certainly be slower than its speed in air. The effect of this is that the path of the light is 'bent' at each boundary between the object and the air (Fig. 6.3). The 'bending' process is called **refraction**. For example, an object at the bottom of a bowl of water always appears closer than it really is. In Fig. 6.5 light passes from object A at the bottom of the water to the surface at C, where it is refracted and then enters the eye. The eye and brain assume that the light has travelled in a straight line and therefore 'see' the object at B.

The amount of refraction at a boundary is expressed by the **refractive index**:

$$\text{refractive index} = \frac{\text{sine} \angle X}{\text{sine} \angle Y} \text{ (Fig. 6.5)} = \frac{\text{speed of light in air}}{\text{speed of light in other medium}}$$

The refractive index of water is 1.33. Rutile titanium dioxide has a higher refractive index than any other transparent material: 2.76. We shall discuss its use as a white pigment in Chapter 8.

Most surfaces involved in reflection or refraction are not flat. The behaviour of light at a curved boundary can be assessed by drawing a tangent to the surface, at the point at which the light strikes it. If it is now imagined that this tangent (or tangential plane) is the surface, the imaginary surface is

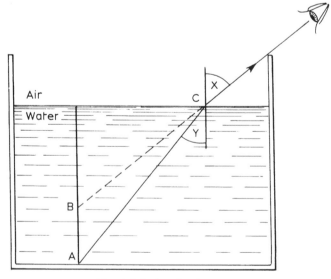

Fig. 6.5 Refraction in water.

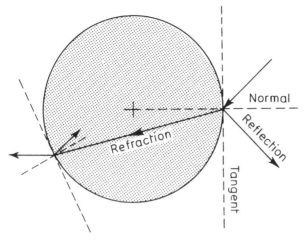

Fig. 6.6 Reflection and refraction at curved surfaces.

thus a flat one and all the laws of reflection and refraction for level surfaces can be applied (Fig. 6.6).

Diffraction

You may have observed that if water waves meet an obstacle of some kind, the waves that pass by the obstacle have the ability to pass around it to some extent, disturbing the still water in the lee of the object. In other words, waves can get around corners. Sound waves can get around the corners of buildings in city streets. The phenomenon is called **diffraction**.

The extent of diffraction depends on the wavelength. In considering the reception of radio waves, the position is slightly complicated by the possibility of reflecting these waves off layers in the upper atmosphere, specifically the ionized Heaviside and Appleton layers, or retransmitting them from satellites. In the daytime, reflection of wavelengths below 10 000 m is inefficient and the 'groundwaves' (those travelling from place to place by the most direct route) are more important. Foreign stations are easily heard on medium and long wavebands (200–2000 m), though one might expect the earth's curvature, hills, etc., to prevent the passage of the waves. Clearly diffraction occurs. Short wave (10–100 m) radio communication is limited to shorter distances. FM radio signals (about 3 m) are affected by hills, though sometimes weak reception is possible on the 'shadow' side of the hill. With UHF television signals (about 0.5 m), hills are more effective in completely preventing reception.

Thus diffraction is not important when the size of the obstacle is much greater than the wavelength and is most important when the size of the obstacle is similar to the wavelength.

Light has such short wavelengths (less than 1 μm) that diffraction is not normally noticed. It is, however, impossible to obtain a shadow with absolutely sharp edges, even with a point source of light. Light, therefore, can pass around obstacles: it can suffer diffraction. A phenomenon produced by the diffraction of light is the halo sometimes seen *closely* surrounding the sun or moon. This is due to diffraction of light by small droplets of water in thin layers of cloud.

These facts are very relevant to paint films. They can explain, for example, why a solution of two transparent resins of different refractive index can dry to give a milky film, if the resins are *incompatible*. Resins that are incompatible are simply resins that will not mix. The solvents help them to mix in solution, but as the liquids evaporate, the molecules of the two types of resin tend to segregate in separate groups. Complete separation into two layers does not occur, because the viscosity is, by this time, too high. Instead, we have a separation into droplets of one resin inside a film of the other.

In this resin film there are hundreds of interfaces or boundaries between the two resins. If the resins differ in refractive index, refraction will occur at every interface. Figure 6.7 shows how light can be returned from a film by successive refractions. Reflection will also occur at each interface and by these means some of the light will be returned towards the surface of the film. It is often forgotten that diffraction will also occur at the edges of the droplets. The net effect of multiple diffractions is extremely complex: it is sufficient to say that diffraction can play an important part in turning the light back towards the film surface.

The overall result is that white light is returned from the film to the eyes of the observer, who concludes that the film is white. The greater the difference in refractive indices and the larger the number of droplets or particles up to a limiting concentration, the more milky and less transparent

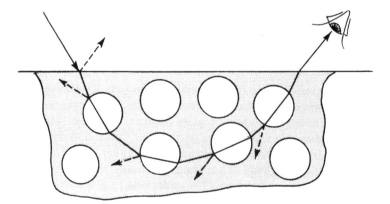

Fig. 6.7 Light returned from a 'white' film.

the film will appear. Solid materials that are transparent in bulk also appear white in powder form, because white light is reflected, refracted and diffracted at the many solid–air interfaces. They will also appear milky or white if dispersed into a clear resin film of different refractive index.

Colour

If two rays of visible light differ slightly in wavelength, they will have different effects on the eye, i.e. they will appear to be different in *colour*. There is a *gradual* change in colour from violet at about 400 nm, through blue, green, yellow and orange to red at about 620 nm. Light at any unit wavelength is almost imperceptibly different from light at the next unit wavelength and one colour merges into the next, so that the number of different colours cannot be counted, and we do not have names for all of these colours. All these wavelengths are present in sunlight as a mixture and the eye registers this mixture of all wavelengths as the colour we call white.

Since the refractive index of a transparent substance varies with the wavelength of light, it is possible to separate the wavelengths. When light passes into glass, each wavelength is bent by a different amount. As the light rays pass out of a glass plate, they are bent back and converge (Fig. 6.3), but if the two sides of the glass are not parallel, as in a prism, the rays diverge (Fig. 6.8). The range of colour produced is called a **spectrum**.

When a coloured object is illuminated by white light, it selectively absorbs some of the wavelengths and transmits others. The eye receives the transmitted wavelengths and summarizes them as one colour. This colour is dominated by the main wavelengths transmitted, which determine its basic **hue**. This hue is modified by undertones of the other wavelengths, which make it an individual shade. Thus there are many red shades; some are bright, some dark; some have

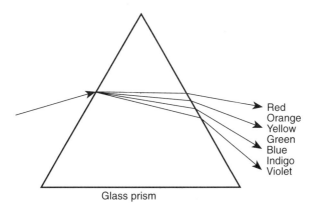

Fig. 6.8 Prism and spectrum.

a yellow undertone (e.g. scarlet) and some a blue one (e.g. crimson). A truly black object absorbs all the wavelengths and a truly white one, none of them.

Colour chemistry

It is interesting to see if there is any connection between the chemical structures and the colours of compounds. Among organic compounds, we find that coloured chemicals usually contain certain groups which are called **chromophores** (colour bearers). These are:

$-N{=}N-$

azo group

$-N\stackrel{\nearrow O}{\searrow O}$

nitro group

$-CH{=}N-$

azomethine group

$\diagdown C{=}O$

carbonyl group

$-\underset{O}{\overset{}{N}}{=}N-$

azoxy group

$\diagdown C{=}S$

thio group

$-N{=}O$

nitroso group

$\diagdown C{=}C \diagup$

ethenyl group

They are usually found attached to aromatic rings, such as benzene, or one of the more complex ring structures, such as

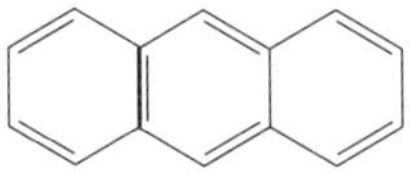

anthracene

Coloured molecules have their colour intensified or modified by certain other groups, known as **auxochromes**:

$-N\stackrel{R}{\diagdown R}$

tertiary
amino group

$-N\stackrel{H}{\diagdown R}$

secondary
amino group

$-NH_2$

primary
amino group

$-OH$

hydroxyl
group

$-OCH_3$

methoxy
group

$-I$ $-Br$ $-Cl$

iodo   bromo   chloro groups

Thus the dye alizarin (red) contains two benzene rings, two keto or carbonyl groups (chromophores) and two hydroxyl groups (auxochromes):

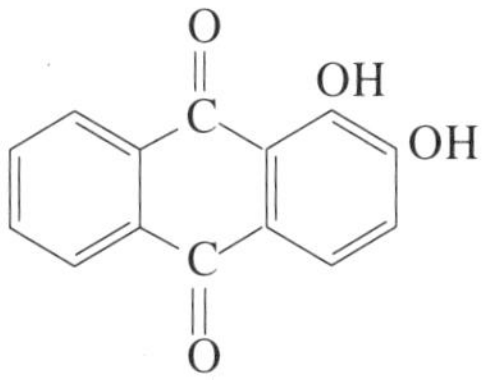

alizarin

It should be noted that all the groups that influence colour contain electronegative atoms, or double bonds, or aromatic rings. All of these attract higher than average densities of electrons, indicating that colour is associated with compounds which have high localized electron distributions. The energy from electromagnetic radiations is, in fact, absorbed by the electrons of a molecule. Colourless compounds absorb only in regions outside the visible spectrum.

Colour in inorganic compounds is again connected with the behaviour of electrons. The elements in the block between groups II and III in the Periodic Table (Fig. 1.1), known as the **transition elements**, contain special electrons which, under the influence of ligands, can absorb energy in the visible region of the spectrum. Ligands are attached groups, ions or molecules, such as the sulphate ion and water molecules in hydrated copper sulphate, $CuSO_4 \cdot 5H_2O$. The exact region in which the absorption takes place is determined by the nature and positioning of the ligands. Thus $CuSO_4 \cdot 5H_2O$ is pale blue, but $CuCl_2 \cdot 2H_2O$ is green.

In other organic compounds the amount of energy that can be absorbed is sometimes sufficient to promote the transfer of a valency electron from one atom or ion to another, e.g. absorption causing

$$Pb^{2+}I_2^- \longrightarrow Pb^+I^- + I$$

is responsible for the golden colour of lead iodide, PbI_2. This type of **charge transfer** is responsible for the intense colour of some inorganic compounds especially when no transition element is present, and it occurs particularly in oxides, sulphides, iodides and bromides.

Colour mixing

Sometimes the eye observes colour which is not to be found anywhere in the spectrum, e.g. grey, brown or purple. Since any colour is the sensation stimulated by a mixture of wavelengths, it will be worthwhile to see what rules govern such mixtures and how new shades are produced. The first point to note is that very different results are obtained by mixing coloured lights on the one hand and coloured pigments on the other. Let us examine these two types of mixing in more details.

Additive mixing

First let us examine the mixing of coloured lights. Suppose we take a white screen into a darkened room and shine coloured lights onto it. Since the screen is white in daylight, it therefore reflects all wavelengths. Thus a yellow beam will give a yellow spot on the screen and a blue beam, shining on another part of the screen, will give a blue spot. If both lights are then focused onto the same part of the screen, we see *white*.

The yellow and blue lights are **complementary** to one another and their mixture in the right proportions is registered by the eye as white. Other complementary pairs can be seen by drawing straight lines through the centre spot of Fig. 6.9 (e.g. red and cyan).

Red, blue and green are called the three **primary** colours. (The blue is in fact a blue-violet.) Secondary colours are produced along the sides of the triangle, by mixing any two primary colours in different proportions, e.g. mixtures of blue and red light vary from purple through magenta to rose-pink as the proportion of red increases. These are not pure colours, and do not occur in the spectrum. Other colours can be made by mixing across the triangle.

Remember that light is energy and the white screen reflects all the light. Thus if two lights of different colour shine onto the same part of the screen, both are reflected and the *amount* of energy reaching the eye is more than would reach it from one light only. Consequently the colour seems brighter. Because the lights are added together to give more light energy and brighter colour, this type of colour mixing is called **additive mixing**.

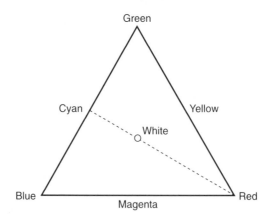

Fig. 6.9 Additive colour mixing diagram.

Subtractive mixing

If we mix a blue pigment and a yellow pigment we do not get white, we get *green*. This is because both pigments **absorb** light. The blue pigment absorbs most of the red, all of the orange and most of the yellow wavelengths from the white light falling on it. The yellow pigment absorbs the remainder of the red, nearly all the blue and almost all the indigo/violet. Looking at the list of the six principal colours of the spectrum, given in the Colour section above, we see that green is the only colour not absorbed by either pigment.

We are assuming that both pigments are extremely fine in particle size and the particles of both are uniformly distributed over the area covered by the mixture. Light falling on the area therefore passes through a double 'filter' and the only wavelengths that can be returned to the eye are those that are absorbed by neither pigment. Note that if both pigments are dull (i.e. absorb strongly), the green will be even duller, because less light energy will reach the eye.

When mixing pigments, blue and yellow are not complementary, but there are complementary colours, shown opposite one another in Figure 6.10, e.g. red and green. Mixtures of complementary colours in the right proportions do not produce white; instead, dark *grey* is produced. They cannot produce white because they do not reflect *enough* light energy. White and black are two extremes: complete reflection and none, maximum light energy reflected back and nil. In between come all the shades of grey, getting darker and

Fig. 6.10 Subtractive colour mixing diagram.

darker as less light is reflected. A neutral grey reflects all the wavelengths of white light, but it does not reflect *enough* light to give the full white colour.

In the pigment circle (Fig. 6.10), a number of pigments are listed. The pigment colours do not correspond to pure colours, but lie between them. If two pigments that are not complementary are mixed, the colour produced will lie between them, on the shorter arc of circle joining them, e.g. red and blue can give violet. If the two pigments are joined by a straight line, the closer that line lies to the centre of the circle, the darker the colour will be.

Because each individual pigment in a mixture subtracts some of the wavelengths and some of the light energy from the light falling on the mixture, this type of colour mixing is called **subtractive mixing**. Strictly speaking, pure subtractive mixing occurs only with dyes, which absorb light, but cannot reflect it. Pigment particles reflect some light from their surfaces. This is why our mixing of complementary colours produced a dark grey and not black. It is the type of mixing that the paint maker must carry out to make a paint with a given colour (Chapter 8).

PART TWO

Applied science

Seven

Paint

first principles

In this chapter, we describe the basic nature of paint, making clear its function, and describe the various routes to attaining the final properties we require.

Wet paint: paint in the can

All objects are most vulnerable at their surfaces. It is the surface of any article that makes continual contact with the air, which may be moist, corroding or oxidizing. The surfaces of objects left in the open bear the brunt of the sun, rain, fog, dew, ice and snow. Under these conditions iron rusts, wood rots (or shrinks and cracks) and road surfaces crack and disintegrate. These, and more sheltered objects, suffer the wear of daily use, scratches, dents and abrasions – at their surfaces. To prevent or to minimize damage, various coatings are applied to these surfaces to protect them. Coatings can also be used to decorate the articles, to add colour and lustre and to smooth out any roughness or irregularities caused by the manufacturing process. Thus the function of any surface coating is twofold: to protect and to decorate.

There are many surface coatings that do this: wallpaper, plastic sheet, chrome and silver plating. No coating material is more versatile than paint, which can be applied to any surface, however awkward its shape or size, by one process or another. Paint is a loosely used word covering a whole variety of materials, with names sometimes more descriptive of their composition or function: enamels, lacquers, varnishes, undercoats, surfacers, primers, sealers, fillers, stoppers and many others. It is essential to grasp at once that these and other less obviously related products, such as plasters, concrete, tars and adhesives, are all formulated on the same basic principles and contain some or all of three main ingredients.

First a **pigment** may be included. Pigments have both decorative and protective properties. The simplest form of paint is whitewash and, when dry, whitewash is nothing more than a pigment – whiting (calcium carbonate) – spread over a surface. It decorates and to some extent it protects, but it rubs off. So most paints contain the second ingredient, the **film-former** or **binder**, which will be a resin or polymer, to bind together the pigment particles and hold them on the surface. If the pigment is left out, the film-former covers and protects the surface, decorating it by giving it gloss or 'sheen'. It is difficult to attach coatings that are not fluid to any but the simplest of surfaces: those that are flat or gently curving. The fluidity of paint permits penetration into the most intricate crevices. It is achieved by dissolving the film-former in a solvent, or by colloidal suspension of both pigment and film-former in a diluent. Thus the third basic ingredient of paint is a liquid. Often the film-former–liquid mixture is called the **vehicle** for the pigment.

If the pigment is omitted, the material is usually called a **varnish**. The term **clearcoat** is used for unpigmented coatings applied over metallic paints. The pigmented varnish – the paint – is sometimes called an enamel, lacquer, finish or topcoat, meaning that it is the last coat to be applied and the one seen when the coated object is examined. **Lacquers** are normally thermo-plastic solution paints or varnishes, but the term is sometimes (confusingly) used to describe all clear woodfinishes. **Enamels** are normally thermosetting paints, hard, with a superficial resemblance to vitreous enamels.

Paint applied before the topcoat is called an **undercoat**. Some undercoats and related finishes may be briefly defined as follows:

- **Fillers** or **stoppers** are materials of high solid content, used to fill holes and deeper irregularities and to provide a level surface for the next coat.
- **Primers** are applied to the filled or unfilled surface, to promote adhesion, to prevent absorption of later coats by porous surfaces and to give corrosion resistance over metals. As their name implies, they prepare the base material for further paint application. Special pigments improve the anti-corrosive properties.
- **Surfacers**, or undercoats in decorative house painting, are highly pig-mented materials containing large quantities of extender (see below). They are easily rubbed smooth with abrasive paper. They provide the body of the paint film, level out minor irregularities in the substrate and must stick well to primer and topcoat.
- **Primer-surfacers** are surfacers that can be applied direct to the object's surface (the substrate), and fulfil both of the functions above in one coat.
- **Sealers** are clear or pigmented materials applied in thin coats to prevent the passage of substances from one coat of paint to another or from the substrate into later coats. They can be required to improve adhesion between coats, where this is otherwise weak.

All these materials are formulated on the principles described above. These are illustrated in Fig. 7.1, which also lists some of the minor ingredients of a paint. Some of the terms in the diagram need a little further explanation at this stage:

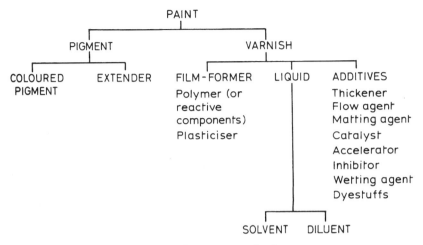

Fig. 7.1 Components of paints.

- **Pigment**. Any fine solid particles that do not dissolve in the varnish. If they do not provide colour they are called **extender** particles. Extenders are much cheaper than prime pigments and can carry out many useful functions, e.g. improvement of adhesion, ease of sanding and film strength.
- **Film-former**. When the coating is dry the film-former is a polymer, but in the wet sample it may be the chemical ingredients, only partly polymeric, which react to form the final dry polymer.
- **Liquid**. Some of the liquids of the paint are often withheld from the paint container and supplied separately as a **thinner**. Users add thinner to adjust the consistency to their requirements.
- **Additives**. Small quantities of substances added to carry out special jobs, such as the improvement of surface appearance.

Application

These are the basic ingredients, but much care in formulation will be required to produce a paint that will be easy to apply under changing conditions and pleasant to look at when dry. It is not within the aims of this book to discuss the methods of application that are available. They are well described elsewhere (see Appendix A). Suffice it to say that the paint may be put on by brushing, by roller, by a whole variety of methods of spraying (compressed air, airless, electrostatic and aerosol spraying), by dipping, electrodeposition, curtain-coating and flow-coating. Either the article is immersed in the paint, the excess of which is then allowed to drain off, or the correct quantity of paint is applied to the article and must not drain, 'run' or 'sag'.

In all cases, any irregularities in the wet film caused by the method of application must flow out to leave a smooth surface.

The first problem is to make the paint easy to handle while it is being applied. Different application methods require paints of different consistency, but in all cases the principle is the same: the greater the content of dissolved polymer in a paint, the more viscous it will be. The weight percentage of involatile material found in the paint is known as the solids content or 'solids' of the paint. Either by adjusting the 'solids' of the paint, or by altering the balance of the types of liquid used, the formulator brings the consistency or viscosity of the paint to that required.

So far so good. The difficulty lies in the next stage. Most methods of application leave some irregularities in the wet film surface: brush marks, spray mottle, roller stipple and so on. These are the methods that apply the correct amount of paint, which must flow at first to remove the irregularities, and then stop flowing, to prevent 'running' or 'sagging' of paint on vertical surfaces. This change in the rate of flow is usually brought about by evaporation of solvent, causing a rise in 'solids' and hence a thickening of the consistency. Flow dwindles until it is scarcely occurring at all. Then the drying mechanism takes over to set the film. A very careful balance of solvents is required to do this satisfactorily and more will be said about this in Chapter 9.

An alternative method, often used in conjunction with solvent evaporation, is to include some material in the paint which gives it abnormal viscosity characteristics, so that it is fluid while being agitated, but thickens up over a period when the agitation stops. This type of paint is said to be 'thixotropic'. More detail is given in Chapter 10.

If the article is allowed to drain after being coated, as in dipping, then it must drain evenly and not so rapidly that the film thickness becomes too low. If possible, thicknesses at the top and bottom of the article must vary only slightly, in spite of the downward drain of paint. Again the same principles must be used to slacken the flow.

With spray application in particular, the position is further complicated by the fact that the paint reaching the surface does not have the same composition as that leaving the spray-gun. The paint is broken up into thousands of fine droplets as it leaves the gun, each droplet presenting a surface – at which evaporation occurs – that is large compared with the droplet's volume. A great deal of liquid can be lost and this must be taken into account in formulating the paint.

It is necessary to point out that this loss of solvent – all of which has been paid for – is only accepted by the paint users on sufferance, as an inevitable consequence of the process. They are constantly looking for new materials which will contain less wasted material: in other words they want paints with higher 'solids'. Increasingly thinners other than water are all seen as hazardous to some degree and undesirable; also legislation is being enacted to reduce the amount of organic material which can be exhausted into the atmosphere, thereby polluting it. Furthermore, solvents come from petroleum, the world stocks of which are limited and must be conserved. For all these reasons, the paint formulator chooses the ingredients to give the highest 'solids' possible.

To keep the paint manageable at high solids, the formulator has to keep the polymer molecular weight down since, as will be seen in Chapter 9, this reduces the viscosity. Alternatively (Chapter 9 again), it is necessary to take the polymer out of solution (if this is practicable and suitable) and make a dispersion or 'emulsion' paint.

Dry film properties

An outline paint formula has been described and it has been shown that application presents problems which influence the formula. After application comes drying, but before this is considered, the properties required from the dry film must be discussed, since these influence the choice of method of drying.

If the dry paint film is to be a useful one, it must stick to the surface beneath, be hard enough and flexible enough for the purpose for which it is used and must retain most of its protective and decorative properties for a long period. The paint should be capable of repair or renovation. Let us now take these requirements separately and see how they are achieved.

Adhesion

We have seen that most molecules or atoms have some attraction for one another. The strength of this attraction varies greatly with the atoms concerned, but all intermolecular attractions have one thing in common: they operate over comparatively short distances (<10 nm) and become weaker the farther apart the atoms are within that short range. At distances above 1 nm their contribution to adhesion is negligible.

Since these attractive forces are the ones that make things stick together, the first problem is to get the molecules within range, i.e. the paint film must 'wet' the surface, displacing air and all the other adsorbed materials. The **critical surface tension** (ν_c) of a smooth solid surface is a numerical measure of the ease of difficulty of wetting it. It is equal to the highest surface tension possessed by any liquid that will spread spontaneously when placed on it. If a paint is to wet the surface, it must have a surface tension equal to or lower than the critical surface tension of the solid. Plastics can have very low critical surface tensions (e.g. polytetrafluoroethene, $PTFE = 18.5 \, \text{nM m}^{-1}$; polyethene = $31 \, \text{nM m}^{-1}$), which limit considerably the choice of solvents for paints to be applied to them (methods of overcoming this are disclosed in Chapter 17).

Clean metals generally have higher critical surface tensions ($>73 \, \text{nM m}^{-1}$), but even the thinnest possible layer of adsorbed oil or grease can dominate the surface (e.g. ν_c for 'clean' tinplate cans has been measured as $31 \, \text{nM m}^{-1}$), leading to wetting problems. Even if the paint wets the contaminated surface, loosely adhering grease, dirt or rust is a menace since,

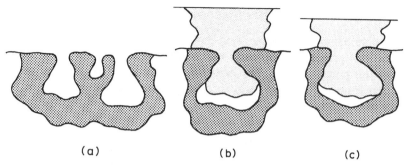

Fig. 7.2 Mechanical adhesion. (a) Cross-section of porous surface, (b) wet paint on surface and (c) some contraction of paint film after drying.

although the paint may stick well to the adsorbed layer, the latter can easily come away from the surface taking the paint film with it. For this reason the cleaning of surfaces before painting is always recommended. The practice of rubbing undercoats with abrasive paper not only provides a level surface for the next coat but also cleans the surface, removing materials difficult to wet or poor in cohesion.

Semi-porous undercoats or surfaces can offer good adhesion by a different mechanism. Providing the liquid paint can displace air from the crevices, the change from liquid to solid will mechanically lock or 'key' the paint on to the surface (Fig. 7.2). If the clean surface is not porous, but has been wetted well, adhesion will depend on the strength of the paint–substrate intermolecular attractions. Sometimes these can be considerably enhanced by the inclusion of particular chemical groups in the molecules of the paint film-former, e.g. carboxyl groups promote adhesion to metals.

Hardness, toughness and durability

These properties are connected in that cross-linked films are generally harder, tougher and more durable than those that are not. High molecular weight linear polymers, however, produce harder, tougher, more durable films than their low molecular weight equivalents. If the molecular weight is high enough, a polymer can be as hard, as tough and as durable in linear form as it can be in cross-linked form.

Flexibility

Hardness often goes with brittleness, and brittle films – particularly those subjected to temperature changes – do not survive long, failing by cracking and then flaking off if adhesion is only mediocre. Flexibility is introduced into cross-linked films by spacing out the cross-links, so that the whole structure becomes a looser, more open cagework of molecules.

Linear polymers are made more flexible by admixture with smaller molecules, which separate the large polymer chains, reduce attractive forces between them and act as a lubricant, allowing the polymer molecules to slide past one another more easily. These smaller molecules are called **plasticizers**.

Some monomers give more flexible polymers than others. They may be copolymerized into brittle polymers to increase flexibility.

Loss of decorative properties due to weathering

Gloss may diminish, the surface ultimately becoming powdery, or the colour may fade or darken. Ultraviolet rays and water, as well as oxygen in the air, gradually attack paint films, breaking chemical bonds and causing polymer molecules to break up into smaller fragments. The surface gradually wears away, roughens and the gloss disappears. Polishing will often rub it smooth again. This type of wear is reduced by choosing resistant polymers and by keeping the proportion of pigments down. If the pigment volume concentration in the film is high, it will only require the breakdown of a small amount of polymer to cause the detachment of surface pigment particles. The powdery surface is said to have 'chalked'.

Colour fastness is dependent on the use of high quality pigments, though complete satisfaction is not guaranteed unless the pigments have been tested with the binder that is to be used in the proposed paint.

Ease of repair and surface renovation; solvent resistance

Paint users often need to be able to make good damage to paint films. The 'damage' may be no more than excessive dirt or dust pick-up during application, spoiling the appearance. It might be shallow scratching. In either case, all that may be needed is light abrasion followed by polishing. To polish well, a surface should be capable of being softened by the heat generated, so that it will partially flow, evening out irregularities. If the polymer is cross-linked, its rigid network of chains of atoms restricts this softening and flow. The best polishing properties are usually shown by linear polymers. Their ability to reflow was demonstrated in the acrylic lacquer process used for cars, in which the lacquer surface was rubbed with very fine emery paper to remove dust and spray mottle. Subjecting the lacquer to heat in an oven reflowed the scratches, so that the surface was bright and glossy again.

Linear polymers also remain soluble in solvents: cross-linked polymers are not soluble. Thus the surface of a linear polymer film can be levelled by light rubbing with a pad soaked in a solvent mixture that will just dissolve the polymer. This process is called 'pulling over' and is used in the furniture industry.

If the paint film is badly damaged, it must be patched. A linear polymer film can be patched by spraying in the damaged area with the same paint.

The solvents redissolve the edges of the existing film, marrying the new area into the old. Cross-linked polymer films will not redissolve and so will not patch in this way. The complete area must be rubbed down and repainted. Any attempted patch will be at best partially successful, with all sorts of trouble likely to occur at the join between the patch and original film.

In property terms, it follows that linear polymer films will have poor resistance to solvents (present in many household articles, e.g. nail varnish). Cross-linked films will be more resistant.

If a paint is to be formulated for a particular purpose, the properties required from the dry film are defined first. These properties and the desired minimum solids of the paint may well determine whether the dry film is to contain a linear or a cross-linked polymer. When the type of polymer film has been selected, the choice of drying mechanism will have been restricted to a narrower area.

Dry paint: how paints dry

In the can of paint the mixture of substances must remain stable for long periods, so the ingredients must not react with one another chemically. Yet the paint must dry on the coated surface. Under some conditions some paints do not dry, which shows that the drying process does not merely consist of evaporation of the liquid. In fact there are three broad mechanisms of drying, two of which involve further chemical reaction.

Drying without chemical reaction

In this case the paint does dry solely by evaporation of liquids. The polymer is already fully formed in the can and, when free of solvent, is relatively hard and not sticky. During the drying process there is no chemical change in the polymer. If it was dissolved in the liquid solvents of the paint, it remains soluble in those solvents. If it was carried as an emulsion or colloid dispersion, it is not soluble in the liquid carrier, but it will be soluble in other solvents because it is a linear polymer. More is said about the sintering of particles of linear polymer into a continuous film in Chapter 11.

Nitrocellulose lacquers and decorative emulsion paints dry by this process. It is sometimes called 'lacquer drying'.

Drying by chemical reaction

There are undoubtedly advantages in producing a dry paint film containing cross-linked polymer as the film-former. Such polymers are, however, insoluble and so cannot be dissolved in solvents to form the vehicle of the paint in the can. We are therefore forced to have linear (or lightly branched) polymers, or even simple chemicals, in the can and to carry out a cross-linking chemical reaction after the paint has been applied. This can be achieved in two principal ways.

Drying by chemical reaction between paint and something outside the paint

Oxygen and water vapour in particular are reactive chemical ingredients of the air. In Chapters 12 and 16 it will be shown how oxygen reacts with drying oils and other unsaturated compounds to produce free radicals and bring about chain growth polymerization. In Chapter 15 we will see how the reaction between water and isocyanates can cause step growth polymerization. Both reactions can give a cross-linked film. In both cases the principle is the same: the air is used as a chemical reactant and is kept apart from the reactive ingredients in the paint by the tightly fitting lid on the can. Assuming that the air is completely excluded, no reaction begins until the paint film is applied, presenting a large surface exposed to the air. As the solvents evaporate, cross-linking begins and the soft, sticky, low molecular weight, linear or branched polymers in the paint are converted to a hard, tough, cross-linked film which will no longer dissolve in the solvents used in the original paint or in any others.

However, if the can lid is not tight fitting, it may cause the formation of a skin, which seals off the paint below from further reaction with the air. Alternatively, the skin may be permeable to the air and the reaction may spread right through the paint, turning it into a cross-linked polymer swollen with solvent. This is called an irreversible gel and, of course, at this point the paint becomes useless.

The drying process is relatively slow at room temperature, chiefly because the reactive ingredients in the air must penetrate into the film before full hardening can occur. The paint hardens more rapidly if the reactive molecules in the paint are large. The effect of cross-linking is to build bigger and bigger molecules. Eventually the whole film could become one molecule. If the building 'bricks' (the original molecules) are large, less 'cement' (cross-links) will be needed to join them together and the whole job (the hardening) will be done faster. However, as we have seen, there are advantages in keeping the molecular size down from the point of view of application and economy. As a compromise, low molecular weight polymer (molecular weight 1000–5000) or large simple molecules are generally used. This leads to moderately high solids in the paint, decidedly a point in favour of cross-linking systems.

Chemical reaction with the air can continue long after the paint film is apparently dry to touch and, in fact, the paint film will change its chemical nature slowly all the time it is in use. This means that the properties of the paint film also change gradually. Glossy house paints derived from drying oils dry by reaction with the air.

The principle of keeping the paint stable by having one reactant outside the can is quite general. Chemicals not normally in the air can be introduced into air in a confined drying chamber. This technique is used in the 'vapour curing' of isocyanate paints (Chapter 15).

Alternatively, the reactive ingredient can be in the substrate before it is coated, or in a previous coating on the substrate. This principle is used in the 'contact process' for curing unsaturated polyester coatings (Chapter 16).

The application of these separation techniques is limited only by the chemistry of individual paint systems and the imagination of paint chemists.

Drying by chemical reaction between ingredients in the paint

Obviously the paint must remain chemically stable in storage, and the reactants must not react until the paint has been applied, yet (in this method) they must all be in the paint. This paradox is resolved either by separating the reactive ingredients in two or more containers and mixing just before use or by choosing ingredients which only react at higher temperatures or when exposed to radiation of some form. The former method produces what is known as a 'two-pack paint'. Two-pack paints are less popular than their ready-mixed equivalents, because measuring is required before mixing and because of the limited period after mixing during which the paint remains usable (the 'pot life'). Sometimes two packs are avoided by so diluting the reactants with solvent that reaction proceeds only very slowly in the can, but much more quickly on a surface once the solvent has gone. Here the paint is not really stable, but a tolerable 'shelf life' is obtained. Whichever method is used, the chemical reactants may be sticky, low molecular weight polymers, or they may be simple chemicals. The reaction produces a cross-linked polymer.

Industrial stoving enamels and anti-corrosive two-pack epoxy coatings dry by chemical reaction of their ingredients.

A summary of the properties associated with the drying mechanisms is given in Table 7.1.

Relative merits

It has been shown that paints are simple in outline, but complex in operation and formulation. How can the best choice of the many ingredients available be made? The choice of pigments and solvents is discussed in the next two chapters. This summary is concerned with the film-former.

Cross-linked or not cross-linked? There are points in favour of both alternatives. The hardest, toughest, most durable and solvent-resistant films contain cross-linked polymers, and these are applied at higher solids. If they are to be matched in all respects except solvent resistance, linear polymers of high molecular weight must be used. For solution polymers,

Table 7.1 Effect of drying process on properties of paints

Method	Mol. wt of film former in can	Solids	Type of polymer film	Ease of polishing, patching and reflow	Rate of dry (no heat)	Minimum drying temperature	Handling and storage	Examples
Evaporation	High	Low, 10–35% (solution)	Linear	Good	Fast	<0°C	Good	Nitrocellulose and other lacquers
		Medium-high, 40–70% (emulsion)	Linear	Not applicable	Fast	~5°C	Good (protect water based from frost)	Decorative emulsion paints; some organosols and plastisols
Reaction between paint and (a) air	Low	Medium to high, 35–100%	Cross-linked	Fair–poor	Slow–moderate	Very slow in cold weather	Cans must be well sealed	Decorative gloss paints; some stoving enamels; one-pack polyurethanes
(b) chemicals added to air	Low				Fast	~5°C	Cans must be well sealed	One-pack polyurethanes
(c) chemicals added to substrate	Low				Fairly fast	10°C	Protect from heat	Unsaturated polyester woodfinishes
Chemical reaction between paint ingredients	Low or very low	Medium to high 30–100%	Cross-linked	Fair–poor	Can be fairly fast	Varies; 10–15°C common	Stoving and radiation curing types stable	Industrial and automotive stoving finishes; radiation curing woodfinish and overprint varnish
							Two-pack or short shelf life	Refinish 2K; acid catalysed polyurethane and polyester woodfinishes

this inevitably means much lower solids at application viscosity. Although higher solids, without loss of molecular weight, can be obtained with emulsion systems, these have their own special formulation problems (Chapter 11).

Against this can be argued the ease of polishing, pulling over or reflowing and the good patching properties of linear polymers. It is also a much simpler, fast method of drying that is not particularly temperature sensitive and produces little or no handling or storage problem. There is no right or wrong here. It simply depends on the job the paint has to do, the drying facilities available, the requirements of the user and whether the difficulties mentioned are slight or severe with the particular resins being suitable for the job.

Eight

Pigmentation

pigments and paint making

The outline formula of a paint has been given in Chapter 7 and the nature of polymers and resins has been described in Chapter 5. In the next three chapters, the other main ingredients of surface coatings will be covered briefly, before particular paints are described in the remainder of this book.

A detailed study of the chemistry of pigments is beyond the scope of an introductory book such as this and is already available elsewhere (see Appendix A). Instead, this chapter reviews some of the key properties of pigments and discusses their selection and methods of incorporating them into paint.

Pigment properties

Any surface coating pigment will need to carry out some (or perhaps, all) of the following tasks:

- to provide colour;
- to provide special effects (e.g. flip, sparkle);
- to obliterate previous colours;
- to improve the strength of the paint film;
- to improve the adhesion of the paint film;
- to improve the durability and weathering properties;
- to increase the protection against corrosion;
- to reduce gloss;
- to modify flow and application properties.

To choose a pigment to carry out a given selection of these nine functions, we must know about the following properties of the pigment:

- tinting strength;
- lightfastness;
- bleeding characteristics;
- hiding power;

- refractive index;
- particle size;
- particle shape;
- specific gravity;
- chemical reactivity;
- thermal stability.

Tinting strength

The majority of paints contain white pigment, which is tinted to the appropriate pastel or mid-shade with coloured pigments. If a lot of coloured pigment is required to achieve the shade, it is said to have poor tinting strength. The tinting strength of a coloured pigment is related to that of some standard pigment of similar hue. If numerical values are given, then

amount of alternative pigment required to produce the shade pigment

= amount of standard pigment required

$$\times \frac{\text{tinting strength of standard pigment}}{\text{tinting strength of alternative pigment}}$$

The tinting strength of a pigment is independent of its hiding power, since the comparison of shades is done at film thicknesses that completely hide the substrate. Relatively transparent pigments can have high tinting strengths.

The phrase 'tinting strength' is sometimes applied to white pigments. A single coloured pigment is used for the comparison of white pigments at a fixed shade.

Lightfastness

To give a good colour initially is not enough. The colour must last, preferably as long as the paint film. Many pigments fade or darken or change shade badly in the light. This is because the ultraviolet rays in sunlight are sufficiently energetic to break certain chemical bonds and thus change molecules. A change in chemical structure means a change in the ability to absorb light in the visible region of the spectrum, with a consequent loss of colour or variation of hue. On the other hand, if the pigment can absorb ultraviolet rays without breakdown, it will protect the binder. The energy is dissipated harmlessly as heat.

Bleeding characteristics

Not all pigments are completely insoluble in all solvents. Colour in the right place is all very well, but when white lettering applied over a red background turns pink, it is not so good.

What happens is that the solvents of the white paint dissolve some of the red pigment in the background coat and carry it into the white overlayer. Organic reds as a group are particularly prone to this fault (though theoretically it can occur with any colour), so it has been given the name 'bleeding'.

Hiding power

Ideally one coat of paint should obliterate any colour. Frequently two are needed, but in any case the total thickness of paint applied should be no more than that necessary for the required degree of protection of the surface and the presentation of a smooth, pleasing appearance. Such thicknesses are surprisingly small, usually 25–100 μm. To obliterate, the pigments used must prevent light from passing through the film to the previous coloured layer and back to the eye of an observer. The pigments do this by absorbing and scattering the light. The **hiding power** of a paint is expressed as the number of square metres covered by 1 litre of paint to produce complete hiding; the hiding power of a pigment is expressed as the number of square metres covered per kilogram of pigment, which has been dispersed in a paint and applied so that it will just hide any previous colour.

Hiding power depends upon the wavelengths and total amount of light that a pigment will absorb, on its refractive index and also on particle size and shape.

Refractive index

Chapter 6 showed how light rays can be bent and how they may be returned to the eye from a paint film. The light rays suffer refraction, diffraction and reflection by transparent particles that have a refractive index differing from that of the film in which they lie. White pigments are transparent in large lumps, but white in powder form, because they have high refractive indices (2.0–2.7). These are greater than the refractive indices of film-formers (1.4–1.6), so the pigments give a film its white colour by the mechanism described for incompatible resins in Chapter 6. Titanium dioxide (TiO_2) pigments, in particular, have excellent hiding power, because their refractive indices are so much higher than those of film-formers and because they have the optimum particle size. Extender pigments are also transparent in bulk and white as powders, but they do not colour paints because their refractive indices scarcely differ from those of film-formers.

Particle size

There is an ideal particle diameter for maximum scattering of light at interfaces and this is approximately equal to the wavelength of the light in the particle. As a rough guide, the optimum diameter is approximately half the wavelength of the light in air, i.e. about 0.2–0.4 μm. Below this size the particle loses scattering power; above it, the number of interfaces in a given weight of pigment decreases. The hiding power of the transparent pigment is reduced. Pigments have particle diameters varying from 0.01 μm (carbon blacks) to approximately 50 μm (some extenders). No sample of pigment contains particles all of an identical size; rather there is a mixture of sizes with an average diameter.

Connected with particle size are **surface area** and **oil absorption**. If a cube of pigment is cut in two, the weight of pigment is the same, the number of particles is doubled, the size of particle is halved and two new surfaces are formed along the cut, in addition to all the original surface of the cube. Thus in any fixed weight of pigment, the smaller the particles, the larger the area of pigment surface.

It is interesting to note that while 1 gram of rutile titanium dioxide white (particle diameter 0.2–0.3 μm) has a surface area of $12\,m^2$, 1 gram of fine silica (particle diameter 0.015–0.2 μm) has a surface area of $190\,m^2$ – about the area of a singles court at tennis!

The surface area is indicated by the oil absorption, which is the minimum weight in grams of a specified raw linseed oil that is required to turn 100 g of pigment into a paste. The oil is added slowly to the pigment with thorough mixing and shearing between the walls of the vessel and a rod. While the oil is in the process of replacing air molecules on the particle surfaces (this is called **wetting** the pigment), the pigment–oil mixture remains a crumbly mass. Pigment, as Fig. 8.1 shows, in the dried powder is in the form of **agglomerates**, with particles packed together, but with air between the individual particles. These need breaking down in dispersion processes and the prime pigment particles separating.

When wetting is as complete as it can be, the next additions of oil will fill the spaces between the particles. As soon as there is a slight oil surplus, the mixture becomes a paste, because the particles are free to flow past one another as a result of the lubricating action of the oil. It can be seen that,

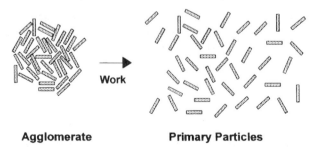

Agglomerate   **Primary Particles**

Fig. 8.1 The pigment dispersion process.

while the oil absorption value is related to the surface area, it depends on how successful the operator is in displacing all the air from the whole surface.

With many pigments the operator will not be at all successful, using such rudimentary equipment. Some surfaces will have more attraction for linseed oil molecules than others. Obviously the greater the area of pigment left unwetted, the lower the oil absorption will appear to be. If the volume of space between the particles varies, as it will do from pigment to pigment according to the particle shape and distribution of sizes, then this will also affect the final figure. At best, oil absorption is a rough guide to surface area, but its merit is that it is easily and quickly measured and gives the paint formulator some idea of the type of pigment that is being dealt with.

The most important part of a pigment particle is its surface. At the surface are those chemical groups that will make contact with the chemical groups in the varnish. These surface groups will determine whether the pigment has any attraction for varnish molecules or not (whether the pigment is easily wetted) and whether the pigment has special attractions for parts of the varnish (e.g. drier molecules, see Chapter 12). This latter feature may be undesirable, since molecules firmly attached to pigment surfaces are not free to move about and carry out their functions in the varnish. If they are molecules of a paint additive, they are bound to be present in small quantities; a pigment of very large surface area could totally absorb the additive, rendering it ineffective. If these groups on the pigment surface attract each other strongly, the pigment particles will tend to cluster, resisting wetting and dispersion and setting up a loosely bound 'structure' of particles in the paint, which will affect its application properties (see Chapter 10). This phenomenon is called **flocculation**. If the chemical nature of the pigment surface is not known, it is a general rule that the larger the surface area, the more active the pigment surface will appear to be.

Particle shape

Particles may be nearly spherical, cubic, nodular (a rounded irregular shape), acicular (needle- or rod-like) or lamellar (plate-like). Since particle shape affects pigment packing, it therefore affects hiding power. Rod-shaped particles can reinforce paint films, like iron bars in concrete, or they may tend to poke through the surface reducing gloss.

Such rough surfaces may help the next coat to stick more easily, so this type of pigment could be useful in an undercoat. Plate-shaped particles tend to overlap one another like tiles on a roof, making it more difficult for water to penetrate the film. Aluminium and mica pigments have this shape. Thus more protection is given with micaceous iron oxide-containing paints, and by heat-resistant aluminium paints. Aluminium and specially treated mica pigments also reflect light from their polished flat faces in metallic paints.

Specific gravity

This is the weight of a substance in grams divided by its net volume in millilitres (excluding, for example, volume occupied by air between pigment particles). For rutile titanium dioxide white, this value is 4.1; for white lead, 6.6. Since pigment is sold to paint makers by the kilogram and is then resold to their customers by the litre, the specific gravity of the pigment is important. An expensive pigment (per kg) may yet prove economical if its specific gravity is low – a little goes a long way. Extender pigments are not only cheap, but have low specific gravities: that is why they are used to increase pigment volume, where the hiding power of the coloured pigment is good enough at low concentrations.

Chemical reactivity

Chemical reactivity can make some pigments unsuitable for some purposes. For example, zinc oxide is amphoteric and should not be used as a white pigment with resins containing a high proportion of acid groups. Soap formation will occur and – because zinc is divalent – this will tend to cross-link the resin, causing excessive viscosity increase on storage. The paint is said to become 'livery' and unusable.

On the other hand, some pigments are included particularly because they are reactive. Excellent examples are the **anti-corrosive pigments**. Especially effective are the chromates of zinc, lead and strontium. When the paint film is permeated by water, these pigments slowly release chromate ions, because of their low but measurable solubilities. The anti-corrosive action of chromates is discussed in some detail in Chapter 17. Another long-established anti-corrosive pigment is red lead, which works well in oil- or alkyd-based paints (Chapter 12). It is thought that this is due to the formation of lead soaps of azelaic acid, $HOOC \cdot (CH_2)_7 \cdot COOH$, one of the aging by-products of unsaturated oils. Both lead and chromate compounds are, however, toxic and the use of lead and chromate pigments is now severely limited; the search continues for less toxic alternatives that are as effective. Nothing with the all-round effectiveness of chromates has been found, but success has been had with zinc phosphate pigments and with silica pigments which release calcium on an ion-exchange basis.

These examples illustrate the advantages of knowing the pigment chemistry. This is especially necessary with the newer (probably organic) pigments, whose precise chemical nature may not be disclosed by the manufacturer. Even with the more traditional pigments, other ingredients may be added by the pigment manufacturer to alter the crystal shape, or to provide a coating on the pigment surface that will make the pigment easier to disperse. Unless precise knowledge can be obtained, the paint formulator is very much in the hands of the pigment manufacturers and must rely on their literature.

Thermal stability

The temperature at which a pigment decomposes or alters its nature (e.g. melts) can be very important if the pigment is required for a paint to be stoved at high temperatures, or if the paint is to be heat resistant.

Where is all this information to be obtained? Partly in general pigment and paint literature (where traditional pigments and well-defined groups of pigments are concerned), partly from the technical literature supplied by the pigment manufacturer and partly by experiment. For a reference to experimental methods for determining pigment properties, see Appendix A.

Pigment types

Whole books have been written on pigments classified by chemical types. Here we have room only to attempt a much broader classification and to say a few words about the implications of the classification. All pigments are either:

- natural or synthetic; and
- organic or inorganic chemicals.

Natural and synthetic pigments

It is unlikely that there is any organic pigment, the naturally occurring form of which is still in industrial use today. Many inorganic pigments, however, are still dug out of the earth's crust, crushed, washed and graded by size. Frequently, there is a synthetic equivalent – i.e. a pigment made from other ingredients by a chemical process – apparently the same chemically, but often different in properties. The differences arise because:

- whereas the natural pigment is available in the crystal form found in nature, the synthetic product may be induced to take a more desirable crystal shape;
- the natural product may be contaminated by some impurity, such as silica, which it is uneconomic to remove; the synthetic product should be pure or nearly so;
- crushing produces a wide range of particles sizes and grading methods may not be able to remove all oversize or undersize particles from desired range; a pigment produced by precipitation under controlled conditions should have a much more uniform size range.

The chief family of pigments in which natural varieties are still of importance is the iron oxide family: ochres, umbers and siennas; red, yellow and black iron oxides.

Organic and inorganic pigments

There are now far more organic pigments than inorganic ones, though some of the newest contain both metallic (inorganic) elements and organic structures. Again, many organic pigments are organic chemicals deposited on an inorganic (e.g. aluminium hydroxide) 'core'. These pigments are called **lakes**. It is difficult to lay down hard and fast rules, as there are exceptions in every case, but for some properties there is a clear-cut advantage for the bulk of one type of pigment over the majority of the other type. These properties are listed in Table 8.1. This list may give the impression that it is best to avoid organic pigments, but their brilliance and clarity are very powerful arguments for using them. The very best organic or organometallic pigments give the highest standard of performance for most uses, e.g. the

Table 8.1 Properties of pigments

Property	Preference	Reasons
Brilliance and clarity of hue	Organic	The most attractive, cleanest colours can only be obtained with organic pigments
White and black paints	Inorganic	The purest white pigment is titanium dioxide and the most jet black, carbon (usually considered inorganic). There are no organic blacks and whites
Non-bleeding	Inorganic	Inorganic compounds have negligible solubilities in organic solvents. Some organics are very insoluble
Lightfastness	Inorganic	The valency bonds in inorganic compounds are generally more stable to UV light than those in organic compounds
Heat stability	Inorganic	Very few organic compounds are stable at or above 300 °C. Some decompose or melt at much lower temperatures
Anti-corrosive action	Inorganic	All anti-corrosive pigments are inorganic
UV absorption	Inorganic	Titanium dioxide blocks harmful UV radiation from the binder and substrate. Fine iron oxides are visibly transparent but again block UV and give protection
Reflective effects	Inorganic	Aluminium and treated mica available in reflective platelet form

phthalocyanine ('Monastral') pigments, the quinacridone ('Cinquasia') pigments, the perylene and perinone pigments, the dioxadine pigments and the high molecular weight azo pigments.

Pigment selection

To sum up, the paint formulator's method of pigment selection is as follows.

- Examine a pattern of the colour to be produced in paint. Estimate the number of different hues that will have to be blended to produce the colour. A suitable pigment has to be found to provide each hue.
- Define the properties required from the pigments.
- Select a suitable pigment or pigments in each hue. In order to obtain pigments with the required properties, consult the following sources where necessary:
  - General paint and pigment literature. There are traditional pigments for particular uses and binders.
  - *The Colour Index* (Society of Dyers and Colourists, Bradford, UK). The properties of pigments are given, together with the names of commercial grades and their manufacturers.
  - Pigment manufacturers' literature. Examples of colour may be given.
- Match the colour (see below) with one or more pigment combinations.
- Test the paints produced.

Dispersion

The next stage is to get the pigment into the paint. The pigment is usually supplied as a powder, in which the granules are actually aggregates of the fine particles produced by the pigment manufacturer. These particles must be dispersed or separated from one another and evenly distributed throughout the paint as a colloidal suspension. For this suspension to have the maximum stability in organic solvents, the surface of each particle should be completely wetted with the varnish; there should be no intervening layers of air or adsorbed water.

Wetting with solvent alone is not enough and pigment dispersions in solvent have poor stability. Each particle in the pigment suspension must be stabilized by polymer chains, anchored to its surface by intermolecular attractions, yet extending out into the varnish because of their attractions for the solvent molecules. When two such particles approach one another, they do not adhere. Contact between them involves the intermingling of polymer chains from the two particles. Locally, in the region between the particles, the concentration of resin molecules is higher than at other points in the paint. This upsets the equilibrium of the system, so solvent molecules diffuse into this region, dilute the concentration and restore equilibrium. The osmotic pressure involved is sufficient to separate the particles.

A resin suitable for dispersion usually contains polar groups (which provide the attraction for pigment surface molecules) and is completely soluble in the solvent mixture of the dispersion. Sometimes, surfactants (Chapter 10) are used to bridge the particle–resin interface and assist wetting. In water, ionic surfactants (e.g. soaps) can provide the pigment surface with an electrical charge. The particles, being of like charge, repel one another and the dispersion is stable.

Dispersion is usually carried out in a **mill**, a machine in which the agglomerates are subjected to the forces of shear and (sometimes) attrition. When **shear** is the dispersing action, the agglomerates are squeezed between two surfaces moving in opposite directions, or in the same direction but at different speeds. It is just like making cocoa, where the powder has to be dispersed in a little milk to form a paste. Dispersion is carried out by a shearing motion between spoon and cup. In paint mills the principle is the same, but the power and degree of dispersion are much greater.

During the manufacture of the pigment, the particles are reduced to their ultimate size by crushing or grinding in a liquid slurry, where **impaction** forces are used in a process known as **comminution**. Where **attrition** or rubbing forces are part of a dispersion process, the conditions are much milder and there is no attempt to fracture individual particles. In the viscous medium of the varnish, agglomerates – not particles – are broken down.

Several types of mills are used, depending on the difficulty experienced in dispersing the particular pigment. The **high speed disperser** (Fig. 8.2) is used

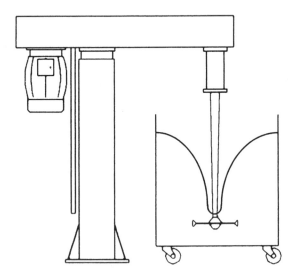

Fig. 8.2 A high speed disperser.

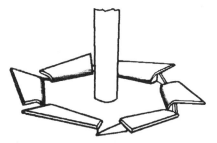

Fig. 8.3 A high speed disperser blade.

for easily dispersed pigments and consists of a horizontal disc with a serrated edge (Fig. 8.3), which rotates are high speed about a vertical axis. The dispersion quality may not however be sufficient for full gloss paints. However, pre-dispersion in a high speed disperser may be the preliminary to faster throughput to a sand or bead mill.

The **ball mill** is a cylinder revolving about its axis, the axis being horizontal and the cylinder partly filled with steel or steatite (porcelain) balls, or pebbles. The speed of rotation is such that the balls continually rise with the motion and then cascade down again, crushing and shearing the pigment. Depending on the method of loading, the ball mill can provide actual grinding or size reduction by impaction. This is rarely required, but its ability to provide extreme conditions means that it can be used for the most difficult pigments, such as carbon black. Since grinding times are long, for economic reasons ball mills are only used for these pigments.

In the **sand mill** (Fig. 8.4) or **bead mill**, the axis of rotation may be horizontal or vertical, the grinding medium is coarse sand or beads (glass, zirconia) and the charge is induced to rotate at higher speeds by revolving discs in a stationary container. The ball mill produces a batch of pigment dispersion; the sand grinder gives a continuous output of dispersed pigment. Sand or bead mills in a variety of configurations are now the most frequently used dispersing machinery.

Roller mills, which have their major use in ink production, consist of a number of horizontal steel rolls placed side-by-side and moving in opposite directions, often at different speeds, with very small clearances in between. In these gaps, the pigment–resin mixture is sheared. The **triple roll mill** is shown in Fig. 8.5. A paste is fed in at D, is lightly sheared between rollers A and B and more severely between B and C. A scraper blade E removes the dispersion. All rollers are independently driven and run at different speeds to increase shear. The mill is particularly suitable for the production of pastes of very high pigment content.

Several other mills are available, which the reader may discover elsewhere (Appendix A). These few examples are quoted to illustrate the principles involved.

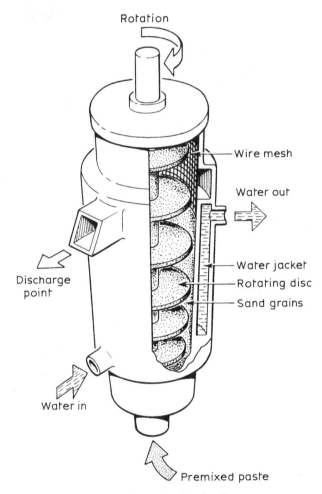

Rotation

Wire mesh

Water out

Water jacket

Rotating disc

Sand grains

Discharge point

Water in

Premixed paste

Fig. 8.4 A sand or bead mill.

It is obvious from the mention of stiff pastes that the whole paint is not charged initially into the mill. In fact, the paint maker aims to put in the maximum amount of pigment and the minimum amount of resin to obtain the largest possible paint yield from the mill. This mixture forms the grinding stage. When the dispersion is complete (after a period varying from 10 min to 48 h according to the materials and machinery involved), the consistency is reduced with further resin solution or solvent. This is the 'let-down' stage and may take up to 2 h. The third or final stage (carried out in a mixing tank) consists of the completion of the formula by addition of the remaining ingredients and any necessary tinting. A breakdown of a possible

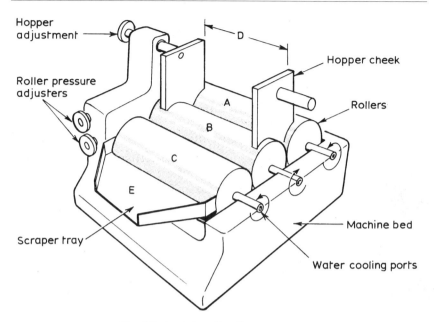

Fig. 8.5 A triple roll mill.

gloss paint preparation using high speed disperser and sand mill looks like this:

	wt%	
Grind stages:		
Pigment	25.0 ⎫	
Resin solution	5.0 ⎬ High speed disperser	
Solvent	8.0 ⎭	
Then add:		
Resin solution	2.0 ⎫ Single pass through sand mill	
Solvent	2.0 ⎭	
Then add:		
Resin solution	50.0 ⎫	
Solvent	5.0 ⎬ Let-down and tint stage	
Additives	3.0 ⎭	
	100.0	

The exact composition of both milling operations is found by experiment, to give the minimum grinding time and the most stable and complete dispersion. Let-down also requires care, as hasty additions in an incorrect order can cause the pigment to reaggregate (flocculate).

The amount of pigment in the formula is that required for the appropriate colour, hiding power, gloss, consistency and durability. As a rough guide, the amount might vary from one-third of the binder weight to double the weight (for a paint, when including extender).

Colour matching

The colour of a paint is usually matched to some colour pattern. It must, at any rate, be matched to the colour of the previous batch of the same paint. If a new shade is to be matched in a paint, the procedure for matching is as follows.

- The main hue and the undertones are observed; also whether the colour is 'clean' or 'dirty'.
- From experience and by application of the principles of subtractive mixing, the pigments likely to give the colour are chosen. The procedure is described under Pigment selection (above).
- Each of these pigments is dispersed separately and each dispersion is converted to a paint. The colour matcher now has several paints, each containing only one pigment. These are known as **colour solutions**.
- The colour solutions are blended, the choice of solutions and proportions being varied, until finally the correct blend to give the colour is obtained. The proportions of pigment in this blend are recorded.
- The final paint is often made by blending colour solutions, but frequently a more stable dispersion is obtained if all the pigments are dispersed simultaneously in the same mill. This is called **co-grinding** the pigments. If the colour is then slightly off shade, it is adjusted by additions of colour solutions or concentrated pigment dispersions.

In larger paint companies, where spectrophotometers or colorimeters and a computer are available, these stages can often be speeded up considerably. The new shade can be measured by the colorimeter and expressed precisely as a group of numerical values. This information can be passed to the computer, which has been previously programmed with the values for numerous pigments and with the 'rules' governing the mixing of these pigments to give the desired colour. The computer will then print out a choice of pigmentations (type and proportions) for achieving the colour.

The formulator makes a selection and will usually find that the colour that is produced to the computer's instructions will be close to the desired colour and that the recipe will require only minor modifications to give the exact shade.

Most colours, even dark ones, require substantial proportions of white pigment. Clean colours require organic pigments and should be matched with the minimum number of pigments. The subtractive effect increases with the number of pigments, reducing the reflected light intensity and hence 'dirtying' the colour. If darkening or 'dirtying' of a colour is required, it can be done with black pigment. The use of black pigment can markedly alter the shade. Thus brown is a mixture of yellow (or red) and black.

Paint making and mixing schemes

While the description of the paint-making process may have given the impression that all paint is essentially made as a batch process, starting from the ingredients and ending with paint, the colour matching description will have given hints that other approaches are possible. While white paints will essentially be based on titanium dioxide pigment, possibly with some extender, coloured paints invariably contain a number of pigments.

In larger factories, colour solutions may be made and stocked in large quantities, providing storage is available for the required range. This is especially applicable to making industrial paints, where this may be the only economical way to make a range of colours and batch sizes, and where a quick and flexible response to customer orders is required.

This mixing concept has been extended into two large paint markets, to give in one case extended customer choice, and in the other, the only possible means of access to the colour range required. For the decorative paint market, concentrated **tinters** have been developed, often capable of use in both solvent and water-borne paints, to offer the customer a wider choice of colours than is possible with a standard range, often limited by shelf storage space in the shop. The shop will be equipped with a dispenser for the tinters, and a can-shaker to ensure complete mixing.

For the car refinish market, it is essential that every colour made can be offered, and here **mixing schemes** use a range of colour solutions, capable of use alone, or in any blend with any other colour. These are also used to produce metallic paints, and the paint dealer or car repairer will have recipes to mix all the colours they require. However, especially in the case of metallic paints, the spray operator may need to be skilled in using solvent and clear (unpigmented) additives on occasion to achieve the desired finish.

Nine

Solvents

chemistry and physics of solvents and diluents

Before we start, it should be explained that the word 'solvent' is used loosely in the paint industry to denote any liquid that is present in, or added to a paint; it tells you nothing about the liquid's ability to dissolve the polymers in the paint. The term 'diluent' is also used, and is correct in that all of the substances discussed below will, when added, reduce the solid content of the paint. The term diluent, however, has a further meaning which will be discussed. In this chapter we will have to use the word solvent in the paint sense and in the scientific sense (as defined in Chapter 1). It is hoped that the meaning intended will be clear from the context and, to minimize the confusion, the word 'liquid' will be used as much as possible, where no description of solvent power is intended.

Chemical types

The following liquids are commonly used in paints to carry the pigment and binder:

- water;
- aliphatic hydrocarbon mixtures, chiefly alkanes (paraffins);
- terpenes;
- aromatic hydrocarbons;
- alcohols;
- esters;
- ketones;
- ethers and ether-alcohols;
- nitroalkanes (nitroparaffins);
- chloroalkanes (chloroparaffins).

Water

Water is the main ingredient of the continuous phase of most emulsion paints (Chapter 11). It is also used alone, or blended with alcohols or ether-alcohols, to dissolve water-soluble resins (e.g. in water-reducible paints) or dyestuffs (e.g. in water stains). It is in fact now the major 'solvent' used in decorative paints, and holds a strong position in paints used in car manufacturing and in can lacquers.

Any type of resin can be made water-soluble. The usual procedure is to incorporate sufficient ionic groups, usually carboxyl groups, into the polymer to give the resin a high acid value (Chapter 12) of 30–50. These groups are then neutralized with a volatile base, such as ammonia or an amine, whereupon the resin becomes a polymeric salt, soluble in water or water–ether-alcohol mixtures. For example, drying oils can be made acidic if they are heat-bodied in the presence of maleic anhydride, reaction taking place between the double bonds in both materials (see Bodied oils, Chapter 12). If the carboxyl groups are not fully neutralized, an emulsion can be produced.

The polymeric part of the salt is negatively charged and can be attracted to and discharged at the anode (Chapter 2) in an electrolytic cell. If the anode is a metallic object, say a car body, then the discharged resin (or paint if it is associated with pigment particles) will deposit on that object and coat it. This is called **anodic electrodeposition**. If the paint film deposited is not porous then, as the film thickness increases, so does its electrical resistance (or insulating power) and thus the deposition rate slackens and deposition eventually stops. This has the effect of automatically controlling film thickness and of encouraging deposition at points which are more thinly coated. From this arise two of the main advantages of electrodeposition: it can be a fully automated process and suitably formulated paints (with 'good' throwing power) can be made to coat all parts of an article of any shape, however awkward, with a uniform paint coating.

Instead of acidic carboxyl groups, resins can be made to include basic, e.g. amine, groups, and these can be neutralized with relatively volatile acids, e.g. acetic or lactic acids, to produce resin salts soluble in water or water–solvent mixtures. Again, incomplete neutralization produces an emulsion. In these systems, the resin is positively charged and can be discharged at a cathode (Chapter 2). This is the basis of **cathodic electrodeposition**.

Even if the polymeric salts are capable of anodic or cathodic electrodeposition, water-borne paints made from them do not have to be applied in this way. The solution or emulsion paints produced can be applied by virtually any technique. There is increasing use of this type of paint in all markets as environmental concerns, legislation, economics and lessening of world petroleum stocks encourage the use of decreasing amounts of volatile organic solvents.

An alternative method of obtaining water solubility, without salt or ion formation, is to incorporate short lengths of naturally water-soluble polymer into the main polymer structure. Often polyethene glycol is used. This is a polyether (Chapter 15) of molecular weight about 1000, which is made from ethylene oxide and ethylene glycol. Between 5 and 20% by weight is incorporated into the resin, usually by esterification. The resin thus modified can be emulsified in water by simple stirring, or can be dissolved in water–ether–alcohol mixtures. Whereas ammonia will evaporate from the paint film eventually, polyethene glycol will not, so the film remains permanently sensitive to water.

The virtues of water are its availability, cheapness, lack of smell, non-toxicity and non-flammability. However, it is not an ideal paint liquid, because of its limited miscibility with other liquids and because the film-formers designed to be dissolved or dispersed in it usually remain permanently sensitive to it. In fact, its abundance in nature makes it any paint film's worst enemy, since it is always around to cause swelling of the film, hydrolysis and substrate corrosion.

Other major problems with water-borne paints concern the rheology of the paint and drying of the paint. Viscosity does not change with solids in the same way as for solvent-borne paints, and this means the use of special techniques. Where drying is concerned, water has five time the latent heat of vaporization (Chapter 1) of organic solvents, and also its rate of evaporation is affected by the relative humidity at the time of drying. Furthermore, water has a decidedly higher surface tension (p. 8) than organic solvents which leads to poorer wetting of surfaces (water $= 73\,\text{mN}\,\text{m}^{-1}$; xylene $= 30\,\text{mN}\,\text{m}^{-1}$). The reduction of surface tension can be a reason for retaining some solvent in water-borne compositions; another reason is to improve the solubility of resins. These difficulties are being understood and overcome, and the use of water is set to continue to increase, while the use of organic solvents will continue to decrease.

Aliphatic hydrocarbons

Aliphatic hydrocarbons are usually supplied as mixtures, because of the difficulty of separating the individual compounds. These are all distillates from petroleum refining (Chapter 3). A boiling range for the mixture is usually quoted, e.g. SBP (special blend of petroleum) spirit No. 3, 98–122 °C. Many mixtures also contain a percentage of aromatic hydrocarbons, e.g. white spirit, 155–195 °C, contains about 15% aromatic, though this may be absent in so-called low odour grades.

Terpenes

Terpenes once commonly used are turpentine, dipentene and pine oil (p. 38). Turpentine varies with grade, but is principally α-pinene; dipentene is mainly limonene; while pine oils are mixtures, chiefly of terpene alcohols. Turpentine was once the main solvent for house paints, but has now been completely replaced by white spirit. Dipentene can be used as an anti-skinning agent (Chapter 10).

Other solvents

The aromatic *hydrocarbons, alcohols, esters* and *ketones* supplied to the paint industry are fairly pure named compounds. Some aromatic mixtures are sold cheaply under proprietary names.

Ethers are not often used, but *ether-alcohols* – which contain both the ether $(-C-O-C-)$ and alcohol $(C-O-H)$ groups – are, e.g.

$CH_3 \cdot O \cdot CH_2 \cdot CHOH \cdot CH_3$ $CH_3(CH_2)_3 \cdot O \cdot CH_2 \cdot CH_2OH$

1-methoxypropan-2-ol   2-butoxyethanol

(propylene glycol monomethyl ether)   (ethylene glycol monobutyl ether)

$HO \cdot CH_2 \cdot CH_2 \cdot O \cdot CH_2 \cdot CH_2 \cdot OH$ $CH_3 \cdot (CH_2)_3 \cdot O \cdot CH_2 \cdot CH_2 \cdot O \cdot CH_2 \cdot CH_2OH$

diethylene glycol (DEG)   DEG monobutyl ether

Their *acetate esters* are also used. Ethylene glycol-based ether-alcohols and their esters are now less used on account of their toxicity, being replaced by the propylene glycol versions.

Nitro- and *chloroalkanes* (or paraffins) are uncommon solvents in paints, the latter because they are somewhat toxic, the former because of the cost and some recent evidence of toxicity.

Solvent properties

The most important properties of a liquid for paints are:

- solvency, i.e. whether it is a solvent or non-solvent for a given film-former; this depends on the film-former and is not an independent property of the liquid;
- viscosity or consistency;
- boiling point and evaporation rate;
- flash point;
- chemical nature;
- toxicity and smell;
- cost.

Let us consider these properties in more detail, referring to Table 9.1, which lists some common solvents together with figures for the first four properties.

Solvency

Polymers dissolve by the mechanism described for inorganic solids in Chapter 1. A solvent for a polymer is a liquid, the molecules of which are strongly attracted by the polymer molecules. The molecules of a non-solvent are only weakly attracted. Simple inorganic and organic solids have fixed solubilities because, above a certain concentration of molecules in the

Table 9.1 Solvent properties

Solvent	Formula	Solvency H-bonding group	Solubility parameter	Viscosity at 20°C (centipoises or mPa s)	Boiling point (°C)	Flash point (closed cup) (°C)
Water	H_2O	III	47.7	1.002	100	None
Aliphatic hydrocarbons						
Cyclohexane	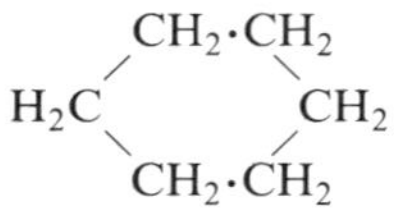 (structure)	I	16.7	0.89[a]	81	3
White spirit		I	15.1	1.09[a]	155–195	33[c]
Odourless white spirit		I	14.1	1.38[a]	180–207	55
Terpenes						
Dipentene	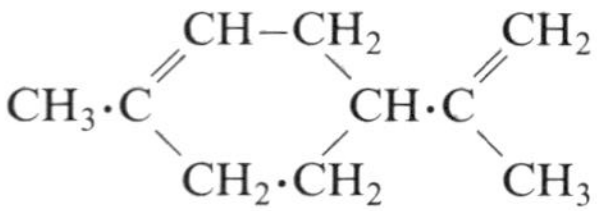 (structure)	I	17.3	0.975[a]	175–190	32[c]
Turpentine	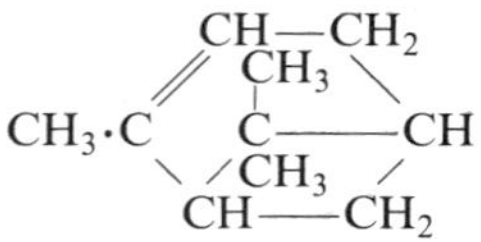 (structure)	I	16.5	1.26[a]	150–170	33
Pine oil		I	17.5	6–26[a]	195–220	74–88[d]
Aromatic hydrocarbons						
Toluene	$C_6H_5 \cdot CH_3$	I	18.2	0.55[a]	111	4
Xylene	$C_6H_4 \cdot (CH_3)_2$	I	18.0	0.586	138–144	27
Styrene	$C_6H_5 \cdot CH{=}CH_2$	I	19.0	0.77[a]	146	31
Vinyl toluene	$CH_3 \cdot C_6H_4 \cdot CH{=}CH_2$	I			164–170	32

Alcohols						
Methanol	CH_3OH	III	29.6	0.547	65	12–14
Ethanol	C_2H_5OH	III	25.9	1.200	78	14
n-Propanol	$CH_3 \cdot (CH_2)_2 \cdot OH$	III	24.3	2.25	97	15
Isopropanol	$(CH_3)_2 \cdot CH \cdot OH$	III	23.5	2.4	82	12
n-Butanol	$CH_3 \cdot (CH_2)_3 \cdot OH$	III	23.3	2.948	118	35
Isobutanol	$CH_3 \cdot CH(CH_3) \cdot CH_2 \cdot OH$	III		3.76	108	25
sec-Butanol	$CH_3 \cdot CH_2 \cdot CHOH \cdot CH_3$	III	22.0	3.15	100	24
Cyclohexanol	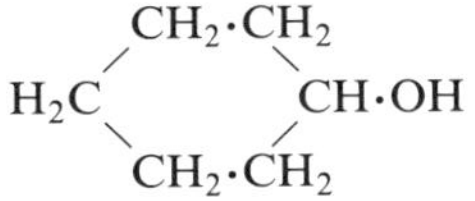 (cyclohexyl–OH structure)	III	23.3	52.7[a]	162	68
Ethylene glycol	$HO \cdot CH_2 \cdot CH_2 \cdot OH$	III	29.0	17[a]	198	111
Glycerol	$CH_2OH \cdot CHOH \cdot CH_2OH$	III	33.7	494[b]	290	160
Esters						
Methyl acetate	$CH_3 \cdot CO \cdot O \cdot CH_3$	II	19.6	0.38[a]	57	–9
Ethyl acetate	$CH_3 \cdot CO \cdot O \cdot C_2H_5$	II	18.6	0.455	77	–4
Butyl acetate	$CH_3 \cdot CO \cdot O \cdot C_4H_9$	II	17.3	0.671[a]	127	23
Methoxypropyl acetate	$CH_3 \cdot CO \cdot O \cdot CH(CH_3) \cdot CH_2 \cdot O \cdot CH_3$	II	18.8	1.2	140–150	46
Ketones						
Acetone	$CH_3 \cdot CO \cdot CH_3$	II	20.4	0.316[a]	56	–17
Methyl ethyl ketone	$CH_3 \cdot CO \cdot C_2H_5$	II	19.0	0.42[a]	80	–4
Methyl isobutyl ketone	$CH_3 \cdot CO \cdot CH_2 \cdot CH(CH_3)_2$	II	17.1	0.546[a]	116	16
Cyclohexanone	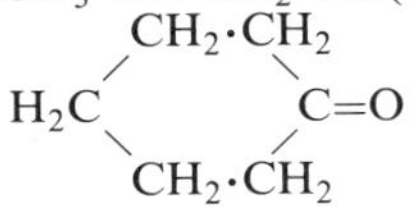 (cyclohexanone structure)	II	20.2	1.94	157	47
Methyl cyclohexanone	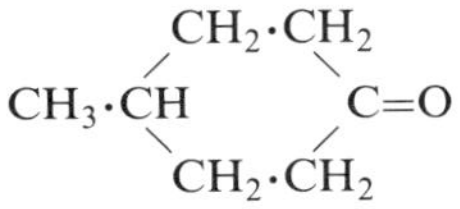 (methyl cyclohexanone structure) and isomers	II	19.0	1.75	165–175	47

Table 9.1 Solvent properties (*continued*)

Solvent	Formula	Solvency		Viscosity at 20 °C (centipoises or mPa s)	Boiling point (°C)	Flash point (closed cup) (°C)
		H-bonding group	Solubility parameter			
Ethers and ether-alcohols						
Diethyl ether	$C_2H_5 \cdot O \cdot C_2H_5$	II	15.1	0.233	35	−40
1-Methoxy propan-2-ol	$CH_3 \cdot O \cdot CH_2 \cdot CHOH \cdot CH_3$	II	20.8	1.65[a]	120	38[d]
1-Ethoxy propan-2-ol	$CH_3 \cdot CH_2 \cdot O \cdot CH_2 \cdot CHOH \cdot CH_3$	II	18.4	1.68[a]	132	43[d]
2-Butoxy ethanol	$C_4H_9 \cdot O \cdot CH_2 \cdot CH_2 \cdot OH$	II	18.2	3.318[a]	171	61
Diethylene glycol (DEG)	$HO \cdot CH_2 \cdot CH_2 \cdot O \cdot CH_2 \cdot CH_2 \cdot OH$	III	18.6	30[a]	245–250	124
DEG monomethyl ether	$CH_3 \cdot O \cdot CH_2 \cdot CH_2 \cdot O \cdot CH_2 \cdot CH_2 \cdot OH$	II	19.6	3.53[a]	194	93[d]
DEG monobutyl ether	$CH_3 \cdot (CH_2)_3 \cdot O \cdot CH_2 \cdot CH_2 \cdot O \cdot CH_2 \cdot CH_2 \cdot OH$	II	20.4	4.74[a]	232	232[d]
Nitroparaffins						
Nitromethane	$CH_3 \cdot NO_2$	I	25.9	0.62	101	35
Nitroethane	$C_2H_5 \cdot NO_2$	I	22.6	0.62	114	28
1-Nitropropane	$CH_3 \cdot CH_2 \cdot CH_2 \cdot NO_2$	I	21.8	0.81	132	34[d]
Chlorinated paraffins						
Methylene chloride	CH_2Cl_2	I	19.8	0.425[a]	41	None
Ethylene dichloride	$CH_2Cl \cdot CH_2Cl$	I	20.0	0.838	84	13
1,1,1-Trichloroethane	$CCl_3 \cdot CH_3$	I	17.5	0.83	72–88	None

[a] at 25 °C; [b] at 26.5 °C; [c] minimum; [d] open cup.

solution, the strong attractions operate often at short range to overcome the energy of movement of the free solid molecules, so that they group together, take up the crystal pattern and crystallize. The concentration of dissolved solid drops below the limiting figure and crystallization ceases. Amorphous polymers, as we have seen, cannot crystallize. Since the forces of attraction between polymer and solvent molecules are as strong as those between the polymer molecules, dissolving the polymer is like mixing a very viscous liquid with a very fluid one. There is no solubility limit; a 'good' solvent is miscible with the polymer in all proportions.

It is not easy to look at the chemical composition of a polymer and predict solvents for it. A general guide is that like dissolves like, but this is limited, sometimes wrong, and gives no indication of what will happen if a liquid mixture is used. There is still no complete explanation of polymer solvency, but this property was rationalized for the paint formulator in 1955 by the American paint chemist, Burrell. He suggested that for every liquid two factors, or parameters, govern the solvency. The first is the hydrogen bonding capacity of the liquid. He classified paint solvents into three groups:

I. weakly hydrogen-bonded liquids (hydrocarbons, chloro- and nitro-paraffins);
II. moderately hydrogen-bonded liquids (ketones, esters, ethers and ether-alcohols);
III. strongly hydrogen-bonded liquids (alcohols and water).

For each solvent, a second parameter is calculated from the latent heat of evaporation, by use of the equations of thermodynamics, and the numerical value of this parameter is a measure of the attractive force between the liquid molecules. This parameter is called the **solubility parameter**.

A polymer has a solubility parameter range for each group of solvents. Polymer parameters may be calculated, but are usually obtained by experiment. A selection of solvents, uniformly covering the whole spectrum of parameters, is assembled. A typical selection is shown in Table 9.2. Attempts are made to dissolve the resin at a technically useful 'solids' level (e.g. 20% for nitrocellulose, 50% for an alkyd) in chosen solvents. If the polymer dissolves in any two solvents in a group, it will dissolve in all the solvents of that group with parameters between those of the chosen pair. Thus the object is to establish the extremes of the range in each group, giving three parameter ranges for the polymer. Table 9.3 gives parameter ranges for some resins.

An untried liquid will dissolve the polymer if its own parameter falls within the polymer's appropriate parameter range. Roughly speaking, the parameter of a mixture of liquids is the average parameter. For example, a polyester resin would not dissolve in *n*-butanol (23.3, group III) or xylene (18.0, group I). However, a 1:4 butanol–xylene mixture dissolved the resin. The mixture's

Table 9.2 Spectrum of solubility parameters

Group I
14.1	Low odour mineral spirits
15.1	*n*-Heptane
16.7	Cyclohexane
17.3	Dipentene
18.2	Toluene
19.0	Trichloroethylene
19.4	Tetralin
20.4	Nitrobenzene
21.8	1-Nitropropane
22.6	Nitroethane
24.3	Acetonitrile
25.9	Nitromethane

Group II
15.1	Diethyl ether
16.3	Methylamyl acetate
17.3	Butyl acetate
18.2	DEG monobutyl ether
19.0	Dibutyl phthalate
20.0	2-Butoxyethanol
21.2	Cyclopentanone
21.8	DEG monomethyl ether
24.7	2,3-Butylene carbonate
27.1	Propylene carbonate
30.0	Ethylene carbonate

Group III
19.4	2-Ethyl hexanol
21.0	*n*-Octanol
22.2	*n*-Amyl alcohol
23.3	*n*-Butanol
24.3	*n*-Propanol
25.9	Ethanol
29.6	Methanol

parameter is roughly calculated thus:

$$\text{butanol contribution} \quad 1 \times 23.3 = 23.3$$

$$\text{xylene contribution} \quad 4 \times 18.0 = \underline{72.0}$$

$$\text{mixture's parameter} \quad = 95.3 \div 5 = 19.1$$

The hydrogen bonding capacity of the mixture corresponds approximately to group II. Note that ethyl acetate (18.6, group II) is a good polyester resin solvent. Where many liquids are involved, the approximation may be too great, but the effect of adding a new liquid to a paint may best be assessed by deciding what effect it will have on the hydrogen bonding

Table 9.3 Solubility parameter ranges of some polymers

Polymer	Chemical type	Parameter range in		
		Group I	Group II	Group III
NC, RS 25 cps/m Pa s (dry)	Nitrocellulose	22.6–25.9	15.9–30.0	29.6
NC, SS $\frac{1}{2}$ sec (dry)	Nitrocellulose	22.6–25.9	15.9–30.0	25.9–29.6
CAB $\frac{1}{2}$ sec	Cellulose acetate butyrate	22.6–25.9	17.3–30.0	25.9–29.6
EC, N-22	Ethyl cellulose	16.5–22.8	15.1–22.4	19.4–29.6
PMM	Polymethyl methacrylate	18.2–25.9	17.3–27.1	0
Acryloid B-72	Acrylic copolymer	21.6–25.9	18.2–27.1	0
Vinylite AYAA	Polyvinyl acetate	18.2–25.9	17.3–30.0	0
45% O.L. linseed glycerol phthalate	Alkyd	14.3–24.3	15.1–22.4	19.4–22.3
30% O.L. soya glycerol phthalate	Alkyd	17.3–25.9	7.1–30.0	0
Beetle 227–8 (dry)	Urea formaldehyde	0	0	19.4–22.3
Uformite MX-61 (dry)	Melamine formaldehyde	17.3–22.6	15.1–22.4	19.4–22.3
Epikote 1001	Epoxy resin	21.6–22.6	18.2–27.1	0
Epikote 1004 and 1007	Epoxy resin	0	17.1–27.1	0
Epikote 1009	Epoxy resin	0	17.1–20.2	0
Epikote 1004 DHC ester	Epoxy ester	17.3–22.6	15.9–20.2	0
Propylene glycol maleate phthalate	Unsaturated polyester	18.8–25.9	16.3–30.0	0

Some of the figures in this table do not appear in Table 9.2. This is because the choice of solvents in Table 9.2 is merely an example. Other solvents can be, and often are, chosen for parameter range measurements.

characteristics of the liquid mixture and whether it will raise or lower the average parameter.

The simple initial concept of Burrell outlined above is still a very useful one for the paint chemist and technologist, and the use of this concept has been refined by others. It is likely that readers will encounter references to the **Hansen 3-D solubility parameter**. The Danish chemist Hansen split the solubility parameter as used above into three components representing dispersion, polar and hydrogen bonding contributions. These are related by the equation

$$\delta^2 = \delta d^2 + \delta p^2 + \delta h^2$$

The solubility parameter of a solvent may be represented as a single point in a three-dimensional graph, where the axes of the graph are the polar, hydrogen bonding and dispersion contributions. The range of solubility parameters for a polymer becomes a **solubility sphere** (Fig. 9.1). To give a sphere, the δd axis is plotted using a scale twice that of the δp and δh parameters.

Just as with the single parameter above, the method may be illustrated using the example of a xylene–butanol mixture. We calculate the dissolving power of individual solvents from the separation between the centre of the resin's solubility sphere and the solvent's value, relative to the calculated sphere radius (R), using the equation for S_{rel}

$$S_{rel} = \frac{\{4(\delta d^s - \delta d^p)^2 + (\delta p^s - \delta p^p)^2 + (\delta h^s - \delta h^p)^2\}^{1/2}}{R}$$

For non-solvents, S_{rel} will be greater than 1, while for good solvents, S_{rel} will be less than 1, and hence the solvent's parameter will be inside the solubility sphere. Thus for an epoxy resin, the centre of the sphere is $\delta d = 19.8$, $\delta p = 10.6$ and $\delta h = 10.3$, with $R = 9.6$. For xylene, $S_{rel} = 1.24$ and n-butanol $= 1.09$, and both would be poor solvents if used alone. However, a blend of 62% xylene and 38% n-butanol has $S_{rel} = 0.80$, and the solubility parameter of this blend lies inside the solubility sphere of the

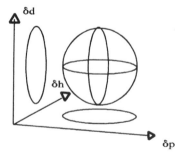

Fig. 9.1 Graphical representation of the Hansen solubility sphere of a polymer.

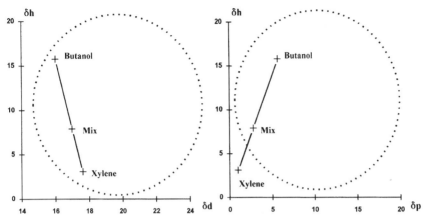

Fig. 9.2 Projections of the solubility sphere of epoxy resin.

polymer and is predicted to dissolve the resin as illustrated (Fig. 9.2). The reader should note that until recently, older units for solubility parameter $(\text{cal cm}^{-3})^{1/2}$ were used. These values are now more properly quoted using $(\text{J cm}^{-3})^{1/2}$ where all solubility parameters are 2.04 times greater. The older units may, however, still be encountered.

The solubility parameter also helps us to understand how different water is, when we compare its value of 47.7 with those of other solvents in Table 9.1; almost all organic solvents have half this value or less. Examination of the Hansen components for water (not shown) would reflect the high values for both the polar and hydrogen bonding components.

Viscosity

Measurement

The viscosity of a simple liquid has been explained and defined in Chapter 1. It is the outward evidence of the internal resistance to flow and can be measured in units called 'poises', or in SI units of 'pascal-seconds'. Both are used in this section of the book; instruments currently in use may still be mainly calibrated in poises. Viscosity can be measured by any method involving either the flow of the liquid, or the movement of some object in the liquid.

In the first category are the simple flow cups, perhaps the most widely used consistency control devices in the paint industry. In these, a given volume of liquid is timed as it falls through a hole of definite dimensions. A viscosity in poise or Pa s is not obtained; the flow time is in itself a sufficient comparative measure of consistency. The consistency for a spraying paint, for example, can be specified as a given flow time in a particular flow cup at an appropriate temperature (usually 25 °C). However, viscosities read directly in viscosity units are obtained easily if the viscometer contains a paddle or disc, mechanically driven to rotate the liquid in an enclosed space. The force tending to twist the stationary part of the apparatus in contact with the liquid is usually measured and the scale is calibrated in poises or Pa s. Comparisons between measurements made in different types of instrument can then be made. Another advantage is that either the rate of shear (proportional to the speed of rotation) or the stress (proportional to the driving force) can be varied, depending on the instrument.

The second category of instrument may alternatively involve a ball falling through a liquid, or rolling in a tube containing the liquid. The time taken for movement over a given distance is measured, but conversion of the answer to poises is easily done. The liquids need to be at least partly transparent to use this type of measurement. Another type, more suitable for very viscous liquids, involves the rise of a bubble in a tube containing the liquid. The time of rise in seconds can be converted to viscosity units. This is typically used for resin solutions and frequently in resin processing. For viscous liquids there is also a method in which the vibration of a rod is damped by the liquid, the effect being related to the viscosity. Results can be obtained in poises or Pa s.

Facts and theory

There are three essential facts concerning the viscosity of a polymer solution. Let us consider them in turn.

• Early in this chapter it was stated that polymers do not have a solubility limit because they cannot crystallize. Nevertheless there are forces of attraction operating between the polymer molecules in solution. Though these forces may be weak compared with those operating in, say, a sodium chloride solution, they operate over a very long length of molecule and there are frequent encounters with other long molecules, enabling those forces to come into play. In addition, there is the possibility of simple mechanical tangling, as with pieces of string. Since both factors have been quoted in Chapter 1 as causes of increasing viscosity, it is not surprising that even comparatively low concentrations of polymer can cause considerable thickening of simple liquids. *As the 'solids' of the solution increase*, the encounters and entanglements between molecules become more frequent and *so the viscosity increases*, e.g. solution viscosities for RS $\frac{1}{2}$ second nitrocellulose (Chapter 11): 12%, 0.1 Pa s; 20%, 1 Pa s; 30%, 10 Pa s. Eventually the solution becomes so viscous that it cannot be used for paints unless further liquid is added. At higher 'solids' still, the solution almost creases to flow and might be mistaken for a solid. So, although there is no limit at which the solution becomes saturated, there is a limit at which the solution becomes too viscous to use. No precise figure can be quoted for this, since it depends upon the use.

- If we have two solutions of the same polymer, e.g. polymethyl methacrylate, in the same solvent at the same level of solids, the more viscous solution will contain polymer molecules of higher molecular weight. Since we have the same total weight of polymer in both solutions, there are fewer polymer molecules present in the high molecular weight polymer solution, but they are longer. The increased opportunity for entanglement reinforced by chemical attraction outweighs the reduced number of molecules. All the attractions between polymer molecules in a useful solution are continuously forming and breaking with the movement of the molecules. If they were permanent, the solution would not flow, since the polymer molecules would form a semi-rigid, reinforcing network in the liquid. However, to double the molecular weight and halve the number of molecules involves making a number of these temporary associations into permanent chemical bonds. The network is not made rigid, but it is a good deal less flexible and less amenable to being deformed in flow than before. *Thus high molecular weight polymers give more viscous solutions*, e.g. 10% solutions of PVA of molecular weight 15 000, 2.5 mPa s; 73 000, 12 mPa s; 160 000, 88 mPa s.

- If we take a given sample of polymer of fixed molecular weight and dissolve it at the same 'solids' in a variety of 'true solvents', *the viscosities of the solutions will be proportional to the viscosities of the original solvents*, e.g. 12% polystyrene in methyl ethyl ketone (0.4 mPa s), 40 mPa s; in ethyl benzene (0.7 mPa s), 160 mPa s; in *o*-dichlorobenzene (1.3 mPa s), 330 mPa s. This is important, because we can reduce the paint viscosity without lowering the solids or polymer molecular weight, simply by changing to a less viscous solvent, if a suitable one is available.

Solvents in the middle 80% of the polymer's parameter range are almost certainly 'good' solvents. Good solvents dissolve the polymer at all concentrations. Some liquids ('poor' solvents) give solutions at certain concentrations, but further dilution precipitates the polymer. In these quasi-solutions, polymer molecules collect in clusters, raising the effective molecular weight of the polymer and giving an abnormally high viscosity.

It is worth noting here that if good solvents in a solution are partly replaced by non-solvents, or if poor solvent is used instead, the viscosity can rise. This happens when the average solubility parameter of the mixture moves to the extreme of the range for the polymer and clustering of polymer molecules occurs. This is the stage prior to precipitation.

Finally, a word about the viscosity of emulsions. If a polymer is fully emulsified in a liquid and no part of it is in solution, then the molecular entanglements and associations do not occur. Consequently the *emulsion viscosity must be independent of the molecular weight of the dispersed*

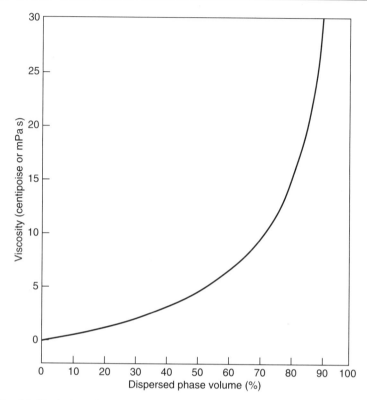

Fig. 9.3 Variation of emulsion viscosity with dispersed phase concentration.

polymer. At low solids the viscosity of the emulsion is the viscosity of the continuous phase liquid. The viscosity of the emulsion rises slightly as the solids rise, because of polymer–particle collisions and weak interparticle attractions, but there is no substantial increase in viscosity until the concentration of particles becomes so high that the particles can scarcely move at all. A typical solids–viscosity curve is shown in Fig. 9.3. Thus the paint formulator can use a high molecular weight polymer at high concentrations only by using it in emulsion or dispersion form.

Boiling point and evaporation rate

We have already seen that the flow of paint on vertical surfaces can be controlled by solvent operation. To get the right solvent balance for doing this satisfactorily, it is necessary to know the relative evaporation rates of solvents available.

In particular we wish to know the evaporation rates of thin films of solvent at normal drying temperatures. These figures are available for many solvents at room temperature (20–25 °C). One system quotes relative evaporation *rates*, with butyl acetate as the reference solvent rated arbitrarily as 100. Thus isopropanol at 200 evaporates twice as quickly. Another system quotes relative evaporation *times*. Diethyl ether is the reference solvent, with the rating of 1.0. *n*-Butanol, with a value of 33, takes 33 times as long to evaporate.

In practice, however, we are not concerned with the evaporation of thin films of single liquids under carefully controlled conditions of temperature, humidity and air movement. We are concerned with the evaporation of mixtures of liquids in the presence of polymers under varying atmospheric conditions. Evaporation rates of solvents in mixtures cannot be predicted from the individual evaporation rates: attractive forces between molecules can delay evaporation and these forces will vary from mixture to mixture.

Attraction for the polymer molecules will also delay evaporation, so good solvents for the dissolved polymer will evaporate more slowly than evaporation rates suggest, while diluents will be least affected. At best, the published evaporation rates are a rough guide on which to base solvent selection and they are only suitable for paints that lose most of their solvent at room temperature. It is necessary to follow this with actual trial of the proposed liquid mixture in the paint. This may need to be repeated several times, with logical replacement of solvents by those evaporating more slowly, to increase flow, or by others evaporating more rapidly, if excess flow (sagging) is occurring.

Because evaporation rates are no more than a guide, many paint chemists feel justified in using as a guide the *boiling point* (BP) of a pure solvent, or the boiling range (BR) of an impure solvent or mixture. The boiling point is a fixed property of a liquid and does not vary with the method of measurement provided that the atmospheric pressure is 760 mm Hg. It is usually true that liquids with low boiling points evaporate more rapidly at room temperature than those with high boiling points. But the boiling point tells you when liquid will *boil*, not *how fast* it will *evaporate* at some lower temperature.

When a pair of liquids have boiling points within 30 °C of one another, it is difficult to say which will evaporate more rapidly at room temperature. Alcohols, in particular, are slower than their boiling points suggest, because they are strongly hydrogen-bonded (group III in Table 9.2).

In spite of this, boiling points are a sufficiently good guide for most paint purposes. Solvents are graded roughly into three groups: 'low boilers' (BP below 100 °C), 'medium boilers' (BP 100–150 °C) and 'high boilers' (BP above 150 °C). Low boilers are used in spraying paints because they evaporate between gun and surface and give the necessary increase in solids and viscosity. High boilers are used to give flow and can be the sole solvents where the paint must be kept fluid for fairly long periods, e.g. where brushed paint must 'marry in' when two areas overlap. Medium boilers can be used in all types of paint to give flow at first, followed by fairly quick set-up, but in spraying paints they necessitate greater care by the user, since the nearer the spray-gun gets to the work, the wetter the applied paint will be – the solvent has less time in which to evaporate.

Most paints contain as much diluent as possible, since diluents are usually aliphatic hydrocarbons, which are much cheaper than the true solvents. The limit is decided for the formulator by the fact that the polymer must be kept in solution in the can and at all stages of application and drying. For this reason, the evaporation rates of diluents and solvents have to be balanced carefully, to make sure that there is always enough true solvent present and, particularly, to ensure that the last molecules to evaporate are solvent molecules. If there is too much diluent at any stage, the polymer precipitates and the film appears milky if it is unpigmented, or low in gloss if it is pigmented.

Flash point

Flash points give an indication of flammability or fire risk. The flash point is the lowest temperature at which enough vapour is given off to form a mixture of air and vapour immediately above the liquid, which can be ignited by a spark or flame under specified conditions. Most countries have regulations concerning the storage, transport and use of products containing the more highly flammable solvents. Highly flammable materials are usually defined as those having a flash point below a certain figure and they may also be required to support combustion in a combustion test. The latter clause is introduced so that paints which are safer to use, but have a measurable – and perhaps low – flash point are not penalized. For example, a water borne paint may contain a small proportion of low boiling alcohol. This could be detected in a flash point test on the paint, yet the paint might not support combustion. The combination of a flash point below 32 °C and combustibility is used in the *UK Highly Flammable Liquids and LPG Regulations 1992*.

Since regulations may vary so much from country to country and even may vary according to use or method of transport, it is usually necessary to know accurately the flash point of any paint. The flash point of a mixture of liquids can be lower than the flash point of any of its constituents, so measurements should always be made on the actual sample. These are usually carried out in 'closed' or 'open' cups of varying design and it is important to state which cup was used. Closed cups give lower flash points.

One further property relating to flammability is the **auto-ignition temperature**. This is the temperature at which ignition of vapour can occur spontaneously at a hot surface without the presence of a separate ignition source. This is generally of concern in processing, particularly in resin making when using certain linear aliphatic hydrocarbons at temperatures above 200 °C.

Chemical nature

It is easy to forget that solvents are chemical and can react with other paint ingredients in some circumstances. For stability in the can, this is undesirable. Examples of the influence of chemical reactivity on solvent selection are given in Chapters 14 and 15.

Toxicity and smell

Some liquids, e.g. benzene, have a cumulative poisonous effect and others can be harmful above certain concentrations in the air. The smell of a liquid may be enough to prevent its use for some purposes. Smell is largely a matter of customer's opinion, but toxicity information is obtainable from the sources quoted in the Appendix.

Safe working conditions are established in the UK by the **Occupational Exposure Limits** (OELs) set by the Health and Safety Executive (HSE), and in the USA by the **Threshold Limit Values** (TLVs) set by the American Conference of Government Industrial Hygienists (ACGIH). OELs and TLVs are concentrations of substances in the working atmosphere which, it is believed, people may be exposed to day after day without adverse effect. They may be quoted as **long term exposure limits** (*time-weighted averages* for a normal working day), or as **short term exposure limits** (maximum concentrations for 10 or 15 min exposure) or absolute **ceiling limits** which should never be exceeded. OELs and TLVs are quoted as parts per million (p.p.m.) or $mg\,m^{-3}$, and are available not only for vapours but also for dusts. The publications of the HSE and the ACGIH also draw attention to absorption through the skin where this is an important additional hazard.

Examples of long term exposure limits for common solvents are 50 p.p.m. for *n*-butanol, 100 for xylene, 200 for methyl ethyl ketone, 400 for ethyl acetate and 1000 for ethanol (1995–96 values). The concentrations of these vapours in the working environment can be measured and, if OELs are being exceeded, it is often possible to solve the problem by changing working methods, e.g. by installing vapour extraction units. Sometimes the alternative of reformulating the paint proves to be the only solution. This is particularly the case when the long term exposure limit for a solvent is very low (say, less than 10 p.p.m.) and its volatility is high.

Pollution concerns, if paints or solvents containers are damaged in transport, also now influence labelling, and certain white spirits require labelling to warn of the possibility of leakage causing environmental damage.

Cost

Apart from water, hydrocarbons (particularly aliphatic) are the cheapest solvents; some esters, less common ketones and nitroparaffins are the most expensive.

Ten

Paint additives

modifying application, curing and appearance

All the main ingredients of a paint shown in the summary on p. 93, Fig. 7.1, have now been discussed in more detail, but one class of ingredient remains: the additives. These are those remarkable materials which, when added to a paint in amounts that can be as little as 0.001% and seldom more than 5%, have a profound influence on the physical and chemical properties of the paint.

Additives affecting viscosity

In the last chapter we dealt with the viscosities of liquids and polymer solutions. Emphasis was laid on the use of instruments which measure the viscosity in poises. Even better are the types which allow either the stress or the rate of shear to be varied.

With pure liquids, such variations are unnecessary, because the viscosity coefficient is a constant. Referring to Newton's equation in Chapter 1, as the stress increases, the rate of shear does so at the same rate and vice versa. This is also true for polymer *solutions* at paint viscosities. If, however, the 'foreign' material in the liquid is also insoluble in it, the equation may be true no longer. Since most paints contain undissolved materials, it becomes important in many instances to quote not only the viscosity in poises and the temperature of measurement, but also *either* the rate of shear *or* the stress at which the measurement was made.

Why does this deviation from constant viscosity occur? An ideal (pigment) particle dispersion, as we have seen in Chapter 8, consists of a completely uniform distribution of isolated individual particles throughout the liquid. In the same chapter we saw how chemical groups at pigment surfaces can exert strong attractions over other suitable bodies in the vicinity. Our ideally distributed particles will of course move, since they are affected by gravity and by collisions with fast-moving liquid molecules, or slow but heavier polymer molecules. During the course of these movements they are bound to encounter other particles. Should suitable parts

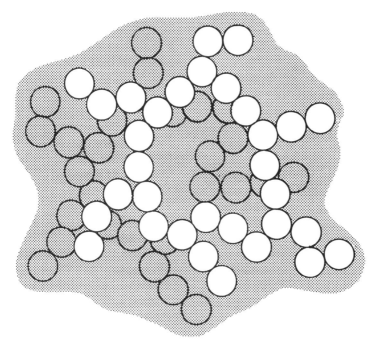

Fig. 10.1 A structure of flocculated particles.

of the surfaces of two particles come into contact, inter-attractions may be strong enough to prevent the particles from separating by their own motion. If this pairing up process continues, quite large groups or clusters may form. A cage-like network or structure may even extend throughout the liquid (Fig. 10.1).

Such an arrangement might give the appearance, in the can, of being very viscous, even jelly-like. But once the paint is stirred, the structure breaks up, the particles separate, the liquid 'gives' under the stirrer and the viscosity appears to drop. Should the stirring stop or slacken, the flocculation (as it is called) will begin again and the viscosity will rise. Thus the paint appears to have two viscosities: a high one when it is still and a low one when it is agitated or sheared. In fact, it has a wide range of viscosities, corresponding to all the rates of shear between zero and some value at which the viscosity becomes a minimum, or to all the stages of partial flocculation between complete flocculation and complete deflocculation.

If flocculation occurs slowly, the viscosity, measured at low rates of shear, increases with time during the rest period after an efficient shearing. When this happens the paint is said to be **thixotropic**. If there is no dependence on time or on the previous treatment of the paint and if the viscosity decreases as the rate of shear increases, then the paint is said to be **pseudo-plastic**. If there is a minimum stress required before any flow can occur at all, the viscosity behaviour is said to be **plastic**. All these types of behaviour are, of course, contrary to Newton's equation and are grouped together under the heading of **non-Newtonian viscosity**.

Most paints show non-Newtonian viscosity to some degree and, if it is marked, it can be most beneficial to the paint. Thus the viscous – even jelly-like – paint in the can does not show hard pigment settlement, applies easily under the shearing action of, for example, a brush or spray-gun, and does not 'sag', because the viscosity rises as soon as the paint is still (or nearly so) on the object being coated. However, non-Newtonian viscosity should be introduced into a paint with caution. While it is true that the paint will not 'sag', it may be extremely difficult to adjust it to obtain reasonable flow-out of application marks as well. Also, if the effect is obtained by pigment flocculation, the perfectly uniform pigment distribution which gives the maximum light scattering and absorption will be lost and hiding power will fall. Let us therefore consider ways and means of achieving non-Newtonian viscosity and discuss their relative merits.

Pigment volume

If the level of pigment in the paint is high, the very bulk of pigment in the paint will cause non-Newtonian viscosity. The close packing of the particles makes some flocculation inevitable and the viscosity will be high simply because the particles impede each other's movements. A further phenomenon, **dilatancy**, can occur at very high pigment volumes. Here, disturbing the arrangement of the particles can pack them into an even tighter mass and, on stirring, the viscosity rises and the paint appears to set solid. When stirring ceases, fluidity apparently returns to the paint. Dilatancy, particularly of pigment settlement in cans of paint, is usually highly undesirable.

Silica and silicates

Dispersion in the paint of a few percent of very fine particle silica, SiO_2 (diameter about $0.015\,\mu m$, $15\,nm$), produces a largely pseudo-plastic effect. The surface forces and very large surface area (e.g. $190–460\,m^2\,g^{-1}$ or $1.3–3.3$ acres per ounce!) are responsible.

Fine particle synthetic aluminium silicates $(Al_2O_3\cdot2SiO_2\cdot2H_2O)$ are also used in paints containing organic solvents. Bentonite is used to impart non-Newtonian behaviour to water paints. It is a complex silicate of the montmorillonite type, containing aluminium, sodium, potassium, iron and magnesium. It has a plate-like particle and will swell to several times its dry volume by adsorption of water between silicate layers.

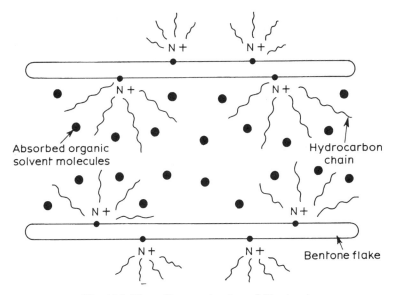

Fig. 10.2 The gelling mechanism of 'Bentone'.

The metal ions in montmorillonites can be exchanged for quaternary ammonium ions, R_4N^+, and when R is a long chain alkyl group, the character of the clay is changed considerably. Organic hydrocarbons are now adsorbed onto the platelet surfaces because of their attraction for the alkyl groups. Thus these treated bentonites, given the trade name 'Bentones', will gel organic liquids or impart non-Newtonian viscosity to paints containing organic solvents. Attractive forces between the particles lead to structure throughout the paint (Fig. 10.2).

Resinous thickeners

Uncrosslinked thickeners, non-aqueous

There are a number of polymers, notably hydrogenated castor oil and its derivatives, polyamides (Chapter 15) and polyamide–oil or polyamide–alkyd reaction products (Chapter 12), which can be used to impart non-Newtonian viscosity to paints based on non-polar (mainly aliphatic hydrocarbon) solvents. The thickening mechanisms of polymeric additives are not fully established, but the resins have in common the following features: borderline solubility in the paints in which they are used and chemical structures involving lengthy soluble non-polar chains (e.g. fatty portions of fatty acids) and polar groups, e.g. −OH, −CONH− and −COOH.

The thickening mechanism is probably similar to that of the particles in the section on silica and silicates (above). A loose network forms, either between very finely divided colloidal particles of resin, via surface forces due to the attractions between polar groups, or between polar portions of large resin molecules. A colloidal particle might have a diameter of 0.01 μm (10 nm) or less. If a single molecule of a polymer of density 1.0 was compactly coiled so as to form a sphere of diameter 10 nm, the molecular weight of the polymer would be 300 000. If a particle of the same size contained solvent molecules, the polymer molecular weight necessary to give that diameter could be much less than 300 000.

Depending on the molecular weight of the polymer in question, such colloidal particles in paint could be clusters of 2–100 molecules. The difference between small polymer particles and large molecules is not great.

Uncrosslinked thickeners, aqueous

Resinous thickeners are also used in aqueous paints. The mechanism of operation is similar, but the polarities of the functioning groups are reversed. Thus solubility is conferred by water-soluble polar chain lengths, such as polyethylene glycol or neutralized polyacrylic acid, and attractions form between the water-insoluble (or hydrophobic) portions of the polymer. The latter are usually concentrated as substantial segments at the ends of the chains (e.g. as in A–B–A block copolymers), or at intervals along the chains. In emulsion paints (Chapter 11) these **associative thickeners** not only form structures between their own molecules, but also anchor onto the hydrophobic portions of the latex polymer or pigment particles, which thus take part in forming a network structure. One important class is the hydrophobically modified ethoxylated urethane (HEUR), based on polyethylene glycol (PEG) and a diisocyanate, and having terminal associating hydrophobic groups, typically C_{12}–C_{18} hydrocarbon chains:

$$R–II–PEG–[II–PEG]_n–II–R$$

where R = associating group, II = diisocyanate constituent. *n* may be 0, 1, 2, etc. The most important conformations that these thickeners can adopt in a latex are loops and links:

where  may be clusters of associating blocks alone or blocks anchored onto pigment or latex.

At low shear rates the ratio between loops and links is constant, but at higher shear rates the links will break faster than they reform, and hence the structure breaks down and the viscosity decreases.

Uncrosslinked thickeners, solvent environment

We should bear in mind the effect of the solvent environment on the thickeners described so far. In non-aqueous paints, polar groups on particle surfaces are even more accessible to small polar molecules (additives or solvents) than to polar groups on other particles or large molecules. Once these small molecules are in place, they prevent particle–particle or molecule–molecule associations by 'blocking' the active sites. Exactly the same is true (with polarities reversed) in aqueous systems.

Alternatively, the addition of polar solvents to hydrocarbon-based paints will improve the solubility of the thickener molecules, and decrease the chance of particles or clusters of molecules forming because of borderline solubility. This will lessen or even remove the effectiveness of the thickener. Again the same principles apply to aqueous coatings: additions of water-miscible organic solvents (e.g. ether-alcohols) increase the solvency of the medium for the hydrophobic portions of the associative thickeners and lessen their tendency to associate.

Microgels

The resinous thickeners above are thermoplastic and all ultimately soluble in one solvent blend or another. As we have seen, their low shear effects can be destroyed by solvent additions. If, however, the polymers are made as latexes or non-aqueous dispersions by emulsion or dispersion polymerization (Chapter 11) and a small amount of cross-linking is introduced via polyfunctional monomer, then insoluble colloidal resin particles are produced. These are called **microgels**.

Microgels thicken paint by forming flocculated network structures. In solvent-borne systems they are not insensitive to solvent selection, but are much less sensitive and are effective in polar organic solvents. Indeed, swelling of the polymer particles by solvent enhances their effect, by increasing particle diameter and hence increasing the apparent gel volume.

A special kind of microgel is the **core–shell** particle (Fig. 10.3). Here a lightly crosslinked polymer core is contained inside an uncrosslinked outer shell, grown or grafted onto it. The outer shell polymer would be soluble in the liquid medium were it not attached to the inner core. Some core–shell microgels are made to function as thickeners in solvent-borne paints, while others are designed for water-based coatings and function in aqueous media; in both cases the structure of the particles is essentially the same. That is, the shell is solvated by the continuous phase, whether it is hydrocarbon or water.

In the case of the aqueous microgels, the solubilizing shell is a copolymer containing water-soluble monomers such as methacrylic acid. The major effect in these microgels is from the greatly enlarged volume of the shell

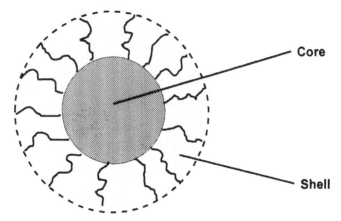

Fig. 10.3 Structure of a core–shell particle.

polymer following neutralization. Their rheological behaviour can be modified by the ratio of core to shell, the concentration of ionized groups and the proportion of cosolvent.

The use of microgels has been particularly successful in solvent-borne and water-borne metallic basecoats used in motor car production. Here the metallic visual effect requires viscosity control during the high film shrinkage that occurs on solvent evaporation, so that the metallic aluminium flakes align parallel to the substrate in the dry film.

Versions which function in aqueous media are also used in automotive refinishing metallic paint to give high viscosity immediately after application to prevent sagging and aluminium flotation. Their use has enabled the development of water-borne car refinishes, which might otherwise only have achieved solvent reduction by the high solids route.

Microgels can also be used in clear coatings without loss of clarity if they are designed to have the same refractive indexes as the film-formers.

Metal chelates

Metal chelates are organometallic compounds in which organic chains are bonded to a central metal atom partly by covalent and partly by coordinate bonds, for example:

$$(HO{\cdot}CH_2{\cdot}CH_2)_2N: \longrightarrow Ti \longleftarrow :N(CH_2{\cdot}CH_2{\cdot}OH)_2$$

with $CH_2{\cdot}CH_2{\cdot}O$ and $O{\cdot}CH(CH_3)_2$ above, and $(CH_3)_2CH{\cdot}O$ and $O{\cdot}CH_2{\cdot}CH_2$ below.

bis (triethanolamine) titanium diisopropoxide

Some chelates of titanium and zirconium have proved especially useful for the thickening of colloid-stabilized aqueous latex paints (Chapter 11). They can induce considerable structure at low shear, leading to non-drip characteristics and good flow with control of runs and sags.

The titanium chelates work best with latexes stabilized by cellulose ether colloids and thickening is thought to occur as the result of hydrogen bonding between hydroxyls in the chelate and in the colloid. An unacceptable irreversible thickening occurs with polyvinyl alcohol (PVA) colloid, due to interchange at the covalent bond between the PVA and the alkoxy group. High surfactant levels interfere with thickening.

The zirconium chelates work with latexes containing sodium carboxymethyl cellulose or cellulose ethers or polyvinyl alcohol, but in each case the pH of the paint must be carefully controlled.

Both types of chelate are effective at or below 1% and the structure formed is thixotropic. If stirred, the paint slowly recovers structure, but never quite to the original value and there is some loss of structure on each repetition of stirring. Methods of manufacture of the paints must be closely controlled.

Choice of additives affecting viscosity

The additives that induce non-Newtonian viscosity in paints should be used as sparingly as possible. The paint formulator will find that it is extremely difficult to make several batches of paint to the same formula and the same viscosity and hence the same application characteristics. Some of the resinous thickeners are themselves difficult to reproduce exactly from batch to batch and the mineral thickeners are difficult to disperse reproducibly. As we have seen, all types are influenced by the ingredients of the paint itself and these too may vary slightly from batch to batch. Such variations may not matter normally, but they can play havoc when coupled with the variations in the additive.

The paint formulator must devise suitable tests which will provide information on the variation in viscosity characteristics and must know how to adjust the paint to give the required characteristics. Tests can include measurements of viscosity at several rates of shear (p. 130). Points plotted on a viscosity–rate of shear graph must fall within a given area. Alternatively, if the paint forms a gel in the can, the strength of the gel in the container can be measured by inserting a razor blade-shaped paddle on a spindle. The paddle is then slowly rotated by a motor. At first the paint and container turn with it against a spring restraining the magnetic platform on which the can stands, but eventually the gel breaks and, at this point, the twisting force or torque is noted and recorded as the gel strength.

The choice of resinous thickener type will depend largely on the solvent system involved. For non-polar organic solvents, linear polymer thickeners are suitable. For polar organic solvents, microgels can be used. For aqueous systems, linear (associative) thickeners, microgels or (if colloids are present) metal chelates can be used. These are all used principally for their modification of application properties. Mineral thickeners are available for all types of solvent. In contrast, their use is principally to control pigment settlement.

Compatibility of the soluble resinous thickener with the main binder needs to be checked and excessive amounts of particulate thickeners can reduce gloss, especially in high gloss paints. Large differences in refractive index between binder and particulate thickeners can cause loss of clarity in clear coatings and should be avoided.

The use of high pigment levels is seldom a *chosen* method of introducing non-Newtonian viscosity. Pigment level is usually dictated by gloss and only matt finishes and undercoats have really high pigment levels.

Additives affecting surface and interfacial tensions

Apart from the defects introduced by the application method, other troubles occur from time to time, often apparently without rhyme or reason. **Popping** – the appearance of holes or craters in the film – is usually due to inadequate drying at room temperature before the paint is stoved, to the presence of too much low (or medium) boiling solvent, to reaction in the paint producing a gas, or to a faulty undercoat. It can be caused by contamination of the undercoat. **Blushing**, the whitening of the surface of a clear film or the loss of gloss of a pigmented one, is due to condensation and subsequent emulsification of water in the film. If the water evaporates after the film has set, it leaves fine air bubbles in the film. Refraction, reflection and diffraction lead to a milky appearance (see the example of two incompatible resins, Chapter 6). These and other defects are put right by changes in drying conditions. One defect, however, usually requires additive to overcome it. This is known as **cissing**.

Cissing is the appearance of small, saucer-like depressions in the film's surface. These are caused by particles or droplets of incompatible material, which either land on the film during drying, or are present in the paint itself. The weight of the particle causes it to sink but, since there is no attraction between paint molecules and molecules in the particle's surface, the surface tension of the paint resists the particle's entry into the liquid film. Immediately under the particle, the paint molecules have no alternative but to be in close proximity to the particle. In the zone surrounding the particle, the liquid 'skin' is depressed by the particle's weight, but attractions from within the paint film pull the surface molecules away from the particle. There is said to be a high **interfacial tension** between paint and particle. The net result, as Fig. 10.4 shows, is a **ciss** mark

To prevent cissing, an additive that reduces interfacial tension is required in the paint. When interfacial tension falls, the particle is wetted by the finish and absorbed into the film. Surface-active agents (**surfactants**, see below) reduce interfacial tension. Alternatively, an agent can be added that will

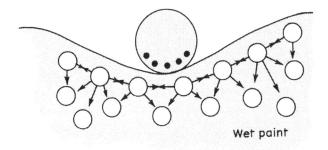

Wet paint

Fig. 10.4 A ciss mark.

reduce the liquid surface tension so much that the interfacial tension also becomes low. Silicone oils (Chapter 13) do this effectively. The silicone oil must be compatible with the finish, or it will itself cause cissing. Very little silicone oil is required, because it finds its way almost entirely to the surface.

Other materials that can prevent cissing are higher molecular weight linear polymers. The mechanism by which they act is obscure.

All such additives are sometimes called **flow agents**.

Surfactants

Surfactants are those chemicals whose molecules have two parts of widely differing polarity and solubility. The soaps, for example, have an ionizing salt 'head' to the molecule and a long, non-polar hydrocarbon 'tail'. A different type are the polyoxyethylene ethers of dodecyl alcohol (dodecanol), e.g. $C_{12}H_{25} \cdot O \cdot (CH_2 \cdot CH_2 \cdot O)_6 \cdot H$. These contain the non-polar dodecyl group $(C_{12}H_{25}-)$, from dodecyl alcohol, at one end of the molecule and the successive polar ether linkages at the other.

Whatever the formula or type, all surfactant molecules have in common these polar and non-polar portions. Possession of the polar portion invariably means the whole molecule has water-dispersibility, and hence surfactants are used in aqueous systems. These polar portions are often referred to as the **hydrophilic** (water-loving) portions, while the non-polar part is the **hydrophobic** (water-fearing) portion.

Thus one end of the surfactant molecule is attracted to polar molecules (e.g. water) and polar surfaces, while the other prefers a non-polar environment. If we have the problem of two materials which will not wet or make chemical contact with one another, surfactants can bridge the gap. In the example already discussed under 'cissing', we will assume that tiny oily particles are dropping onto the surface of a particularly polar resin solution. There is incompatibility and no wetting. If a suitable surfactant can be found with an aliphatic 'tail' attracted to the particle surface and a polar portion attracted to the resin molecules, chemical contact from particle to paint will be established, interfacial tension will decrease and cissing will not occur. The surfactant provides a bridge across the gap between the two unlike molecules.

Wherever this 'bridging' is required, surfactants can be used. If pigment surfaces have poor attraction for binder molecules, surfactants can assist dispersion. When two liquids will not mix, surfactants will stabilize droplets of one liquid in the other, i.e. they will emulsify (Chapter 11). Different surfactants are required for different systems and different applications: the nature and proportions of the two parts of the molecule will vary from use to use. A principal use is as dispersion aids.

Additives affecting appearance

The nature of gloss has been described in Chapter 6. A surface that is sufficiently smooth will be glossy. To reduce gloss we must roughen the surface and break up its smooth outline. In pigmented finishes this is usually achieved by increasing the number of pigment particles present in the paint and hence at the surface of the paint. The protruding particles break up the smooth outline. In varnishes or clear wood finishes, we cannot do this and reduction in gloss (to a 'satin' sheen at least) is usually obtained with additives.

Either a few percent of fine particle silica are used, or else an insoluble wax is dispersed in the finish and floats to the surface during drying. The silica, being an extender, is of course transparent when wetted by the finish. Its high oil absorption makes a small percentage as effective as a much larger quantity of ordinary pigment. However, in many cases the silica, like the wax, tends to float to the surface. It does so as large aggregates containing many particles. The air trapped in the unwetted spaces between the particles reduces the density of the aggregate, allowing it to rise to the surface. When wax is used, only very small quantities are needed and this method gives a very smooth feel to the finish. Polyethene and polypropylene waxes are particularly efficient and, having relatively high molecular weights, do not impair the heat resistance of the coating.

Pigmented coloured finishes frequently change their colour slightly on drying, owing to some migration of pigment to or from the paint surface. The fine particle extenders described above as affecting gloss, and also as affecting viscosity, are effective in controlling this problem known as **floating** or, if severe, **flooding**.

The last classes of additives affecting appearance are light stabilizers and optical whiteners. Paints are susceptible to loss of colour through pigment fading, and to breakdown of polymer in the binder. A prime cause of these is the effect of incident radiation, especially ultraviolet radiation. **UV absorbers** have the physical effect of absorbing UV radiation, thus preventing it from reaching and attacking the binder. These are now used in combination with **hindered amine light scavengers** (HALS). These capture free radicals produced by the action of radiation, which would otherwise cause polymer degradation.

Optical whiteners function by absorbing some UV and reflecting this as blue light, thus countering any yellowness developed in the binder.

Additives affecting chemical reaction

These are the items described as 'activators' (or 'catalysts'), 'accelerators', 'driers', 'inhibitors' (or 'retarder solutions'). Detailed examples will be given later; for the moment it can be said that they fall into two categories:

• those that initiate the drying reaction (e.g. 'activators');
• those that affect the rate of drying (the rest).

An **activator** may be one portion of a two-pack paint, one reactant of the resin-forming ingredients. In this instance the activator can hardly be called an additive. The true additive activator is a chemical which, when added as a minor ingredient, sparks off the chemical reaction in the paint. Such an additive is usually a chemical which decomposes to give free radicals which, in turn, initiate an addition polymerization (see Peroxides, Chapter 5).

Driers (Chapter 12) and **accelerators** are true catalysts in the chemical sense, since they can speed up the chemical reaction responsible for drying, without being consumed in the process. An **inhibitor** (Chapter 5) slows down the chemical reaction, usually by reacting with free radicals, thus preventing them from initiating addition polymerization. Inhibitors are therefore chemically changed in the process. **Retarder solutions** are simply solutions of inhibitor. Retarder can also mean a lacquer thinner containing high boiling solvents. These retard drying simply by being slow to evaporate.

Anti-skinning agents (or **anti-oxidants**) are mild inhibitors of oxidative drying (Chapter 12) and are often included in paints drying by that mechanism, in order to improve stability in the can. They are usually fairly volatile and evaporate once the paint is applied. **Moisture scavengers** are included to inhibit reaction between paint ingredients – e.g. isocyanates or zinc or aluminium pigment – and water, which may have entered the paint during manufacture. Such a reaction might prematurely gel the coating in the container or produce gas under pressure. Calcium oxide has been recommended for zinc-rich coatings and other moisture scavengers are described in Chapter 15.

Additives affecting living microorganisms

Paints are susceptible to deterioration brought about by microorganisms. These include particularly bacteria, yeasts and fungi.

All these types of organism can cause deterioration of liquid paint in the can, but most reported incidents are caused by bacteria. Bacteria can cause gassing, viscosity reduction and colour drift in latex paints (p. 161). They can enter the paint via infected intermediates and raw materials (including water) or unsterile equipment. The best defence against infection lies in high standards of plant hygiene with regular sterilization of equipment. Nevertheless, the inclusion of an **in-can biocide** at low concentration in latex and paint is a wise precaution. Commercial biocides for paint are often mixtures of complex organic chemicals, which together give protection against a wide range of bacterial types. Many effective biocides have functioned by releasing formaldehyde and, with current concerns about toxicity, this class is now much less used.

Once a paint film has been applied it is exposed to a wide variety of airborne spores from yeasts, fungi and algae. If they find nutrients in the paint or in dirt or debris adhering to it, plus moist conditions ideal for growth, they will multiply forming unsightly colonies. Algae also require light for growth and so are found mainly on outdoor paint. As the mildew, moulds and algae multiply, film damage eventually results, leading to loss of protective properties. To prevent this **fungicides** and **algicides** are included in the paint formulation. Again a range of complex organic compounds are effective, as are various organometallic compounds (especially tin complexes). No one compound is completely effective against all species and the widest protection is given by a mixture. Selection is based not only on biocidal effectiveness, but also on solubility, stability in the can and long life in the film. Great care is now taken to use these compounds only when necessary, and at the minimum level that is effective.

Algae are also found among the large number of plant and animal organisms that cause fouling of ships' hulls below the water line. These organisms, which also include barnacles, tubeworms, mussels, polyzoa, hydroids, ascidians, sponges and sea anemones, settle on the hull and adhere strongly when they ship is stationary or moving very slowly. Once established, they colonize the surface, causing resistance to movement. Reductions of speed by even half a knot ($\sim 1\,\mathrm{km\,h^{-1}}$) lead to large extra costs, arising from lengthened journeys or higher fuel consumption. To prevent such growths, ships' bottoms are coated with anti-fouling paints containing so-called **anti-fouling additives**. These are biocides which leach slowly from the film and are effective when released at only a few micrograms per square centimetre per day. In the paint, concentrations are usually much higher than normal additive levels. Many 'additives' are in fact pigments, including the traditional copper compounds, cuprous oxide and cuprous thiocyanate. They are used at high concentration, either in a slowly dissolving film-former (e.g. rosin), or in an insoluble matrix (e.g. chlorinated rubber). Against Enteromorpha seaweed (algae) organometallic compounds (often based on tin) are more effective and are used as non-pigmentary solid solutions in, for example, acrylic binders. Even more ingenious is the direct modification of the acrylic binder by copolymerization (Chapter 11) of a tin acrylate or methacrylate (e.g. a trialkyl tin acrylate). The surface of the polymer is gradually eroded away releasing the tin, so that the coating is kept smooth by a 'self-polishing' action. There is now concern over the use of tin compounds, since the build-up of residual tin found in marine life in estuaries signalled that their continued use is causing environmental damage.

Eleven

Lacquers, emulsion paints and non-aqueous dispersions
paints drying by evaporation

This somewhat curious grouping needs an explanation. Although lacquers and emulsion paints involve completely different technologies, they are grouped together here because the leading examples of these types of paints dry by the same mechanism as defined in Chapter 7: lacquer dry. Non-aqueous dispersions must be dealt with alongside aqueous emulsions, since they are complementary.

Lacquers

Lacquers have found widespread use for general industrial finishing (for paper, textiles, plastics and metal), for the original finishing and refinishing of motor-cars and for finishing furniture, though these are now diminished. Most of the advantages of lacquers over cross-linked finishes are described in Chapter 7, but the principal reason for their continued use in parts of industry today lies in their ability to harden quickly at all practical temperatures and particularly where oven heating is not available. They are particularly suitable where the object to be coated would be deformed at modest stoving temperatures, e.g. wood, thermoplastics. They have a big advantage over the so-called 'cold-curing' cross-linking finishes, in that they are supplied in one pack and present no 'shelf-life' or 'pot-life' problems. They also dry faster than paints drying by an oxidative mechanism.

What is a lacquer? It is a finish, clear or pigmented, which consists primarily of a hard linear polymer in solution. It dries by simple evaporation of solvents (lacquer dry). It is thus possible to make a lacquer from any soluble linear polymer, such as chlorinated rubber, which is used in chemical-resistant lacquers, but here we have space to discuss only two widely used types: acrylic and nitrocellulose lacquers. Since the nature and properties of a lacquer are largely determined by the main polymer, the chemistry of these two families of polymers will be discussed next.

Acrylic polymers

Acrylic polymers are a family of entirely synthetic chain growth polymers, whose monomers are mainly esters of the unsaturated acids:

$$CH_2=CH-\overset{\overset{\displaystyle O}{\|}}{C}-O-H \quad \text{and} \quad CH_2=\overset{\overset{\displaystyle CH_3}{|}}{C}—\overset{\overset{\displaystyle O}{\|}}{C}-O-H$$

acrylic acid   methacrylic acid

Since any alcohol can be used to esterify the acids, a wide variety of esters are available. Each ester monomer, when polymerized alone, gives a different homopolymer. The number of monomers in a copolymer is not restricted, though some combinations will not copolymerize. The proportion of each monomer may be varied widely. Thus the variety of copolymers possible is boundless.

The acrylate homopolymers are softer and more flexible than the corresponding methacrylates. The latter begin with the hard, tough polymethyl methacrylate (PMMA), better known as the plastic with the trade names 'Perspex' and 'Diakon' (UK) and 'Lucite' and 'Plexiglas' (USA). Methacrylate esters of higher alcohols give softer, more flexible polymers and whereas PMMA is insoluble in aliphatic hydrocarbons and softens above 150 °C, polybutyl methacrylate is soluble and softens above 35 °C. Polyisobutyl methacrylate softens above 70 °C, showing the effect of a branched alkyl group. Polylauryl ($C_{12}H_{25}-$) methacrylate is syrupy, but higher alkyl groups lead to hard waxes, which soften at temperatures increasing with the number of carbon atoms in the alkyl group. The non-waxy polymers become more flexible as the size of the alkyl side group increases, since the polymer chains are kept farther and farther apart, but the waxy polymers are partially crystalline, because the alkyl groups are long enough to align parallel to one another.

Copolymers usually contain a blend of 'hard' and 'soft' monomers, e.g. methyl methacrylate and ethyl acrylate, to give the required properties for the particular purpose. Some of the acid monomer may be included, or the basic amide, e.g. acrylamide, $CH_2=CH-CO\cdot NH_2$. A hydroxyl group may be introduced by copolymerizing with, for example, ethylene glycol monomethacrylate (or hydroxyethyl methacrylate).

$$CH_2=\overset{\overset{\displaystyle CH_3}{|}}{C}\cdot CO\cdot O\cdot CH_2\cdot CH_2\cdot OH$$

Cellulose polymers

These polymers are not wholly synthetic, since they are based upon cellulose. This is not made, but found widely in nature, where it forms about half of all the cell wall material of wood and plants. Cotton is almost pure cellulose, and wood pulp is another source.

The cellulose molecule consists of a large number of rings of atoms joined as shown:

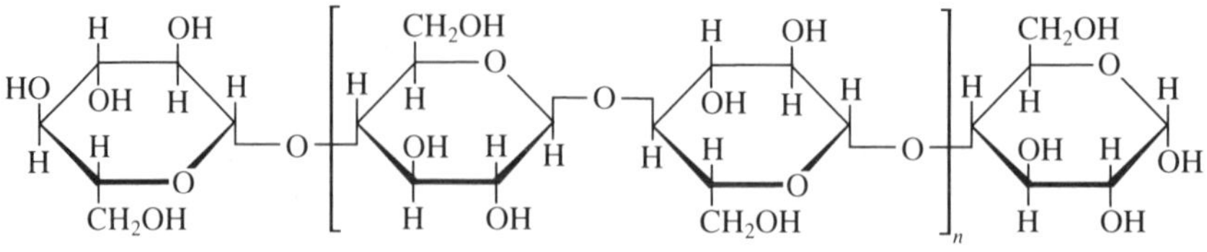

At each corner of each hexagon (except where an oxygen atom is shown) a carbon atom should be imagined. It is omitted here to simplify the diagram.

The repeating unit of the polymer, cellobiose (inside the brackets), may be thought of as two glucose molecules linked by the eliminations of water:

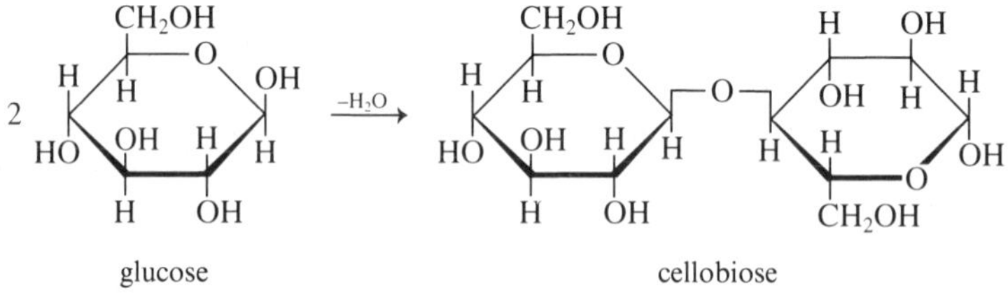

glucose   cellobiose

Glucose is a sugar, a member of a family of compounds called **carbohydrates** because their simple formula, $C_x(H_2O)_y$, appears to be composed of carbon and water (hydrate).

The large number of hydroxyl groups and ether linkages make cellulose sensitive to water, but the molecules are of such immense size (molecular weight 300 000–500 000) and are held together by hydrogen bonding along their lengths to such an extent that they are not dissolved by water or other normal solvents, and can show such uniformity of arrangement of the chains that the fibres are in part crystalline.

The three hydroxyl groups per glucose ring provide the means of converting cellulose to polymers soluble in organic solvents. They may either be esterified by organic or inorganic acids, or etherified with suitable alcohols.

In this way a number of polymers useful in paint and particularly lacquers have been produced:

- ethyl cellulose, ether of ethyl alcohol;
- cellulose acetate, ester of acetic acid;
- cellulose acetate butyrate (CAB), ester of acetic and butyric acids;
- cellulose nitrate (nitrocellulose, NC), ester of nitric acid.

(A further modified cellulose, hydroxy ethyl cellulose, will be mentioned later as a thickener for emulsion paints.)

Nitrocellulose is, of course, chemically the wrong name for the polymer, since it contains nitrate ($C–O\cdot NO_2$) groups and not nitro ($C–NO_2$) groups. Nitration can be stopped at any stage by dilution with water and is usually taken to a nitrogen content of 10.5–12.3% (or an average of 1.8–2.4 nitrate groups per glucose unit). The molecular weight of the product is too high for paint usage, so the molecules are split by hydrolysis with very dilute acids, aided by heat and pressure:

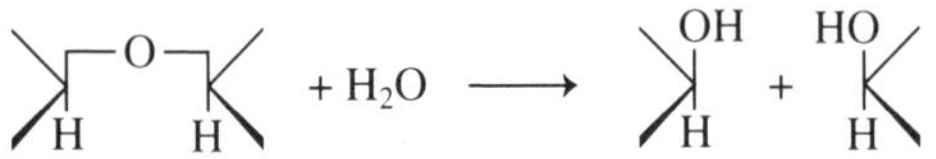

Products of molecular weight varying from 50 000 to 300 000 are produced. Water – which would not be tolerated in a lacquer – is then displaced by an alcohol. NC must be supplied damped by a liquid; if dry, it is classified as an explosive.

Many grades of NC, differing in nitrogen content and molecular weight, are available. Nitrogen content controls solubility:

- 11.8–12.2% N   Dissolves in esters, ketones and ether-alcohols.
- 11.2–11.8% N   Dissolves in mixtures of ethanol and esters or toluene.
- 10.5–11.2% N   Dissolves in ethanol.

Molecular weight determines solution viscosity. No international grading system has been agreed upon. Examples of four systems are given in Table 11.1.

Lacquer film-formers

The acrylic polymer which offers the best all-round properties for a metal-coating lacquer exposed to the weather is the hardest acrylic polymer, polymethyl methacrylate (molecular weight 80 000–150 000). It is hard, clear, scarcely coloured, very little affected by ultraviolet light, insoluble in commercial petrols and resistant to acids and alkalis. Marked deviations from the homopolymer reduce one or more of these properties. It is true that a wide variety of acrylic copolymers are used to make excellent lacquers for many purposes, but in discussing acrylic lacquers we will now confine our attention to lacquers of polymethyl methacrylate.

Table 11.1 Grading of nitrocellulose

Nitrogen content %	Conc. of NC in g/100 ml 95% aqueous acetone	Viscosity of this solution in Pa.s at 20 °C	ICI grade (UK)	Hercules grade (USA)	Wolff Walsrode grade (Germany)
11.8–12.2	40	0.8–1.3	DHX8–13	RS $\frac{1}{4}$ sec	E400
11.8–12.2	40	3–5	DHX30–50	RS $\frac{1}{2}$ sec	E560
11.8–12.2	20	2.5–4.5	DHL25–45	RS 5–6 sec	E840
10.5–11.2	40	0.8–1.3	DLX8–13	SS $\frac{1}{4}$ sec	A500

Polymethyl methacrylate, like nitrocellulose, is a fairly brittle polymer and both must be plasticized for paint uses. It is sometimes convenient to increase flexibility by blending with more flexible resins and other gains – such as improved adhesion – can be obtained. This technique is often used with NC, though with PMMA the film properties are usually downgraded by it and the number of compatible resins that can be used is limited. In fact, it is generally true that the best plasticizers are solvents for the polymer, which have boiling points high enough to ensure little or no evaporation, either at normal atmospheric temperatures, or during moderate stoving schedules. This sort of boiling point (250 °C and above) implies fairly large molecules and liquids more viscous than ordinary solvents, but the substances are simple chemicals, not polymers. The plasticized polymer can be thought of as a high solids solution in plasticizer. We have already seen that less viscous solvents give less viscous solutions. For this reason, less viscous plasticizers give less viscous (or more flexible) plasticized polymer films.

Since group II solvents dissolve most resins, it is not surprising that most plasticizers are esters. Esters of this sort of molecular weight are easily made in wide variety, but ketones are not.

The simplest plasticizer for NC is dibutyl phthalate and for PMMA, butyl benzyl phthalate:

dibutyl phthalate
BP 325 °C

butyl benzyl phthalate
Boiling range 200–288 °C
at 20 mmHg pressure

Compatible resinous plasticizers used with NC include non-drying alkyds (Chapter 12), acrylic resins and many natural resins and derivatives, such as dammar and ester gum. Alkyd plasticizers are today the most usual. Natural resins are mixtures of relatively low molecular weight chemicals, often of uncertain chemistry (see rosin, Chapter 12). Ester gums are produced by reacting polyhydric alcohols, e.g. glycerol, with rosin. An increase in molecular weight occurs, since more than one molecule of monobasic rosin acid will react with one molecule of the alcohol. Replacement of the carboxyl group by an ester linkage increases the range of solvents that may be used.

Resins used with PMMA include NC and CAB, other acrylic polymers and some vinyl polymers. The use of compatible resins with molecular weights in the medium to low range (1000–30 000) can increase substantially the solids of a lacquer at application viscosity.

So far we have seen that our lacquer consists of

- pigment (if required);
- the linear polymer;
- plasticizer;
- compatible resins or polymer (if required).

To this we need add only

- solvents;
- additives

and the lacquer is complete. The liquids that are true solvents for NC and PMMA can be seen from Tables 9.1 and 9.3. Non-solvents are also frequently included to reduce cost, but the average solubility parameter of the liquid mixture must lie within the range for the polymer at all stages, from the lacquer in the can to the almost dry film on the surface being coated. Additives depend on the particular requirements of the lacquer. To complete the picture, the formulae of two typical lacquers are given in Table 11.2. It will be readily seen that they both conform to the same general outline given above.

Hot spray

One more point should be mentioned regarding lacquers. Users, who are otherwise satisfied with the properties of lacquers, frequently complain about their low spraying 'solids'. One way of obtaining higher solids is to apply by hot spray. Here the lacquer is heated on its way to the spray-gun and emerges from the nozzle at 70–90 °C. Since the viscosity of a liquid falls with rising temperature (a result of the increased energy and hence mobility of the molecules), it is possible to start with paint at higher solids and at a viscosity too great for normal spray application. There are two other advantages. Some non-solvents at room temperature become solvents at hot-spray temperatures. Also, cooling between the gun and object provides a viscosity rise, which can prevent sagging. Between 25 and 50% higher solids can be obtained by the hot-spray technique.

Emulsion paints

An emulsion has two parts: the **dispersed phase** (the droplets) and the **continuous phase** (the liquid in which the droplets are dispersed). In Chapter 9 it was explained that the viscosity of an emulsion is little greater than that of the continuous phase. It is not influenced by dispersed phase polymer molecular weight and is influenced by dispersed polymer concentration only at high concentrations. So, if the film-former can be emulsified, then, in principle, the paint can be supplied as an emulsion at much higher

Table 11.2 Formulations of nitrocellulose glossy wood finish and acrylic motor car lacquer

	Nitrocellulose glossy wood finish	wt.
Polymer	NC DHX 3–5 (or RS $\frac{1}{8}$ sec), dry weight	12.3
Plasticizer	Methyl cyclohexanyl phthalate	0.9
	Blown castor oil	0.9
Compatible resin	Ester gum	13.3
	Hard maleic resin ('Cellolyn' 501)	8.6
Solvents	Butyl acetate	16.6
	Ethyl acetate	2.4
	Methyl cyclohexanone	1.9
	n-Butanol (NC damping liquid)	5.7
	Ethanol	5.3
	Toluene	29.6
Additive to prevent 'bloom'	2% Hydroquinone in 2:1 toluene–acetone	2.5
		100.0

Ready-for-spraying at 36% solids

	Grey acrylic motor car lacquer	wt.	
Pigment	Rutile titanium dioxide	11.87	Sand mill grind stage
	Lampblack	0.12	
	Methoxy propyl acetate	1.98	
	Toluene	6.01	
Polymer	'Paraloid'[a] A21LV	4.00	2nd stage
	'Paraloid' A21LV	61.34	
Plasticizer	Butyl benzyl phthalate	8.40	
Solvents	Methoxypropyl acetate	1.48	
	Methyl ethyl ketone	2.40	
	Toluene	2.40	
		100.00	
Thinner	Acetone	39.5	
	Toluene	43.4	
	Methoxypropyl acetate	17.1	
		100.0	

Thin to 11 s No. 4 Ford cup, for spraying at approx. 22% solids

[a] 'Paraloid' A21LV is an acrylic copolymer, largely polymethyl methacrylate, dissolved at 30% solids in

Toluene	50.0
Methyl ethyl ketone	40.0
n-Butanol	10.0
	100.0

solids than would be the case for a solvent-borne paint. This should be true for any paint, but particularly so if the polymer has a high molecular weight as in a lacquer.

Problems

There are some practical drawbacks to this idea. The first concerns the method of film formation. When a thin film of polymer solution is applied to a surface, evaporation occurs and the polymer molecules are gradually brought closer together as their concentration rises. Ultimately, the polymer molecules are closely packed, forming a uniform and continuous film. A polymer emulsion similarly treated starts to lose continuous phase by evaporation. The polymer droplets (or particles) become more closely packed until they are touching all their neighbours. At this point we have a discontinuous film of polymer, containing some liquid in the voids (or spaces) between the particles. It is now necessary for the particle boundaries to merge, so that ideally the outlines of the particles can no longer be seen, and the film becomes completely continuous.

This merging of particles will not take place unless the polymer molecules in the particles have freedom to movement. Either the glass transition temperature (T_g) of the particle must be below room temperature, or heat must be applied. In the former event this will mean that the final film will be relatively soft, unless the droplets contain a solvent that evaporates later, or unless the polymer can harden by chemical reaction, e.g. by oxidative drying. In practice then, there are three possible types of paint:

- a stoving emulsion paint containing a linear polymer;
- an emulsion paint containing a linear polymer with a low glass temperature, or one made mobile by plasticizer or solvent;
- an emulsion paint containing low molecular weight material that will dry by chemical reaction after film formation, either at room temperature or above.

The tendency to **coalescence** (as the merging process is called) can be increased by introducing into the paint a high boiling solvent that is miscible with the continuous phase. Some of this **coalescing solvent** may be left in the voids during the early stages of evaporation, but penetrates the particles during the later stages, giving them greater fluidity. Some time after film formation is complete, it evaporates altogether.

A second difficulty is pigmentation. Either the pigment must be dispersed in the binder and then the whole emulsified, or it must be dispersed in the continuous phase and mixed with the polymer emulsion later. Often only the latter alternative is available and the dispersion must be carried out with the aid of surfactants. Alternatively it can be achieved using a second resin soluble or dispersible in the continuous phase. Having been dispersed, the pigment particles must be integrated into the film during coalescence. If a surfactant was used for dispersion, the film-forming resin must flow over the whole pigment surface and wet it during film formation. If a secondary resin was used, this resin must now merge readily with the main film-former, with which it must be fully compatible. Often, the very mechanism of film formation flocculates well dispersed pigment particles. Failure to integrate the pigment properly can lead to low gloss and film porosity.

The third difficulty concerns application. Emulsions are intrinsically difficult to apply by any method (except perhaps spraying), since they have the viscosity of a solvent (the continuous phase). They are too fluid for brushing or dipping, for example. It therefore becomes necessary to raise the viscosity, either with thickener additives or by use of resinous materials soluble in the continuous phase. The latter convert the continuous phase to a polymer solution, thus increasing its viscosity and therefore that of the paint. In either case the material added to increase the viscosity must be compatible in the paint or poor gloss and film integration will result.

The emulsion

The emulsion can be made in either (1) water or (2) some other non-solvent for the polymer. Emulsions based on (2) are called **non-aqueous dispersions** (NADs). If aqueous emulsions are evaporated to dryness without coalescence of the polymer particles, these may be redispersed in some organic non-solvent and the redispersion is then specifically referred to as an **organosol**. However, if the polymer dispersion is produced by direct polymerization of monomers in organic non-solvent, then it is called an NAD. Methods of making emulsions, both aqueous and non-aqueous, will now be considered.

In the section on Dispersion in Chapter 8, we considered how to make stable colloidal dispersions of solid and found that, for stability, it was necessary to keep the particles apart. We learned that aqueous pigment dispersions could be stabilized by adsorbed surfactant molecules, which ionized in the water to produce an electrical charge barrier around the particle (**ionic stabilization**). This could also be done by using polymer molecules, anchored strongly to the particle, but also extending out into the solvent, in which they are soluble. These polymer molecules provide a steric barrier around the particle and this method of stabilization is called **steric stabilization**. Exactly the same techniques are used to stabilize emulsions.

Emulsions can be made in two ways.

Mechanical means

The film-former or, if it is too viscous, a solution of the film-former, can be dispersed by vigorous agitation. The polymer may be added to the water, the water to the polymer or both may be added to the mixing vessel simultaneously. The stabilizing surfactant may be introduced mixed with one component or the other. The surfactant may be **anionic**, because in water the active portion is an anion, e.g.

$$R \cdot COONa \longrightarrow R \cdot COO^- + Na^+$$
active portion

or **cationic**, e.g.

$$R_4N \cdot Br \longrightarrow R_4N^+ + Br^-$$

or **non-ionic**, e.g. polyoxyethylene ethers of dodecyl alcohol (see Surfactants, Chapter 10). Thus the emulsion particles may be negatively or positively charged, or neutral.

Stability is improved by **protective colloids**, which are polymeric materials with highly polar and non-polar features in their molecular structures. They provide additional steric stabilization. If the protective colloids are not truly compatible with the film-former, gloss will be reduced and the film weakened. Frequently the colloids in water paints (e.g. water-soluble cellulose derivatives) are affected by microorganisms, so fungicides and bactericides may be included to prevent deterioration.

Emulsions made by mechanical means, using surfactants and colloids, are again being used in decorative paints today; this technique, formerly used in oil-bound distempers, and once based on drying oil emulsions, is now a route to water-based alkyd paints. Alternatively, the resins may be internally modified to emulsify in water, and a particularly well-established use is in electrodeposition paints (p. 119).

Emulsion polymerization

Chain growth polymerization has been described in Chapter 5. The assumption there was that the monomers were polymerized in bulk or in solution. Aqueous emulsion polymerization begins with a true solution of water-soluble monomers and initiator, or with the 'solubilization' of insoluble monomers and initiator using surfactants. Surfactant molecules can cluster in water, with their fatty organic portions in the centre of the cluster and their water-seeking portions on the outside. Such an arrangement is called a **micelle**. The centre of the micelle is a refuge for insoluble organic monomer molecules, which are attracted by the organic 'tails' of the surfactant molecules. Thus the micelle becomes swollen with monomer, yet the micelle size remains so small that the micelles are not seen by eye and the monomer appears to have been 'solubilized'. Reaction is started by heat, which causes the decomposition of a water-soluble initiator, e.g. a persulphate, to free radicals. Polymerization to insoluble, emulsified polymer occurs within some of the micelles, which are 'fed' with monomer by adsorption from other micelles. Further surfactant is attracted to and held at the growing particle surface, stabilizing it. Eventually polymerization goes to completion within the polymer/monomer emulsion droplets.

Mention has already been made in Chapter 10 of microgels, which are special swellable emulsion polymers used for their ability to alter the rheology of paints. The special feature of many microgels is that they are multi-layered in a form known as core–shell (p. 142). These emulsion polymers are made by multi-stage processes so that first of all the inner core is made which is often cross-linked, following by an outer shell or **mantle** layer.

In the case of aqueous microgels, this mantle is acrylic polymer containing hydroxyl and carboxyl groups. It is essential that it is joined or grafted to the inner cross-linked core. (Grafting means the linking of one end of a polymer chain to another, as illustrated on p. 164). On neutralization the mantle swells, and because of interactions between particles, imparts non-Newtonian behaviour to the wet coating.

Core–shell emulsion polymers may also be made by the same multi-stage process such that a harder inner core is surrounded by a softer outer shell. This type of emulsion will have its ability to coalesce determined by the nature of the shell polymer.

A further specialized form of emulsion polymer is that made with air or pigment inclusions, or possibly both. The use of such polymers has developed from earlier use of hard polymer beads as spacers to obtain better distribution of expensive titanium dioxide in emulsion paints. As described in Chapter 6, opacity is achieved through scatter from areas of unlike refractive index, and this can be achieved by air inclusions in the film. A number of materials are available, made by emulsion processes where the emulsion polymer contains the inclusions mentioned to provide additional opacity.

Many, though not all, polymers made by emulsion polymerization would be accepted as solids in the dry state. In colloid terms, the word 'emulsion' implies a dispersion of one liquid in another, so that many people have claimed that these products should not be called polymer emulsions. In practice, then, they are also called **polymer dispersions** and the aqueous dispersions are called **latices** (or **latexes**). The paints made from them, however, are usually called emulsion paints.

Household emulsion paints

In Europe the most popular paints of this type are based on vinyl acetate copolymer or acrylic latices. The former are prepared by the emulsion polymerization of vinyl acetate, $CH_3CO \cdot O \cdot CH = CH_2$, and are stabilized by a combination of surfactants and protective colloids. Although polyvinyl acetate (PVA) is a relatively soft polymer, like polymethyl acrylate, its glass temperature is above room temperature. Household paints must coalesce to form a film at room temperature, so PVA must have its glass temperature lowered, either by addition of a lacquer plasticizer, such as dibutyl phthalate, or most usually by copolymerization with other monomers, e.g. ethyl acrylate, 2-ethyl hexyl acrylate or vinyl Versatate. ['Versatic acids' are highly branched carboxylic acids made by Shell Chemicals. 'Versatic' and 'Versatate' are trade names and not chemical names.]

In acrylic latices, the 'hard' monomer is methyl methacrylate and the plasticizing monomer an acrylate, such as butyl acrylate or one of the acrylate comonomers mentioned above. Acrylic latices usually contain copolymers of acrylic or methacrylic acid as colloids and thickeners, these being solubilized by neutralization with base. Whatever the type of latex, coalescing solvents are also normally added to improve film formation. These may or may not be water miscible and include alcohols, glycols, ether-alcohols, ether-alcohol esters and even hydrocarbons, all of high boiled point.

Pigment is dispersed in the continuous phase with suitable surfactants, additives and a water-soluble thickener, e.g. hydroxyethyl ($HO \cdot CH_2 \cdot CH_2 -$) cellulose. The dispersion and latex are carefully blended with efficient stirring to form the paint. It is most stable if just alkaline. The additives include fungicides, to prevent mould growth feeding on the cellulosic or other colloids in the dry film, and biocides, to prevent bacterial degradation of colloid and thickener molecules in the paint can. The bacteria, or enzymes produced by them, rapidly reduce molecular weights of these molecules, leading to a dramatic decline in viscosity and malodorous by-products. Susceptibility to bacterial attack is dependent not only on latex, colloid and thickener chemistry, but also on pH and whether or not other chemicals with a biocidal action, such as certain coalescing solvents or traces of unpolymerized monomer, are present.

Emulsion paint is easy to apply and dries quickly without unpleasant smells. Brushes and rollers can be cleaned in water. However, it has not been possible, to date, to produce a satisfactory full gloss version. To get a smooth surface of high gloss, fine particle latices must be used and relatively large quantities of polymer in solution. Such paints give less than perfect flow on application. In addition, it is hard to keep a water-based coating fluid for long enough to give good merging of brushed overlaps, and application over porous substrates can lead to low gloss. Acrylic latices and vinylidene chloride ($CH_2 = C \cdot Cl_2$) copolymer latices are used in water-based gloss finishes.

Due to the presence of water-sensitive materials and sometimes (in some matt finishes) very high concentrations of pigment and extender, emulsion paint films retain sensitivity and porosity to water vapour, though this is not as bad as might be feared and many emulsion paints can be washed with safety. Over very porous surfaces, loss of plasticizer or coalescing solvent into the substrate can cause poor coalescence.

For many years styrene–butadiene ($CH_2 = CH - CH = CH_2$) copolymers were used in the USA in cheap latex coatings for house interiors. More recent developments have been the formulation of primers and undercoats based on latices with good adhesion for wood, the use of ethylene–vinyl chloride–vinyl acetate and ethylene–vinyl acetate copolymer latices in house paints, and the introduction of 'solid' emulsion paints. **Solid emulsion paint** is in fact paint structured to have a very high low shear viscosity, so that it appears like a stiff jelly in the container. Because of this the paint can be supplied in flat plastic containers which act as ready-for-use roller trays that can be sealed with a push-on lid after use. Under the high shear of the roller, the structure breaks down, allowing satisfactory application without splashing. Structure of this type is achieved by using higher than normal levels of metal chelate thickener, coupled with latices based on appropriate colloids (p. 142).

Typical emulsion paint formulae are given in Table 11.3.

Industrial emulsion paints

Based on the principles described in the earlier part of the chapter, various emulsion paints for industrial uses have been formulated and adopted commercially, especially where legal pressures exist to reduce the levels of solvent emitted into the atmosphere from factories. Many of these lacquers dry at room temperature and differ from household paints principally in the formulation of the resins, often an acrylic copolymer, to give a particular combination of properties in the dry coating. Since drying time at room temperature can be considerably prolonged at high relative humidities, force-drying or stoving is often used to achieve consistent production rates. Emulsion primers or undercoats are found more often than higher gloss high quality topcoats, since flow and levelling are more easily controlled in varying humidity and minor application defects do not spoil the end result. Excellent primers of high corrosion resistance can be formulated on copolymers of vinyl chloride, vinylidene chloride and acrylate esters.

Other industrial emulsion paints are based upon polymers with reactive groups, such as hydroxyl-functional acrylics, reacting with water-soluble amino resins (Chapter 13) and they are normally stoved. A technique for passivating aluminium flakes in water has been developed and emulsions of this type are used with aluminium and other pigments to apply the metallic basecoats of basecoat-clear high quality finishes to motor cars, both in the manufacturing stage and in refinishing after damage repair. These have already been referred to in the previous chapter, describing the use of aqueous microgels for rheology control.

An unusual form of industrial emulsion coating, the **autodeposition** or **autophoretic** process, requires a special mention. In this process the emulsion particles are negatively charged and the coating is maintained at a very low (5–6%) solids content in an acidic continuous phase (pH 2.5–3.5). Extremely clean metal parts are dipped into the coating bath, whereupon acid attack

Table 11.3 Formulations of 'silk' gloss emulsion paint, acrylic water-borne undercoat and coil coating plastisol

'Silk' gloss emulsion paint

Component	wt%	Function
Rutile titanium dioxide	20.00	Pigment
Calcium carbonate	7.00	Extender
Sodium polyphosphate	1.00	Dispersant
Sodium carboxymethyl cellulose	0.30	Thickener
Biocide	0.50	Bactericide
Antifoam	0.30	Antifoam
Texanol ester-alcohol	2.00	Coalescing solvent
Vinyl acetate–vinyl 'Versatate' latex (55% solids)	42.00	Film former
Ammonia (0.880)	0.03	Alkalinity control
Water	26.87	
	100.00	

Ready for brushing at 52% solids.
Tamol 731 is a surfactant from Rohm & Haas.
Texanol is 2,2,4-trimethylpentanediol-1,3 monoisobutyrate and is a trademark of Eastman Chemical.

Acrylic water-borne undercoat

Component	wt%	Function
Rutile titanium dioxide	25.0	Pigment
Calcium carbonate	2.50	Extender
Talc	12.00	Extender
Tamol 731	2.50	Dispersant
Hydroxyethyl cellulose	0.25	Thickener
Biocide	0.50	Bactericide
Antifoam	0.20	Antifoam
Texanol ester-alcohol	1.00	Coalescing solvent
Methacrylate–ethyl acrylate–acrylic acid latex (50% solids)	26.00	Film former
Ammonia (0.880)	0.20	Alkalinity control
Water	29.85	
	100.00	

Ready for brushing at 55% solids

Coil coating plastisol

Component	wt%	Function
Rutile titanium dioxide	6.00	Pigment
PVC powder	60.50	Polymer
Dioctyl phthalate	15.10	Plasticizer
Dioctyl adipate	4.90	Plasticizer
Epoxy plasticizer	3.00	Stabilizers
Ba–Cd–Zn stabilizer solution	1.80	Stabilizers
White spirit	8.70[a]	Solvent
	100.00	

Ready for reverse roller application at 90% solids.
Stove 30–60 s to a peak metal temperature of c. 200 °C
[a] Or to application viscosity.

on the metal takes place. From steel, Fe^{2+} ion are produced and are rapidly converted to Fe^{3+} by oxidizing agents in the bath. The negatively charged paint particles are destabilized by these multivalent cations, and the destabilized particles flocculate and deposit on the metal surface. Coalescence is deliberately avoided by formulation design, so that a porous film forms, through which acid and cations can diffuse, bringing about more coating deposition. When sufficient coating has been deposited, it is rinsed and dipped in dilute chromic acid (which considerably enhances the corrosion resistance) before it is finally force-dried. Because coating forms wherever metal ions dissolve, it coats all recesses and hidden areas and has good 'throwing power'. Thus it can be considered as performing an electrocoat function without the aid of electricity. Since the composition of the bath is continually changing as it is used, careful analytical monitoring and chemical control is required and this can involve discarding some of the bath contents, as well as making additions to the bath.

Non-aqueous dispersions, organosols and plastisols

The methods for making NAD chain growth polymers are similar to methods for making emulsion polymers. The continuous phase can be varied quite widely and, by choosing non-solvents of appropriate boiling point, the temperature of the reaction can be controlled by allowing the continuous phase to reflux. The monomers are initially soluble in the continuous phase, but the polymer produced is not. As it precipitates as fine particles, these are stabilized by graft or other copolymer species. The particles grow by absorbing further monomer, which polymerizes within the particles.

NADs are always sterically stabilized, so the surfactant is polymeric. Frequently graft copolymers are used. These contain polar polymer chains, which anchor well onto the dispersed particles and which are chemically bonded (or 'grafted') onto non-polar polymer chains which are soluble in the continuous phase:

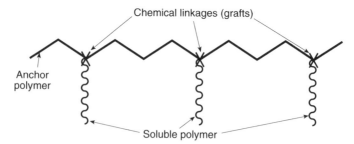

Microgel particles (introduced in Chapter 10) were originally developed by the NAD process before their manufacture was extended to aqueous media.

The principle again is that the most suitable structures are core–shell, so that they have a cross-linked inner part with a joined or grafted outer part soluble in hydrocarbon.

It is also possible to produce step growth polymers in NAD by mechanically emulsifying the liquid or dispersing the solid ingredients in an aliphatic hydrocarbon of very high boiling point, before carrying out the step growth polymerization.

NAD-based coatings can be formulated using the principles already described for aqueous emulsion paints, but bearing in mind that dispersants, thickeners and the like must be suitable for a non-aqueous continuous phase. Using microgel NAD as additive where necessary, the most successful application of NADs has been in metallic automotive topcoats and basecoats as already described. The use of NAD microgels blended with long oil decorative alkyd in decorative paints and clear wood coatings has also been successful, where they help preserve long-term elasticity in the coating.

However, there are two types of coating in this overall grouping which are based on polymer made originally by aqueous emulsion polymerization. The emulsions are spray-dried and the polymer powder so produced is then redispersed in non-aqueous liquids to produce useful coatings. These coatings are called **organosols** and **plastisols**.

Organosols

Organosols made from redispersed PVC homopolymer particles have long been used as coatings for the interiors of cans, because of their extensibility and excellent resistance to permeation by the contents of the can. While PVC powders are low in cost, for some years a much more expensive polymer has been used in an organosol to produce high quality coatings for steel and aluminium sheet (coated originally as a coil of strip metal, hence the coatings are called **coil coatings**). That polymer is polyvinylidene fluoride (PVdF):

$$\cdots -\underset{\underset{F}{|}}{\overset{\overset{F}{|}}{C}}-\underset{\underset{H}{|}}{\overset{\overset{H}{|}}{C}}-\underset{\underset{F}{|}}{\overset{\overset{F}{|}}{C}}-\underset{\underset{H}{|}}{\overset{\overset{H}{|}}{C}}- \cdots$$

It is made by emulsion polymerization, spray-dried and sold as a fine particle powder. Pigment and fluoropolymer particles are separately dispersed in a solution of an acrylic copolymer compatible with PVdF, the two dispersions being blended to give the final paint. After application of paint, the coil is heated to 240–260 °C in 30–60 s, when the PVdF particles melt and lose their identify, forming a blend with the acrylic.

PVdF will not absorb UV light from sunlight and so is not degraded by it. If inorganic pigments of equal quality are used with it, the resultant finish, which is also tough and flexible, lasts for 20 years or more outdoors, even in tropical climates, and fully justifies its extra cost. The coated metal sheet is used for the cladding and roofing of buildings.

Plastisols

A plastisol may be regarded as an organosol in which the continuous phase is almost entirely liquid plasticizer (small amounts of solvent are used for viscosity adjustment). PVC plastisols are made from PVC powder, adipate and phthalate ester plasticizers and minor amounts of epoxy-type resin in solution to aid pigment dispersion and to help (with other additives) to keep the polymer stable to heat and oxidation. The resultant coating is nearly solvent-free and so can be applied in thick films (100–250 μm) and stoved without disruption by escaping solvent. The plasticizer penetrates the particles, aiding sintering as the metal substrate reaches c. 200 °C in 30–60 s.

These coatings are applied to galvanized steel on a coil coating line, as described above for PVdF. They are very tough and flexible and extremely resistant to building site damage. Hence the sheet is used for cladding and roofing. However, this polymer is subject to slow degradation by UV light and the coatings give a life expectancy of 10–15 years in temperate climates.

A plastisol coil coating formulation is shown in Table 11.3.

Twelve

Oil and alkyd paints
paints drying through oxidation

The paints considered in this chapter all dry by the second mechanism discussed in Chapter 7 and, in particular, by the same form of that mechanism. This is known as **oxidative drying** and occurs when the film-former becomes cross-linked as a result of a chemical reaction with oxygen in the atmosphere.

Before proceeding with the chemistry of the oxidative drying process, the reader would be well advised to turn back to p. 44 and re-read the section on oils. You should know the general formula of an oil and, in particular, the arrangement of double bonds in conjugated and non-conjugated fatty acids. To understand the mechanism of drying fully, you should also re-familiarize yourself with the free radical polymerization mechanism discussed in Chapter 5.

Oxidative drying

Such are the complexities of the action of oxygen on drying oils that the mechanisms have not been completely unravelled in over five decades of research. The outline that follows is therefore the fullest explanation based on the evidence available, but not a completely proven one.

The facts are as follows.

- There is almost always a delay or **induction period** before drying commences.
- Presence or addition of certain substances (anti-oxidants) can further delay or even inhibit drying.
- During drying there is a considerable uptake of oxygen. Linseed oil, containing suitable driers (see below), can react with nearly 40% of its own weight, ultimately retaining about half that in the dried film.
- In non-conjugated oils, oxygen uptake is accompanied by an increase in hydroperoxide content.
- Without driers and at 25 °C, a thin film of linseed oil (non-conjugated) requires 120 h to dry. However, the rate of drying is considerably accelerated

by the addition of catalytic amounts of certain metal soaps (driers), e.g. the same linseed oil dried in $2\frac{1}{4}$ h.

- Volatile decomposition products are produced in quantity during the drying process and continue to be produced throughout the life of the film. Drying and further chemical change during the aging of the film are additionally affected by exposure to natural ultraviolet light.

The following additional facts concern conjugated oils.

- Oils containing fatty acids with conjugated double bonds, dry much faster than those with non-conjugated unsaturation.
- Compared to linseed oil, a film of tung oil (conjugated) requires 48–72 h to dry without added driers; with driers added, the same tung oil dried in $1\frac{1}{4}$ h.
- In conjugated oils, hydroperoxide formation is not substantial until after the film has set.

The term **autoxidation** may be encountered in describing these oxidative drying processes. This simply means that atmospheric (gaseous) oxygen is involved and that conditions are mild, compared with other reactions termed oxidations where conditions can involve heat and reactants other than oxygen (p. 23).

Drying mechanisms

An explanation of the drying mechanism must be consistent with the above facts. These facts suggest that differences in mechanism exist between conjugated and non-conjugated oils. Better evidence exists to explain the drying mechanisms of non-conjugated oils and, since most oxidative coatings now contain these, they will be explained first and in greater detail.

While what follows is given in a factual form, the practical difficulties in gaining full understanding are large, and a considerable body of early work was carried out by painstaking methods, but without the aid of modern instrumentation. These difficulties require us to recognize that:

- oils are complex mixtures of fatty acids, and occur with natural inhibitors present, and hence ultra-pure materials for reliable experimentation are extremely difficult to obtain;
- transport of oxygen into the film and release of by-products from the film are necessary, and hence reaction is dependent on the availability of surface, and may be variable through the depth of a typical drying film;
- free radicals if involved are extremely reactive and short lived, and their involvement is of necessity deduced by indirect means.

Non-conjugated oils

As you will recall, the non-conjugated oils we refer to contain linoleic and linolenic acids. All oils contain some oleic acid.

The drying mechanism for this type of oil involves the formation of hydroperoxides and their subsequent reaction in various ways. There are different paths with possibilities of both cross-linking and scission (that is, the splitting off of parts of chains).

The initial reaction between oxygen (or free radicals) and non-conjugated oils occurs without loss of unsaturation. The first reaction products in the case of linoleic acid are hydroperoxides formed at either the 9 or 13 position. This involves a shift in one of the double bonds, with that double bond also changing from *cis* to *trans*.

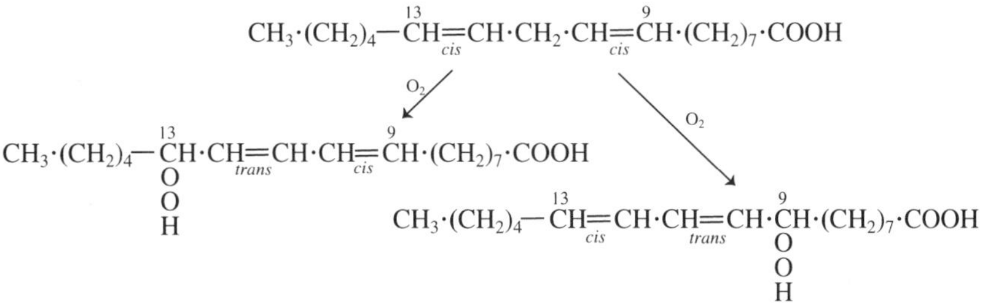

$$CH_3 \cdot (CH_2)_4 - \overset{13}{CH} = CH \cdot CH_2 \cdot CH \underset{cis}{=} \overset{9}{CH} \cdot (CH_2)_7 \cdot COOH$$

$$CH_3 \cdot (CH_2)_4 - \overset{13}{CH} \cdot CH \underset{trans}{=} CH \cdot CH \underset{cis}{=} \overset{9}{CH} \cdot (CH_2)_7 \cdot COOH$$
$$\begin{matrix} O \\ O \\ H \end{matrix}$$

$$CH_3 \cdot (CH_2)_4 - \overset{13}{CH} \underset{cis}{=} CH \cdot CH \underset{trans}{=} CH \cdot \overset{9}{CH} \cdot (CH_2)_7 \cdot COOH$$
$$\begin{matrix} O \\ O \\ H \end{matrix}$$

One mechanism offered in explanation for this shows involvement of one of the existing double bonds and the CH_2 adjacent to the double bond in what is known as the **allylic** position (denoted by *):

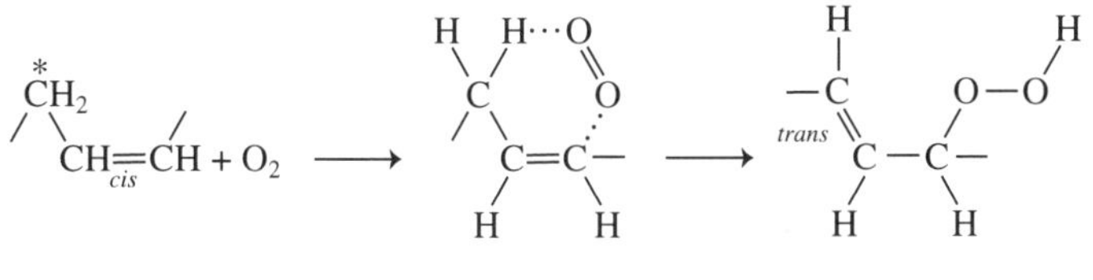

Hydroperoxides are decomposed by driers to give free radicals as described in the next section on driers. Free radicals are also obtained when hydroperoxides are decomposed by heat and light, as already mentioned (p. 65):

$$RC-O-O-H \longrightarrow RC-O \cdot + \cdot O-H$$

Radicals, once present in a system, will remove a hydrogen atom from any allylic carbon; this new radical may then rearrange to a conjugated form. Allylic CH_2 groups positioned between two double bonds are the most readily attacked. Oxygen can add to these radical sites to form peroxy radicals. These radicals in turn may remove hydrogen atoms from other molecules in a form of chain reaction, generating further radical sites for oxygen attachment. In this manner an increased concentration of hydroperoxide is built up. By the same mechanism, linolenic acid has four likely hydroperoxide structures.

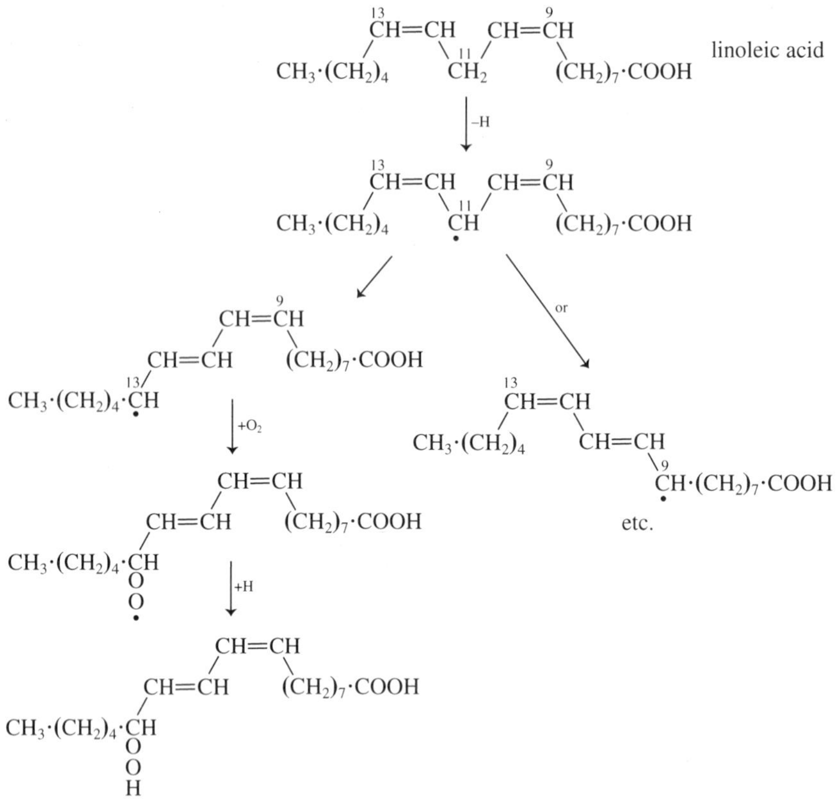

$$\underset{CH_3\cdot(CH_2)_4}{\overset{13}{CH}=CH}\quad\overset{}{\underset{CH_2}{}}\quad\overset{9}{CH}=CH\quad(CH_2)_7\cdot COOH \qquad \text{linoleic acid}$$

```
        13                    9
        CH=CH   CH=CH
       /      \  11 /     \
CH3·(CH2)4    CH2        (CH2)7·COOH          linoleic acid

                  | -H
                  ↓

        13                    9
        CH=CH   CH=CH
       /      \  11 /     \
CH3·(CH2)4    CH         (CH2)7·COOH
               •
          ↙                         ↘  or

            9
            CH=CH
           /      \
      CH=CH      (CH2)7·COOH              13
    13/                                    CH=CH
CH3·(CH2)4·CH                             /      \
           •       | +O2       CH3·(CH2)4    CH=CH
                   ↓                               \ 9
            CH=CH                                   CH·(CH2)7·COOH
           /      \                                 •
      CH=CH      (CH2)7·COOH              etc.
    /
CH3·(CH2)4·CH
           O
           O
           •       | +H
                   ↓
            CH=CH
           /      \
      CH=CH      (CH2)7·COOH
    /
CH3·(CH2)4·CH
           O
           O
           H
```

It is now normal to think of autoxidation reactions generally as a sequence of reactions commencing with a radical formed by abstraction of a hydrogen at a vulnerable position on the molecule under attack. The free radical nature of these reactions offers explanation for both the induction period and the effect of anti-oxidants. Any induction period may be explained by the necessity to build up an initial hydroperoxide concentration and to remove naturally present antioxidants by reaction before any chain reaction can commence. **Anti-oxidants** are structurally very similar to and function essentially in the same manner as free radical inhibitors. The reaction of inhibitor illustrated on p. 67 is shown involving a peroxy radical.

During the initial stages of drying, the chief method of cross-linking is by direct combination of free radical sites on different oil molecules (Fig. 12.1). Peroxy, ether and carbon–carbon links can be formed, and chains can be linked as shown below, with ether or C–C links most numerous.

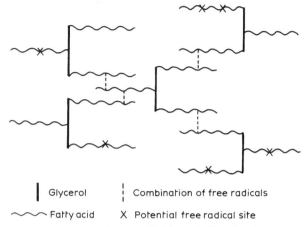

| Glycerol | Combination of free radicals |

~~~ Fatty acid   X Potential free radical site

Fig. 12.1 The cross-linking of oil molecules.

$$RC-O\cdot + \cdot O-CR \longrightarrow RC-O-O-CR$$
$$RC-O-O\cdot + \cdot CR \longrightarrow RC-O-O-CR$$
$$RC-O-O\cdot + \cdot O-O-CR \longrightarrow RC-O-O-CR + O_2$$

$\left.\begin{array}{c}\\\\\\\end{array}\right\}$ peroxy linkage

$$RC-O\cdot + \cdot CR \longrightarrow RC-O-CR \qquad \text{ether linkage}$$
$$RC\cdot + \cdot CR \longrightarrow RC-CR \qquad \text{C-C linkage}$$

Ether links are most likely in room temperature cure, while C–C links are most numerous in stoved films. While peroxy links are not found in high concentration, they are stable when built into polymeric structures.

Free radicals may also undergo decomposition or **scission** reactions. Hexanal is a principal oxidation by-product of linoleate oxidation, one of the causes of the acrid smell during drying.

$$CH_3\cdot(CH_2)_4-\underset{\underset{H}{\overset{O}{\underset{|}{O}}}}{CH}\cdot CH=CH\cdot CH=CH\cdot(CH_2)_7\cdot COOH$$

$\downarrow$ linoleic acid hydroperoxide

$$CH_3\cdot(CH_2)_4-\underset{\overset{|}{O}}{CH}\cdot CH=CH\cdot CH=CH\cdot(CH_2)_7\cdot COOH + \cdot OH$$

$\downarrow$

$$CH_3\cdot(CH_2)_4-CHO + \cdot CH=CH\cdot CH=CH\cdot(CH_2)_7\cdot COOH$$
hexanal

Aldehydes, ketones and carboxylic acids are all known to be decomposition products.

Conjugated oils

The drying of conjugated oils is less well understood. Reaction between oxygen (or free radicals) and conjugated oils occurs with loss of unsaturation, and the first reaction is possibly attack by oxygen on conjugated double bonds by direct addition, to form free radicals with two unsatisfied valencies (diradicals):

$$-\text{CH}=\text{CH}-\text{CH}=\text{CH}-\text{CH}=\text{CH}- + \text{O}_2 \longrightarrow -\text{CH}=\text{CH}-\text{CH}=\text{CH}-\overset{\cdot}{\text{C}}\text{H}-\text{CH}-$$

(with numbering 6 5 4 3 2 1 under the left side, and 2 1 over the right side, and a chain $\underset{\overset{|}{\text{O}\cdot}}{\overset{|}{\text{O}}}$)

These double bonds may rearrange as with the non-conjugated oils to give radical sites in the 1,4 and 1,6 positions as well.

Cross-linking occurs when diradicals attack double bonds in other oil molecules. What is known is that the cross-links from conjugated oils are mostly C–C bonds. This contributes to the higher durability of conjugated oil films.

Since the diradicals formed as above still leave conjugated double bonds, a possible maximum functionality of eleostearic acid is four (counted as radical sites), and that of the triglyceride 12. It is this higher functionality which accounts for the faster dry of tung oil.

Synthetic alternatives

It is useful to mention here that one other group of compounds is known which takes part in oxidative reactions; these are compounds containing the **allyl ether** group, which we will encounter further in Chapter 16. This forms hydroperoxides at its allylic position; in this case reaction is encouraged by its position between a double bond and an ether oxygen.

$$\text{R}-\text{CH}_2-\text{O}-\text{CH}_2-\text{CH}=\text{CH}_2 \xrightarrow{\text{O}_2} \text{R}-\text{CH}_2-\text{O}-\underset{\overset{|}{\text{OOH}}}{\text{CH}}-\text{CH}=\text{CH}_2$$

$$\xrightarrow{\text{Co}} \text{R}-\text{CH}_2-\text{O}-\underset{\overset{|}{\text{O}\cdot}}{\text{CH}}-\text{CH}=\text{CH}_2$$

Hydroperoxide breakdown with cobalt drier occurs followed by dimerization, just as with oils. One reason why these compounds are not fully suitable as alternatives for drying oils is that they can produce the decomposition product acrolein, which is toxic and has a particularly objectionable smell.

$$\text{R}-\text{CH}_2-\text{O}-\underset{\overset{|}{\text{O}\cdot}}{\text{CH}}-\text{CH}=\text{CH}_2 \longrightarrow \text{R}-\text{CH}_2-\text{O}\cdot + \text{OCH}-\text{CH}=\text{CH}_2$$

Acrolein

Summary

The mechanisms above explain the drying rates and properties found and also indicate differences in dry between conjugated oils and non-conjugated oils.

- In non-conjugated oils and in allyl ethers, hydroperoxides are the primary products and free radicals are released by hydroperoxide decomposition. In contrast, oxygen attack has been considered to produce free radicals directly in conjugated oils.
- In the initial stages of drying, allyl groups and non-conjugated oils yield only one free radical site at every point of oxygen attack, while conjugated oils yield two free radical sites. For fatty acids of both types, there is one point of attack for every double bond is excess of one. Thus, if the distribution of double and triple unsaturation is similar in both types of oils, conjugated oils have a higher initial functionality.
- The conjugated oils contain high proportions of acids with triple unsaturation (Chapter 3), and therefore have more positions susceptible to oxygen attack.

Driers

So far the accelerating action of the driers has not been explained. A drier is a metal soap with an acid portion that confers solubility in the oil medium. Synthetic acids, e.g. octanoic (2-ethyl hexoic, $C_4H_9 \cdot CH(C_2H_5) \cdot COOH$), are normally used; naphthenic acids (derivatives of cyclopentane and cyclohexane) formerly used are now rare. Driers are typical additive materials, being present in quantities generally less than 1%.

Primary driers are true catalysts and contain metals of variable valency, the lower valency being the more stable one, yet capable of oxidation to the higher valency by the products of the drying process. Cobalt and manganese are the principal **oxidative driers**. The metals act in two ways:

- they promote the uptake of oxygen;
- they catalyse the decomposition of peroxide in free radicals.

The **secondary** and **through driers** are of a group of metals, including lead, zirconium, calcium and cerium, which assist the drying of the lower layers of the paint film by mechanisms which are not fully understood, but probably involve interactions of the metal with carboxyl (and perhaps hydroxyl) groups in the film-former. Because lead is a cumulative poison, permissible levels of lead in household paints have been reduced so much that paint manufacturers now formulate paints of this type without the use of any added lead at all.

How driers promote oxygen uptake is uncertain, but the decomposition of hydroperoxide by primary driers occurs as follows:

(a) $\quad Co^{++} + R-O-O-H \longrightarrow Co^{+++} + RO\cdot + OH^- \left.\begin{array}{l} \\ \\ \end{array}\right\}$ oxidation

$\qquad$ (cobaltous) $\qquad\qquad\qquad$ (cobaltic) $\qquad\qquad\qquad\qquad\qquad$ of cobalt

$\qquad\qquad\qquad\qquad$ or $\longrightarrow Co^{+++} + RO^- + \cdot OH \left.\begin{array}{l}\\\end{array}\right\}$

(b) $Co^{+++} + R-O-O-H \longrightarrow Co^{++} + RO_2\cdot + H^+$ $\left.\begin{array}{l} \\ \\ \\ \end{array}\right\}$ reduction
$\quad Co^{+++} + OH^- \longrightarrow Co^{++} + \cdot OH$ $\qquad$ of cobalt
$\quad Co^{+++} + RO^- \longrightarrow Co^{++} + RO\cdot$

(c) $\qquad\qquad$ overall
$\qquad\qquad 2R-O-O-H \longrightarrow RO_2\cdot + RO\cdot + H_2O$

The cobalt is oxidized and reduced over and over again.

Cobalt leads to more rapid drying at the surface of the film than in the lower layers. If the surface dries first, it will buckle or 'shrivel' when the rest of the film contracts as it hardens. For this reason, secondary driers are usually added with cobalt to accelerate the drying of the bulk of the film. The presence of calcium accelerates the loss of unsaturation in the film. When the levels of the two types of drier are 'balanced', shrivelling does not occur. Manganese is less pronounced in its surface bias, but is also used with secondary driers.

In the late 1970s the aluminium alkoxide derivatives were introduced as through driers. These compounds may have the general formula

$$\begin{array}{c} R' \\ | \\ Al-OR \\ | \\ R'' \end{array}$$

in which the substituent groups $-OR$, $R'-$ and $R''-$ can react with hydrogen atoms, particularly in carboxyl and (less readily) in hydroxyl groups of alkyds, to bond several alkyd molecules to one atom of aluminium and eliminate volatile by-products from the drying film (ROH, HR' and HR''). Thus cross-linking through the film is achieved. Careful formulation (particularly adjustment of the carboxyl:aluminium ratio to 1.0 or less) is required to ensure stability in the can. In thick films the improved through dry that these driers can bring has been especially beneficial. Improved durability and yellowing resistance are found. Zirconium functions in a similar manner, and together these are termed **coordination driers**.

In typical air-drying decorative paints, the usual combination now is a mixture of cobalt, calcium and zirconium. Aluminium is used in special circumstances and with more difficult pigment combinations. For the newer developments of high solids coatings (see later in this chapter), which are longer in oil length and lower in molecular weight, the same combination is used, but here the aluminium drier described above often has a special advantage, and is used in place of zirconium. For emulsified alkyds, again cobalt, calcium and zirconium are most often used. Problems of formulation include ensuring that the driers are in the dispersed phase, and the pH has been found important in ensuring this.

In stoving paints it is usually satisfactory to include low levels (0.01–0.05% metal on binder) of a single primary drier. Cobalt, manganese, cerium and iron (which only acts as a drier at higher temperatures) are used. Unacceptable discoloration in pale colours can be a hazard, particularly with iron and cerium.

Bodied oils

In Chapter 7 the faster rate of drying which can be obtained by beginning with large molecules requiring fewer cross-links was discussed. To speed up the drying of oils, it is logical to polymerize a few oil molecules together before making the paint, stopping the polymerization process before the viscosity of the oil rises excessively. This is called '**bodying**' the oil.

A well recognized approach is used to make **stand oils** or **heat-bodied** oils. Conjugated oils are bodied or polymerized rapidly at 200–260 °C, while non-conjugated oils body more slowly at 260–290 °C. Stand oils contain mainly carbon–carbon linkages. In conjugated oils these are formed by the reaction known as the **Diels–Alder** reaction:

$$
\begin{array}{c}
\text{HC} \overset{\text{CH}}{\underset{\text{CH}}{\diagup\diagdown}} + \begin{array}{c}\text{CH}\\ \| \\ \text{CH} \\ | \\ \text{CH} \\ \| \\ \text{CH} \\ | \\ \text{CH}_2\end{array} \longrightarrow
\begin{array}{c}\text{HC} \overset{\text{CH}}{\diagup} \text{CH} \\ \| \qquad | \\ \text{HC} \diagdown \diagup \text{CH} \\ \qquad \text{CH} \diagdown \text{CH}=\text{CH}-\text{CH}_2-\end{array}
\end{array}
$$

For this reaction it is necessary that one molecule should contain two conjugated double bonds and the other molecule a double bond. Isomerization to conjugated form, followed by Diels–Alder reaction, accounts for part of the thermal polymerization of non-conjugated oils, but the dominant reaction is thought to arise from the transfer of a hydrogen atom from one molecule to another, in a reaction known as the **ene** reaction:

$$
\begin{array}{c}
-\text{CH}=\text{CH}-\text{CH}_2-\text{CH}=\text{CH}- \\
+ \\
-\text{CH}=\text{CH}-\text{CH}_2-\text{CH}=\text{CH}-
\end{array}
\longrightarrow
\begin{array}{c}
-\text{CH}=\text{CH}-\overset{\cdot}{\text{C}}\text{H}-\text{CH}=\text{CH}- \\
+ \\
-\overset{\cdot}{\text{C}}\text{H}-\text{CH}_2-\text{CH}_2-\text{CH}=\text{CH}-
\end{array}
\longrightarrow
$$

$$
\begin{array}{c}
-\text{CH}=\text{CH}-\text{CH}-\text{CH}=\text{CH}- \\
| \\
-\text{CH}-\text{CH}_2-\text{CH}_2-\text{CH}=\text{CH}-
\end{array}
$$

Here only one double bond is lost.

An older way of providing bodying is to carry out oxidative drying in the bulk oil by bubbling air through it at 75–120 °C. The product is called **blown oil**. Alternatively, driers may be formed in the oil by heating it to 230–270 °C with metal oxides or salts, which react with fatty acids from the oil. The addition of soluble driers can be carried out at a lower temperature (95–120 °C), accompanied by 'blowing' with air. These products are called **boiled oils** (although in fact the oils do not reach boiling point). Nowadays the name 'boiled oil' is given to mixtures of oil and driers blended in the cold, even though no heating has occurred and the oil molecular weight is not in any way increased.

Blown and boiled oils contain principally carbon–oxygen–carbon links between molecules. In contrast, since the weaker ether and peroxide links are excluded, stand oils are more durable and lighter in colour. The 'body' gives fuller films with higher gloss.

Oil paints and varnishes

Paints containing oil as sole film-former are scarcely ever used as finishes because of poor gloss, soft films and inferior water resistance, but where used, the oils were usually highly pigmented to minimize these disadvantages and found outlets as wood and steel primers. The chief oil used was linseed; oils with no triple unsaturation in their fatty acids are too slow drying. Conjugated oils give wrinkled, frosted films on drying, but are usefully mixed with linseed stand oil to increase the rate of drying and water resistance. Nowadays oils are replaced in primers by alkyds, and are formulated without the need for lead driers.

It is interesting to note that oil paints were the first 100% solids finishes. The oil is both solvent and film-former and virtually non-volatile. Modern resins of higher molecular weight have led to lower solids, while improving paint performance. Paints combining the present standard of paint properties and very high solids are general targets of research in the paint industry.

Oils are still used in **oleoresinous** paints and varnishes, and in inks. In these the oil is either mixed with, or heat-bodied in the presence of, some other resin. The resin may react with the oil to give larger molecules containing fatty acid. Reaction occurs with rosin and its derivatives, phenolic resins and petroleum resins. Other resins, such as terpene resins and coumarone–indene resins, do not react with the oil, but heat helps to dissolve the resin and causes the oil to body. The oils used are those found suitable for oil paints.

An outline of the resins mentioned follows.

Rosin consists principally of abietic acid and its isomers.

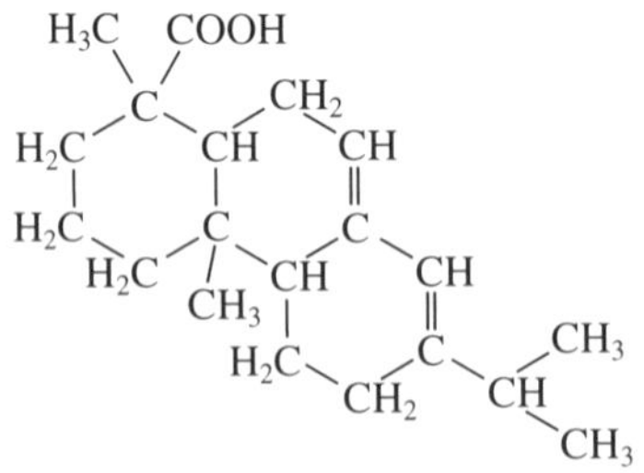

abietic acid

Petroleum resins are produced by the polymerization of the mixture of unsaturated hydrocarbons found in petroleum. As many of these compounds contain more than one double bond, varying amounts of unsaturation can be obtained in the resins.

Terpene resins are made by polymerization of terpenes. These lose all unsaturation in the process.

Coumarone–indene resins are made by polymerization or copolymerization of coumarone and indene:

coumarone   indene

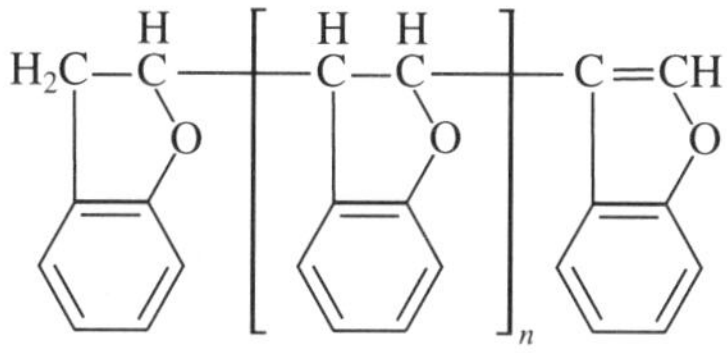

polycoumarone

The resins possess only slight unsaturation, since the attached rings are aromatic.

Phenolic resins. A variety of resins is possible, but in each case, the formation of a **phenolic condensate** is the first step:

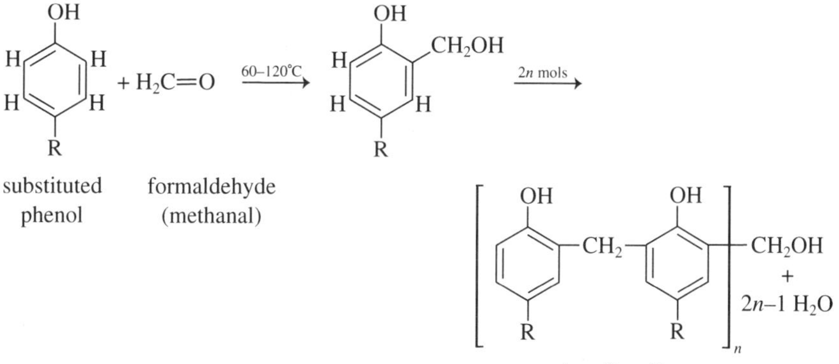

substituted   formaldehyde
phenol   (methanal)

$(n = 2$ to $4)$

With 0.5–1.0 molecules of formaldehyde to one of phenol, an acid catalyst is used. As can be seen, a short-chain linear polymer is produced. The group R— may be hydrogen, an alkyl group or even an aromatic substituent. It need not necessarily be *para-* to the phenolic hydroxyl, but the lightest colour, best solubility in oils and greatest reactivity are obtained with the substituent in this position. This type of phenolic resin is called a **novolac**.

With higher proportions of formaldehyde and the *para*-position vacant, branching is possible. An alkaline catalyst is used to obtain a slower rate of reaction, thus making it easier to stop the reaction before cross-linking occurs. This type of phenolic resin is called a **resole**.

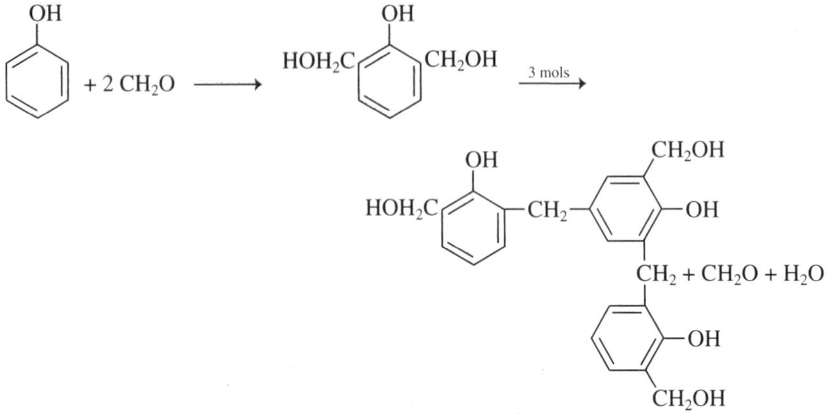

If the *para*-substituent group R— (p. 177) is an alkyl group of about four or more carbon atoms (e.g. tertiary butyl, $(CH_3)_3C-$), the product is soluble in oils; otherwise, it is soluble in alcohols, ether-alcohols and alcohol–aromatic hydrocarbon mixtures. If the resin is made from a phenol blocked in the *para*-position, with a high formaldehyde level and using an alkaline catalyst, it will react when heated with unsaturated materials such as drying oils or rosin to produce a **modified phenolic resin**. The mechanism is thought to be of the type

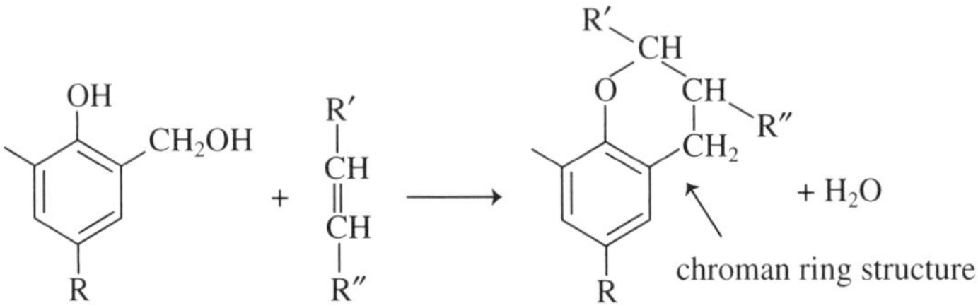

chroman ring structure

The reaction with rosin is used to modify phenolic resins and provides an alternative route to solubility in oils. The rosin carboxylic group is afterwards esterified with a polyhydric alcohol, such as pentaerythritol, to increase the resin molecular weight even further.

The methylol group in phenolic resins can be butylated, as in nitrogen resins (Chapter 13), and this improves compatibility with epoxy resins.

The main use for phenolic resins is in oleoresinous varnishes. Apart from the chemical-resistant finishes produced with epoxy resins (Chapter 14), other outlets include modification of alkyds to improve resistance to water and alkali, combination with urea–formaldehyde resins (Chapter 13) to provide metal coatings and combination with polyvinyl formal or butyral resins to produce wire enamels.

The effect of the resin, whatever its type, is to improve the drying time of the paint, since the resins are hard in their normal state and only require solvent evaporation for drying. It should be noted that the addition of resin means that solvent (white spirit, turpentine or possibly aromatic hydrocarbons) will now be required to reduce the viscosity of the paint for application, though the low molecular weights of the oils, resins and oleo-resinous products allow medium–high solids at application viscosity.

Other properties are altered by the resins. They harden the film and improve the gloss, though flexibility is reduced. Phenolic, coumarone–indene, petroleum, terpene and other resins free from acid or ester groups give improved water and chemical resistance. Phenolic resins improve the outdoor durability; unfortunately they give poor initial colour and aggravate the yellowing of the film as it ages. Yellowing is marked when the oil used contains appreciable amounts of linolenic acid (e.g. linseed oil). Manganese driers can also aggravate yellowing.

The weight ratio of oil:resin is called the **oil length**. Between three and five parts of oil to one of resin is described as 'long oil', $1\frac{1}{2}$–3:1 as 'medium oil' and $\frac{1}{2}$–$1\frac{1}{2}$:1 as 'short oil'. As the oil length increases, the properties contributed by the resin become less and less important. For example, long oil varnishes are the slowest drying, but have the best outdoor durability, because the oil contributes the necessary flexibility.

Alkyd resins

Incorporation into alkyd resins provides yet another method of improving the drying potential and film-forming properties of drying oils. In principle, alkyds need not be made from oils; they are step growth polymers of dibasic acids and dihydric alcohols. The name 'alkyd' derives from its ingredients: *alc*ohols and *ac*ids. An oil-free resin could be made from ethylene glycol (ethane diol) and phthalic anhydride (benzene 1,2-dicarboxylic anhydride):

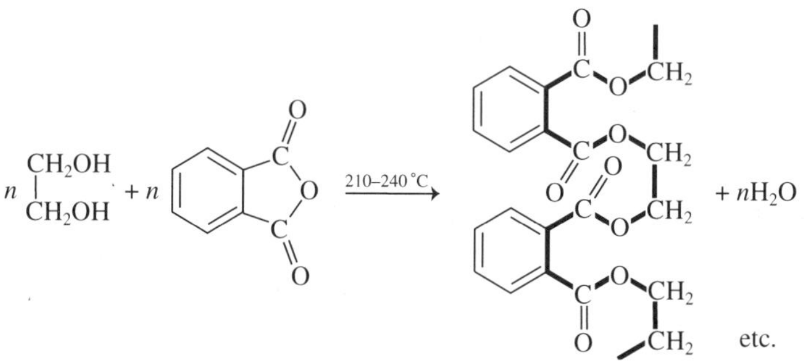

The polymer is quite linear (unbranched), as can been seen by following the thickened valency lines. It is a polyester resin, because the polymer chain is held together by a series of ester linkages; in this form it is referred to as a

saturated or **oil-free polyester** because it contains no aliphatic double bonds in either backbone or side chains. To include an oil fatty acid we would use a monoglyceride instead of ethylene glycol. This will leave our alcohol a hydroxyl functionality of two (being the mono fatty acid ester of glycerol), e.g. linseed oil monoglyceride (containing linolenic acid):

$CH_2{\cdot}O{\cdot}CO{\cdot}(CH_2)_7{\cdot}CH{=}CH{\cdot}CH_2{\cdot}CH{=}CH{\cdot}CH_2{\cdot}CH{=}CH{\cdot}CH_2{\cdot}CH_3$
|
$CHOH$
|
CH_2OH

A portion of the alkyd molecule appears like this:

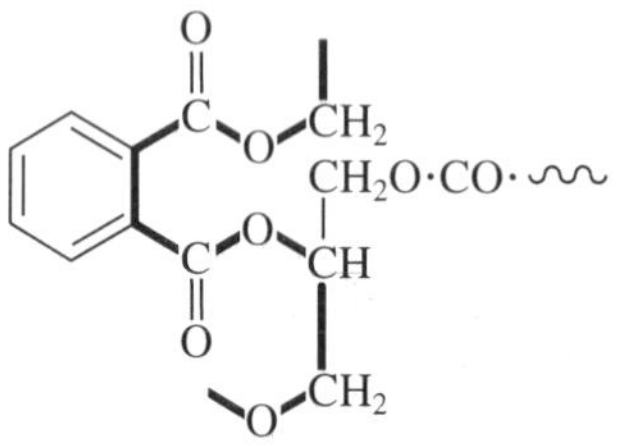

HO·CO·∿∿

fatty acid

In fact, alkyds are not normally made from monoglycerides prepared separately, but from oils included at the start of processing. The oil is converted to a monoglyceride by heating with glycerol and a catalyst before the phthalic anhydride is added:

$CH_2O{\cdot}CO{\cdot}$∿∿		CH_2OH		$CH_2O{\cdot}CO{\cdot}$∿∿		
n $CHO{\cdot}CO{\cdot}$∿∿	+	$2n$ $CHOH$	$\xrightarrow[\text{catalyst}]{220\text{–}240\ ^\circ C}$	$3n$ $CHOH$	$\xrightarrow[\text{addition of phthalic anhydride}]{150\text{–}240\ ^\circ C}$	alkyd resin
$CH_2O{\cdot}CO{\cdot}$∿∿		CH_2OH		CH_2OH		
triglyceride (oil)		glycerol (propane triol)		monoglyceride		

Monoglyceride is formed, because some of the ester linkages in the original oil are broken and the fatty acid molecules produced react with glycerol hydroxyl groups. When this ‘redistribution’ of fatty acid has gone as far as it can, the average molecule will be monoglyceride, though there will still be some oil, some glycerol and some diglyceride.

This redistribution is also carried out with other polyols such as pentaerythritol, which is commonly used in long oil alkyds. It is customary still to refer to this as a **monoglyceride stage** since the oil yields monoglyceride, even though another polyol is also involved whose partial ester is not a glyceride.

Since the oil would body oxidatively in the presence of oxygen, air is displaced from the reaction vessel by inert gas (e.g. nitrogen), which continues to blanket the liquid for the remainder of the process. It is usually the practice to include a solvent for the alkyd (typically xylol) as up to 8% of the charge. The solvent accelerates reaction by distilling over into a separator with the water produced. Being immiscible with water, the solvent separates on condensation and is returned to the reaction vessel. Better colour and batch-to-batch reproducibility are obtained than by just allowing the water to boil off.

The course of the reaction can be followed by (1) the drop in acid content, since the $-COOH$ group disappears for every ester link formed, (2) the rise in viscosity and also less reliably by (3) the volume of water produced. The acid content is measured by titration of resin solution with alcoholic alkali solution and the **acid value** (AV) or **acid number** (AN) of the resin is the number of milligrams of potassium hydroxide required to neutralize 1 g of solid resin. AV values of commercial alkyds are usually between 5 and 30. Water evolution is never sensitive enough to monitor the reaction properly, and the first two tests are those normally carried out together.

Ingredients

Many acids and alcohols may be used, but those with more than two functional groups must be kept at a low concentration, so that branching rather than cross-linking occurs, or the groups in excess of two must be neutralized, e.g. hydroxyl with fatty acid. It should be noted that although glycerol has three hydroxyl groups, the secondary hydroxyl group is much less reactive than the two primary groups at 150–180 °C. The secondary group becomes reactive at 200–260 °C. Examples of polybasic acids and polyhydric alcohols used are given below:

Acids

Phthalic anhydride ⎫
Isophthalic acid ⎬ Chapter 4

Succinic acid, $HOOC \cdot (CH_2)_2 \cdot COOH$

Adipic acid, $HOOC \cdot (CH_2)_4 \cdot COOH$

Sebacic acid, $HOOC \cdot (CH_2)_6 \cdot COOH$

Terephthalic acid,

$$HOOC \langle \bigcirc \rangle COOH$$

Trimellitic anhydride,

Alcohols

Ethylene glycol ⎫
Glycerol ⎬ Chapter 3
Pentaerythritol ⎭

Diethylene glycol Chapter 9

1,2-Propylene glycol,
$$CH_3 \cdot CHOH \cdot CH_2OH$$

Trimethylol propane,
$$CH_3CH_2 - \overset{\displaystyle CH_2OH}{\underset{\displaystyle CH_2OH}{\overset{|}{\underset{|}{C}}}} - CH_2OH$$

neo-Pentylglycol,
$$HOCH_2 - \overset{\displaystyle CH_3}{\underset{\displaystyle CH_3}{\overset{|}{\underset{|}{C}}}} - CH_2OH$$

Of these, phthalic anhydride as the acid component, and glycerol and penta-erythritol as polyols, are most commonly used. Monobasic acids, such as benzoic acid, can be used to stop chain growth and hence to limit molecular weight.

Drying or non-drying oils can be used, according to the type of alkyd required. The non-drying oils, such as castor oil, give the plasticizing alkyds frequently used in lacquers. The drying and semi-drying oils give drying alkyds. Alkyd molecular weights are of the order of 1000–5000. For a 58% linseed oil glycerol phthalate alkyd (see below), this allows about 2–10 fatty acid chains per average molecule. Thus a drying alkyd will dry faster than the corresponding oil, because of the higher functionality of the molecule and its greater size. Whereas the choice of oil for oil and oleoresinous paints is limited, air-drying alkyds are mainly made from semi-drying oils, such as soya bean oil. These have acceptable drying times because of the increased fatty acid functionality.

Alkyds, like oleoresinous varnishes, have an oil length and this is expressed as a percentage: the number of grams of oil contained in 100 g of final resin. Below 45% oil it is a short oil alkyd, 45–60% medium oil and above 60% long oil. Long oil alkyds are soluble in aliphatic hydrocarbons, such as white spirit, while short oil alkyds require aromatic hydrocarbons (Tables 9.1 and 9.3).

Alkyds allow great scope for variation. Changing the polybasic acids and polyhydric alcohols alone can vary the properties, since increasing the number of carbon atoms between the carboxyl groups in the acids (or the hydroxyl groups in the alcohols) leads to more flexible resins. The extent of branching is controlled by acid/alcohol functionality. The alkyd may dry either by lacquer dry or oxidatively, according to the oil used. The rate of dry and colour retention will vary with the oil and oil length. The latter controls solubility, cross-linking potential and compatibility with other resins.

Modifications are possible to give more unusual properties to the resin. Inclusion of polyethylene glycol as polyol produces an alkyd which is water dispersible. One such alkyd has been used in white spirit-based gloss house paint. This paint can be washed from brushes using a weak detergent solution in water without requiring the use of organic solvent.

Another modification involves treating prepared long oil alkyd with polyamide resin (p. 212) while still hot. Providing this is done carefully, a thixotropic alkyd is obtained which may be used as a rheology control additive, as described in Chapter 10.

The extent and speed of lacquer dry can be increased by modification of the alkyd with polystyrene, poly(vinyl toluene) or methacrylate polymer chains. These may be introduced by carrying out the chain growth polymerization of the vinyl or acrylic monomer in the presence of (1) the completed alkyd, (2) the monoglyceride or (3) the oil. The reaction with the fatty acid portions proceeds most readily if the double bonds are conjugated. With non-conjugated fatty acids, there is some grafting (Chapter 11) of the chain growth polymer onto the active methylene groups, but much of the polymer formed is unattached to the alkyd. Dehydrated castor oil (DCO) is the oil most usual in these alkyds, since it contains a significant amount of the more reactive conjugated acid. Some loss of unsaturation occurs when the copolymerization is successful. This means that the modified alkyd dries more rapidly at first by lacquer dry, but forms fewer cross-links than the unmodified alkyd and so is less solvent resistant and more prone to swelling and wrinkling when recoated with the same paint.

Alkyds are used in stoving and air-drying finishes, alone and with other resins. Their scope is so great that it will be necessary to confine the following brief discussion on paints to those containing alkyds alone. Their use with nitrogen resins in stoving finishes is covered in Chapter 13.

Alkyd finishes

Air-drying finishes

Household air-drying finishes are based on long oil alkyds, driers, pigment, white spirit and additives. Industrial air-drying finishes are frequently based on styrenated or vinyl toluene-modified alkyds. In household paints, soya bean oil and tall oil alkyds are most common and used because they yellow less than the linseed oil alkyds once more commonly used. Alkyds give better durability out of doors than oleoresinous varnishes. Phthalic anhydride is responsible for their good resistance to degradation by ultra-violet light. The water and alkali resistance of alkyds is not as good as that of the best oleoresinous varnishes, but alkyds have better colour and gloss retention and are less prone to fail eventually by cracking. These are important features in a house paint.

The relatively low molecular weights of alkyd resins allow medium–high solids at brushing viscosity (about 5 poises or 0.5 Pa s). Films dry to touch in about 1–4 h in warm weather, but take longer when it is cold. Once air is allowed into the paint can, drying occurs at the surface forming a skin: it does not usually go deeper. Alkyd films harden slowly over many days. Those designed to last outdoors are soft and easily marked, but very flexible.

With requirements now to reduce solvent emissions from paint application on environmental grounds, two approaches are being applied to decorative formulations. The first method is to reduce solvent content by making the paint at higher solids. For this to be possible, the alkyd used must be of lower molecular weight and, to retain functionality of fatty acid chains and drying potential, must also be of very long oil length. These alkyds are made from isophthalic acid, rather than orthophthalic anhydride, since this gives a narrower molecular weight (MW) distribution, eliminating lower MW fractions with poor drying potential, and higher MW fractions which would reduce the solids.

The other approach is to dissolve or to emulsify the alkyd in water. Emulsification can be assisted by adding a surfactant, and also by making the alkyd partly water-soluble. Typically non-ionic surfactants are used. The alkyd may have an increased acid value and be partly amine neutralized, or may include water-soluble components (Chapter 9). The latter is now more usual. These emulsions are made by mechanical means (p. 158).

Fully water-soluble air-drying finishes are prepared as above by neutralization of higher acid value alkyd resins. Made using shorter oil alkyds, they find industrial use, and typically they contain some water-miscible solvent to improve application viscosity and flow.

Stoving finishes

A few stoving finishes contain alkyd alone; more usually the finish is hardened by blending the alkyds with nitrogen resins (Chapter 13). Short–medium oil length alkyds are used, probably with some proportion of aromatic solvent present. Some stoving primers and one-coat paints are based on alkyd alone, typically medium oil length linseed oil alkyd, possibly rosin-modified. Water-based industrial stoving alkyds are prepared as above by partial neutralization of higher acid value alkyd resins (Table 12.1).

Examples of coatings based on alkyds, oil and oleoresinous film-formers are given in Tables 12.2–12.5.

Table 12.1 Formulation of glossy water-thinnable air-drying alkyd paint for agricultural machinery

		wt%	
Resin	Worléesol 65	10.80	
Neutralizing agents	DMAMP 80	0.63	
	Ammonia (25%)	0.10	
Solvent	Water	10.00	
Pigments	Paliogen red L3880 HD	4.60	Bead mill
	Sicotrans red L2915 D	0.85	grind stage
	Rutile titanium dioxide	0.25	
Additive	Activ 8 (Vanderbilt)	0.10	
Driers	Cobalt complex (8% Co)	0.25	
	Manganese complex (9% Mn)	0.30	
Resin	Worléesol 65	23.30	
Neutralizing agent	Ammonia (25%)	0.90	
Solvents	Water	45.42	
	2-Butoxy ethanol	0.90	
	Propylene glycol *n*-butyl ether	0.90	
Additives	Surface wetting improver		
	Worlée-Add 330	0.50	
	Flow/Marr improver		
	Worlée-Add 327	0.20	
		100.00	

The pH should be adjusted to 8.2–8.5 with ammonia if necessary. Application may be by spray or dipping. Paint should be thinned as required with water:propylene glycol *n*-butyl ether 8:2. The paint has a dust free air drying time of 30 min, and tack-free time of 3 h.
Worléesol 65 is an air-drying short oil alkyd at acid value 35 70% solids in 2-butoxy ethanol–*sec*-butanol–propylene glycol *n*-butyl ether 1:1:1, made by Worlée-Chemie GmbH.
DMAMP 80 is dimethyl amino methyl propanol supplied by Lehmann and Voss.
Based on Worlée-Chemie formulation No. 553/1.

Table 12.2 Formulation of alkyd gloss finish for decorative use

		wt%
Pigment	Rutile titanium dioxide	27.0
Resin	68% OL tall oil fatty acid–pentaerythritol alkyd at 75% solids in white spirit	55.0
Thixotropic resin	60% OL soyabean oil–pentaerythritol alkyd polyamide modified at 60% solids in white spirit	5.0
Driers	Cobalt octoate (10% Co)	0.2
	Zirconium complex (6% Zr)	0.5
	Calcium octoate (5% Ca)	1.7
Solvent	White spirit	10.6
		100.0

Ready for brushing at 72.5% solids.   OL = oil length.

Table 12.3 Formulation of oleoresinous varnish for exterior use

		wt%
Resins	Modified drying oil	16.0
	Oil-soluble, non-heat-reactive phenolic resin	38.4
Solvents	White spirit	42.0
	Isopropyl alcohol	2.4
Driers	Cobalt octoate (6% Co)	0.3
	Manganese octoate (6%Mn)	0.1
	Zirconium complex (6% Zr)	0.8
		100.0

Ready for brushing at 55% solids.

Table 12.4 Formulation of wood primer (solvent-borne)

		wt%
Pigment	Rutile titanium dioxide	16.30
Extender	Magnesium silicate	34.60
Resins	Long oil soya beam oil–pentaerythritol alkyd at 70% solids in white spirit	18.20
	Medium oil isophthalic alkyd at 50% solids in aromatic hydrocarbon blend, BR 160–180 °C	12.10
Additives		
Pigment dispersant	Soya lecithin	0.25
Anti-settling aid	Bentone 38	0.25
Anti-skinning aid	Methyl ethyl ketoxime	0.15
Driers	Cobalt octoate (10% Co)	0.15
	Zirconium complex (6% Zr)	0.40
Solvents	Alcohol	0.10
	White spirit	17.50
		100.00

Ready for brushing at 70% solids.   Bentone 38 is a treated clay from Rheox Inc.

Table 12.5 Formulation of decorative gloss paint based on emulsified alkyd

		wt%	
Rutile titanium dioxide Kronos 2190	Pigment	24.00	
Borchigen DFN (33%)	Dispersing aids	3.60	Dispersed in bead mill
Tamol 731 (25%)		0.50	
Antifoam Agitan 295	Additive	0.10	
Water	Diluent	7.25	
Then add to			
Uradil AZ516Z (60%)	Resin emulsion	50.00	
Thickener Borchigel DP40 (12.5%)	Additives	11.70	
Levelling agent Fluorad FC129 (10%)		1.20	
VX71	Driers	1.65	
		100.00	

Paint properties: solids 54.9%, pigment volume 16.3%, viscosity 3 Pa s.
Uradil AZ516Z is a 63% OL tall oil alkyd emulsion provided at 60% in water. It is amine and solvent free, and supplied by DSM Resins bv.
VX71 contains 0.9% cobalt, 3.6% calcium and 5.4% strontium, and is supplied by Durham Chemicals.
Borchigen DNF and Borchigel DP40 are supplied by Borcher AG.
Tamol 731 is a product of Rohm and Haas.
Agitan 295 is supplied by Münzing Chemie GmbH, and Fluorad FC129 by 3M.

Thirteen

Thermosetting alkyd, polyester and acrylic paints

paints based on nitrogen resins

The finishes in this and subsequent chapters dry by the third mechanism described in Chapter 7, that is to say, ingredients entirely within the paint undergo a chemical reaction on the surface of the article being coated to produce a cross-linked polymer film. The particular reaction in this chapter is the condensation reaction between methylol groups ($-CH_2OH$) attached directly to a nitrogen atom. For this reason, the family of resins involved are described here as 'nitrogen resins', although they are more widely known as **amino resins**. That name is misleading, since many of the ingredients are amides, not amines.

The principal coatings resins in this group are the urea–formaldehyde (UF), melamine–formaldehyde (MF) and the acrylamide– or methacrylamide–formaldehyde copolymer resins. The reaction concerned normally proceeds rapidly only when assisted by heating or acid catalyst. Thus the finishes in which nitrogen resins are found are principally stoving finishes, though the acid-catalysed cold-curing finishes are particularly important in the woodfinish market. The chief characteristic that nitrogen resins bring to finishes is hardness, although another virtue is their relatively good colour. Let us now consider their chemistry.

Nitrogen resins

What follows must be qualified by saying that the complete chemistry of the nitrogen resins has not been established beyond doubt. The account presented here is therefore a simplified outline, but one which nevertheless helps the paint and resin formulator to understand the processes involved.

Urea–formaldehyde

UF resins get their name from their main ingredients: urea and formaldehyde. Both are soluble in water; urea, a white crystalline solid melting at 133 °C, to the extent of 80 g in 100 g of water, while formaldehyde is usually supplied as the 37% aqueous solution, 'formalin', or the solid, paraformaldehyde. 'Paraform' is the step growth homopolymer produced by evaporating formalin:

$$_{n+1}CH_2{=}O \ + \ _{n+1}H_2O \longrightarrow \left[CH_2 {<}^{OH}_{OH} \right]_{n+1} \longrightarrow$$

$$HO{\cdot}CH_2{\cdot}(OCH_2)_n{\cdot}OH + nH_2O$$

It yields formaldehyde gas when heated to 180–200 °C. Reactions between urea and formaldehyde proceed as follows:

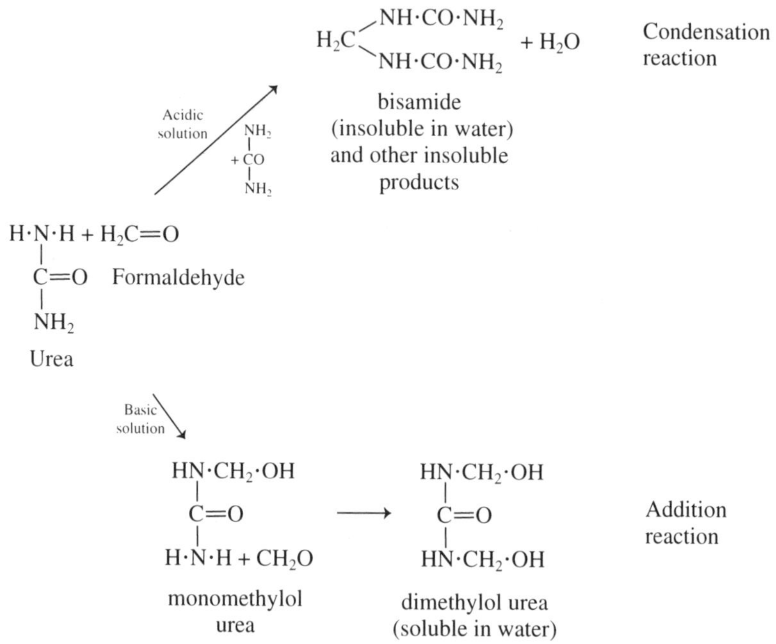

$$H_2C {<}^{NH{\cdot}CO{\cdot}NH_2}_{NH{\cdot}CO{\cdot}NH_2} \ + H_2O \qquad \text{Condensation reaction}$$

bisamide
(insoluble in water)
and other insoluble
products

Acidic
solution
$$\begin{matrix} NH_2 \\ | \\ +CO \\ | \\ NH_2 \end{matrix}$$

$$H{\cdot}N{\cdot}H + H_2C{=}O$$
$$\begin{matrix} | \\ C{=}O \quad \text{Formaldehyde} \\ | \\ NH_2 \end{matrix}$$
Urea

Basic
solution

$$\begin{matrix} HN{\cdot}CH_2{\cdot}OH \\ | \\ C{=}O \\ | \\ H{\cdot}N{\cdot}H + CH_2O \end{matrix} \longrightarrow \begin{matrix} HN{\cdot}CH_2{\cdot}OH \\ | \\ C{=}O \\ | \\ HN{\cdot}CH_2{\cdot}OH \end{matrix} \qquad \text{Addition reaction}$$

monomethylol   dimethylol urea
urea   (soluble in water)

The methylol group gets its name because it consists of a *methyl* group in which one hydrogen atom is replaced by an alcoho*l*ic hydroxyl. Note particularly that acid conditions give a condensation reaction, while basic conditions give an addition reaction. It is the latter route that interests us at present, because preparation of water-soluble dimethylol urea gives possibilities of subsequent conversion to a water-soluble resin by a switch to acid conditions:

$$HN{\cdot}CH_2OH + HN{\cdot}CH_2 - \xrightarrow[\text{solution}]{\text{acid}} HN{\cdot}CH_2{\cdot}N{\cdot}CH_2 - + H_2O \qquad \begin{matrix}\text{Condensation} \\ \text{reaction} \\ \text{(polymerization)}\end{matrix}$$

The reaction is shown linking two dimethylol urea units, but continues through the other groups present to form polymer.

Dimethylol urea has a functionality of four (two $-CH_2OH$, two $HN-$), so resin formation occurs with ultimate cross-linking. The high concentration of hydroxyl groups present in the linear version of the resin means poor solubility in organic solvents. This situation can be remedied by reaction of the hydroxyl with a suitable alcohol in an etherification reaction.

$$HN \cdot CH_2OH + HO \cdot R \xrightarrow[\text{solution}]{\text{acid}} HN \cdot CH_2 \cdot O \cdot R + H_2O$$

Condensation reaction (etherification)

ether link

If the alcohol $R \cdot OH$ is ethyl alcohol, the product is soluble in ethyl alcohol, but if it is butyl alcohol, the product is soluble in aromatic hydrocarbons (or their mixtures with alcohols).

Thus to make a resin suitable for non-aqueous paints, urea, formalin and butanol are placed in the reaction vessel and heating is commenced with the mixture slightly alkaline. When sufficient of the mono- and dimethylol ureas have been formed, the mixture is made acid. The two condensation reactions described above – polymerization and etherification – now proceed in direct competition. The nature of the final resin will depend upon the proportions of the three ingredients, the amount and type of acid and the temperature and time of reaction. The proportions might be 1 mol urea:2 mol formaldehyde:1 mol butanol. An idealized reaction scheme would be:

$$n \overset{\displaystyle NH_2}{\underset{\displaystyle NH_2}{C}=O} + 2nCH_2O + nC_4H_9OH$$

$\downarrow$ basic

$$n \overset{\displaystyle HN \cdot CH_2OH}{\underset{\displaystyle HN \cdot CH_2OH}{C}=O} + nC_4H_9OH \xrightarrow{\text{acidic}} n \overset{\displaystyle HN \cdot CH_2 \cdot O \cdot C_4H_9}{\underset{\displaystyle HN \cdot CH_2OH}{C}=O} \longrightarrow$$

$$\begin{array}{l} HN \cdot CH_2 \cdot O \cdot C_4H_9 \\ | \\ C=O \\ | \\ HN \cdot CH_2 \cdot N \cdot CH_2 \cdot O \cdot C_4H_9 + H_2O \\ * \\ C=O \\ | \\ HN \cdot CH_2 \cdot N \cdot CH_2 \cdot O \cdot C_4H_9 \\ * \\ C=O + H_2O \\ | \\ HN \cdot CH_2- \text{ etc.} \\ * \end{array}$$

It is most unlikely that only a linear polymer will be formed, because the reactive

$$\underset{*}{\overset{|}{HN}-}$$

groups in the growing polymer molecule can react with $-CH_2OH$ groups in other molecules to produce branching. Rings could be formed:

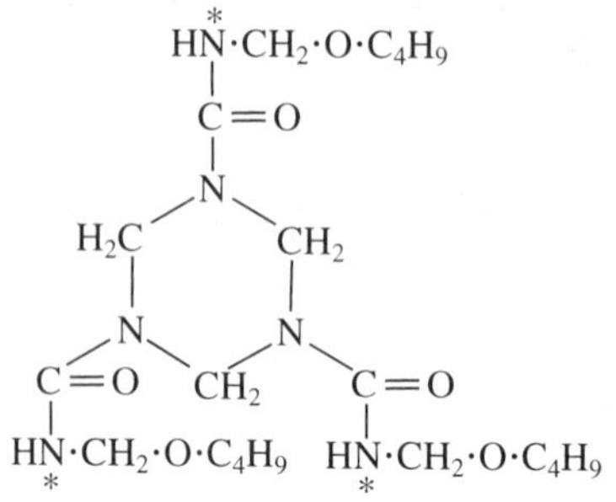

$$
\begin{array}{c}
\overset{*}{H}N \cdot CH_2 \cdot O \cdot C_4H_9 \\
|\\
C=O \\
|\\
N \\
H_2C \diagdown \quad \diagup CH_2 \\
|\qquad\qquad | \\
N \qquad\quad N \\
C=O \quad CH_2 \quad C=O \\
|\qquad\qquad\qquad | \\
\overset{}{H}N\cdot CH_2\cdot O\cdot C_4H_9 \;\; \overset{}{H}N\cdot CH_2\cdot O\cdot C_4H_9 \\
*\qquad\qquad\qquad *
\end{array}
$$

with chains leading from the rings via reaction at *. Consequently, the reaction is stopped by cooling the mixture and neutralizing the acid, while there are still appreciable numbers of

$$
\begin{array}{c}
| \\
HN-
\end{array}
$$

and $-CH_2OH$ groups unreacted. The resin may later be 'cured' to a cross-linked, insoluble state by allowing the reaction to proceed further on the coated article:

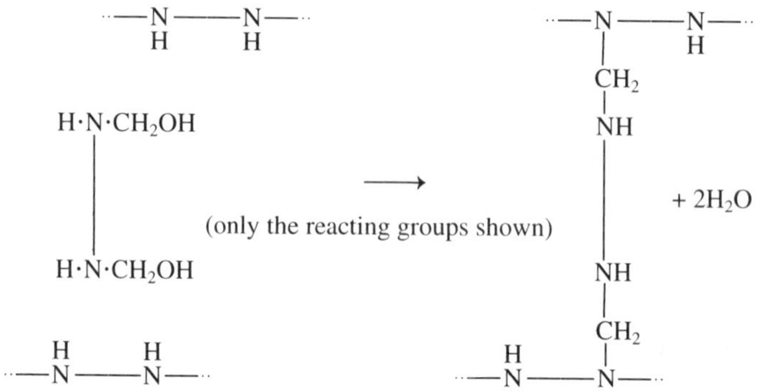

$$
\begin{array}{ccc}
\cdots\!-N\!-\!\!-\!\!-N\!-\!\cdots & & \cdots\!-N\!-\!\!-\!\!-N\!-\!\cdots \\
\;\;\;H\quad\;\; H & & \;\;\;\;|\qquad\;\; H \\
& & \;\;\;CH_2 \\
H\cdot N\cdot CH_2OH & & \;\;\;NH \\
|& \longrightarrow & \;\;\;\;| \qquad +\,2H_2O\\
| & \text{(only the reacting groups shown)} & \;\;\;| \\
H\cdot N\cdot CH_2OH & & \;\;\;NH \\
| & & \;\;\;CH_2 \\
\;\;\;H\quad\;\; H & & \;\;\;\;H\quad\;\; | \\
\cdots\!-N\!-\!\!-\!\!-N\!-\!\cdots & & \cdots\!-N\!-\!\!-\!\!-N\!-\!\cdots
\end{array}
$$

In all this the butylated hydroxyls are not entirely inactive, since butanol is one of the products of the cross-linking reaction. The reaction may be

$$
\begin{array}{ccc}
|\qquad\qquad\qquad | & & |\qquad\quad | \\
N\cdot CH_2\cdot O\cdot C_4H_9 + H\cdot N & \longrightarrow & N\cdot CH_2\cdot N + C_4H_9OH \\
|\qquad\qquad\qquad | & & |\qquad\quad |
\end{array}
$$

but it occurs less readily than the reactions involving non-etherified hydroxyl, since one of the effects of increasing the extent of butylation is to decrease reactivity and hence the curing rate of the resin. This is also reduced by choosing an etherifying alcohol containing more carbon atoms (e.g. nonanol, $C_9H_{19}OH$). However, both of these changes serve to increase the solubility of the resin in organic solvents and to improve its compatibility with other resins. A balance between reactivity and solubility is required in coatings resins. Those most frequently used are *n*- and *iso*-butylated resins.

Melamine–formaldehyde resins

Melamine is a white, crystalline solid melting at 354 °C and only slightly soluble in water. It is made as follows:

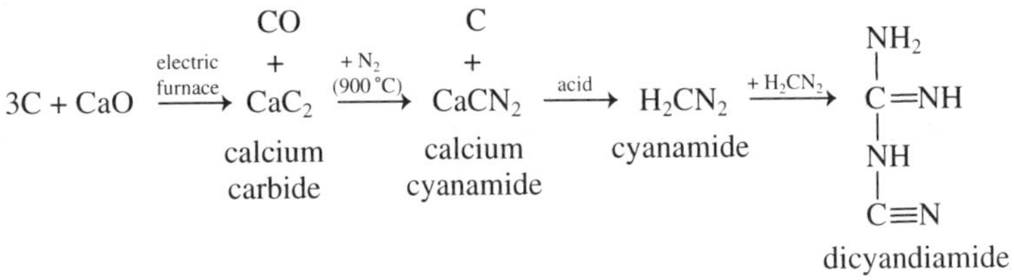

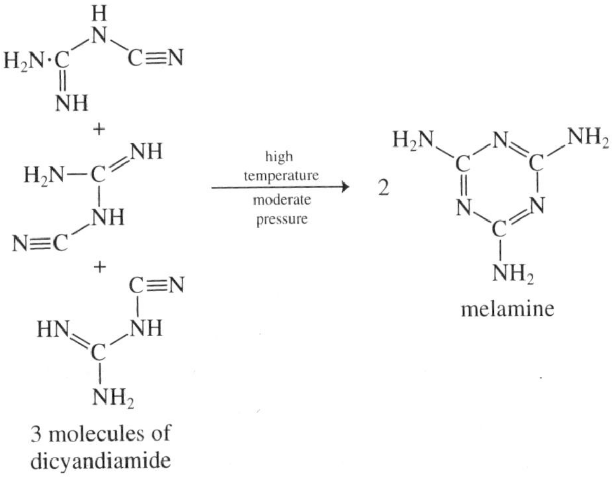

The primary amino groups of melamine react with formaldehyde as do those in urea, but with a difference. Both hydrogen atoms may be converted to methylol groups, making a maximum of six for the molecule. If melamine, formaldehyde and butanol are reacted together under acid conditions, a butylated MF resin can be obtained:

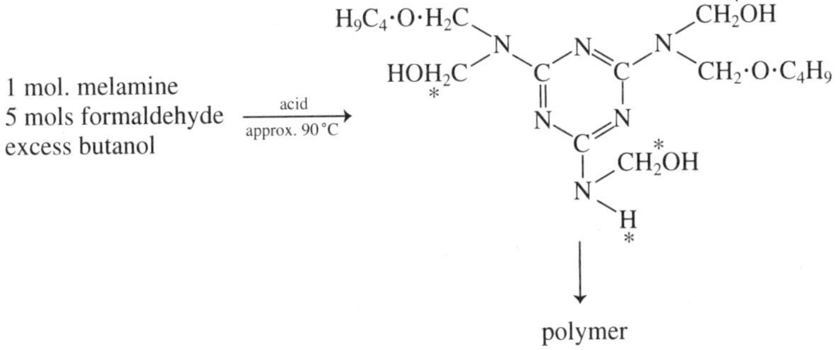

The reactive groups marked * undergo the condensation reactions shown for UF resins, as well as etherification by condensation between methylol groups:

$$-CH_2OH + HOCH_2- \longrightarrow -CH_2 \cdot O \cdot CH_2- + H_2O$$

The polymerization is stopped short of cross-linking which, with further heating, will occur rapidly due to the high functionality. These butylated MF resins have been used for a wide range of industrial stoving finishes.

If melamine is converted to methylol melamine with formaldehyde and then acidified and etherified with a very large excess of methanol, polymerization can be largely avoided and the end product is essentially hexamethoxymethyl melamine (HMMM):

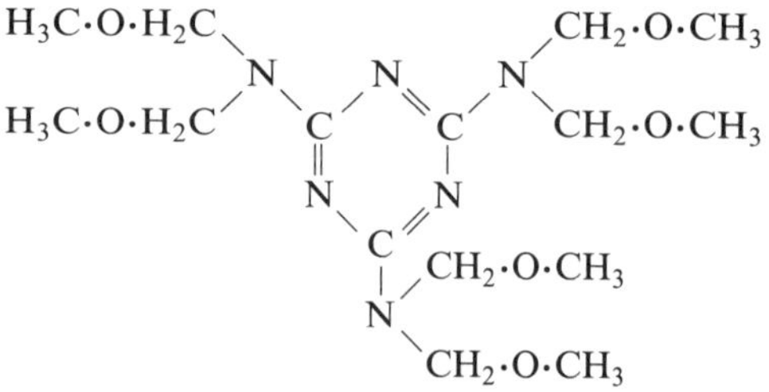

HMMM 'resins' generally contain an average of not less than 5.5 methylol groups per molecule of melamine and not less than 5.5 methoxy ether groups. They are soluble in all common organic solvents except aliphatic hydrocarbons and also dissolve in water alone, or mixtures of water and small amounts of an organic solvent (e.g. ethanol). They have very low molecular weights and so can be used in coatings of high solids and low viscosity. Greater reactivity can be obtained by going for less than full methylolation and/or less than full etherification by methanol. Some condensation inevitably occurs, with an increase in molecular weight and viscosity. Such resins are then methylated melamines rather than true HMMMs.

Without the addition of acid catalysts or presence of carboxyl groups, HMMM resins require high temperatures for reaction, but with acids present, reaction occurs at temperatures from room temperature to 150–175 °C. The methoxymethyl groups react with hydroxyl, carboxyl and amide groups. HMMM molecules react preferentially with the other resins rather than with each other and, although six groups are theoretically available for reaction, in practice only three normally react.

Thus HMMM resins are very suitable for a wide variety of finishes, but particularly for water-based or high solids content finishes.

Acrylamide-based acrylic nitrogen resins

The two long-established nitrogen resins described above have the following features in common:

- In the monomer, $-NH_2$ is attached to carbon, this carbon atom being adjacent to an atom or group with a greater attraction for electrons, e.g.

$$H_2N-\underset{\underset{O}{\|}}{C}-$$

in urea, where the oxygen is strongly electron-attracting and

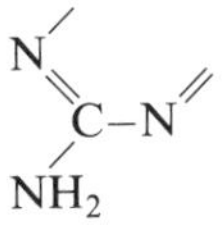

$$\underset{NH_2}{\overset{N}{\diagdown}}\underset{}{\overset{}{C}}-N$$

in melamine, where the nitrogen is electron-attracting. The $-NH_2$ group is capable of methylolation and the methylol groups are capable of etherification.

- The monomer must have a functionality of three or more if the resin produced is to be cross-linked ultimately.

Such features have been assembled in other monomers to give useful nitrogen resins. Cost considerations have prevented most of these from commanding wide usage. The same features and cross-linking mechanism can be built into acrylic polymers. The polymer may be a copolymer of any suitable acrylic or vinyl monomers, but must contain copolymerized amide (acrylamide or methacrylamide). The amide group is thus attached to the polymer chain

$$-\underset{\underset{NH_2}{|}}{\underset{\underset{C=O}{|}}{CH}}-CH_2- \qquad \text{acrylamide incorporated in polymer}$$

and bears a marked resemblance to urea.

The $-NH_2$ group can be methylolated with formaldehyde and the methylol group etherified (e.g. with butanol). Sufficient amide is copolymerized to give the average polymer molecule a functionality well in excess of three. Cross-linking occurs in the same way as the other nitrogen resins, copolymerized acid (e.g. methacrylic acid) acting as a catalyst for both etherification and cross-linking.

Typical resin recipes (before reaction with formaldehyde) might be:

1. Styrene 82.5%, acrylamide 15%, methacrylic acid 2.5% by weight.
2. Styrene 38.5%, ethyl acrylate 44%, acrylamide 15%, methacrylic acid 2.5%.
3. Methyl methacrylate 25%, ethyl acrylate 60%, acrylamide 15%.

Hardness decreases in the order (1) > (2) > (3). Flexibility increases with decreasing hardness.

There are other ways of producing cross-linked acrylic resins (see below and Chapters 14–16). This route is described here because it makes use of nitrogen resin chemistry.

Paints based on nitrogen resins

UF and MF finishes

While the normal cross-linking condensation reaction of UF and MF resins proceeds at temperatures of 90–180 °C, with higher levels of stronger acid it can also occur at temperatures above 15 °C, and hence two types of finish can exist:

- the one-pack stoving finish;
- the two-pack cold-curing finish.

Stoving finishes

These finishes have dominated the markets for metal coatings applied in the factory. With the resin properties already described – hardness and good colour – would come the disadvantages of brittleness and poor adhesion, if these resins were used on their own. It is therefore necessary to plasticize the resins and this is usually done by blending with compatible alkyds, saturated polyesters or acrylic resins. Between 5 and 40% nitrogen resin is used in the blend. Short or medium oil alkyds, based on drying or non-drying oils, are used in a wide range of general purpose finishes.

While polyesters were very briefly mentioned in the previous chapter, there we were concerned with the oil modification that changes a polyester into an alkyd. Polyesters are sometimes referred to as oil-free alkyds, but this is somewhat misleading. Polyesters may use, for example, any of the materials from the table given in the previous chapter (see p. 181). However, while alkyd formulation principally concerns the choice of oil/fatty acid used, and the amount, for polyesters we concern ourselves very much more with the choice of polyester components rather than relying principally on glycerol, pentaerythritol and phthalic anhydride. In fact, in formulating our polyester we are frequently looking for a superior performance and choose our ingredients accordingly.

One factor in the selection of ingredients is the susceptibility to hydrolysis of the polyester backbone. This is strongly influenced by the choice of polyol. Hydrolysis occurs more readily if the polyol contains a hydrogen atom or atoms in the carbon atom which is in a position β to the hydroxyl group, thus

$$HO-CH_2-\overset{*}{C}H_2-$$
$$\alpha\beta$$

Comparing the structures of the diols below,

$$\begin{array}{ccc}
\overset{H^* \; H^*}{\underset{H^* \; H^*}{HO-\overset{|}{\underset{|}{C}}-\overset{|}{\underset{|}{C}}-OH}} & \overset{CH_3 \quad\; CH_3}{\underset{H^* \; OH \; CH_3}{CH_3-\overset{|}{\underset{|}{C}}-CH-\overset{|}{\underset{|}{C}}-CH_2OH}} & \overset{CH_3}{\underset{CH_3}{HOCH_2-\overset{|}{\underset{|}{C}}-CH_2OH}} \\
\text{ethylene glycol} & \text{trimethylpentane diol} & \text{neopentyl glycol}
\end{array}$$

we can see that two of the structures have these atoms, while in the other they are absent.

Likewise, in considering susceptibility to heat and light breakdown, it is also this atom which is split away with the acid group which is created,

$$R_1-\overset{\overset{\displaystyle O}{\|}}{C}-O-CH_2-\overset{\overset{\displaystyle H^*}{|}}{\underset{\underset{\displaystyle H}{|}}{C}}-R_2 \xrightarrow[\text{or light}]{\text{heat}} R_1-\overset{\overset{\displaystyle O}{\|}}{C}-OH^* + CH_2=CH-R_2$$

and overall, of the three polyols above, neopentyl glycol is recognized as giving superior durability performance.

Specially formulated saturated polyesters are used where long-term outdoor durability is required (e.g. pre-painted strip steel for cladding industrial buildings) or good chemical and corrosion resistance are needed (e.g. on washing machines). Maximum durability is obtained with polyesters made from *iso*- rather than *ortho*-phthalic acid and polyols such as neo-pentyl glycol and trimethylol propane. Here they are used principally with MF resins, especially of the HMMM type, rather than UF resins, for the reasons given below. Polyesters will be encountered used with other curing reactions in the chapters that follow.

Even more durable polyesters are those reacted further to form **silicone-modified polyester resins**. Silicones are semi-organic compounds of the general type

$$-\overset{\overset{\displaystyle R}{|}}{\underset{\underset{\displaystyle R}{|}}{Si}}-O-\overset{\overset{\displaystyle R}{|}}{\underset{\underset{\displaystyle R}{|}}{Si}}-O-\overset{\overset{\displaystyle R}{|}}{\underset{\underset{\displaystyle R}{|}}{Si}}-$$

where R may be a hydrogen atom or some group containing carbon and hydrogen atoms. While the lower molecular weight silicones are oils used sparingly as additives (p. 145), the higher molecular weight silicones are resins. Since the Si–O and Si–C bonds are much more resistant to heat than C–C bonds, these resins may be used as film-formers in heat-resistant coatings. In the context of this chapter, if –R in the general formula is a methoxy ($-O \cdot CH_3$) or hydroxyl group, reaction can occur with hydroxyl or carboxyl groups on a polyester. These modified polyesters, again cross-linked with MF resin, are used in coil coatings and have high resistance to degradation by UV in sunlight.

Acrylic resins used with MF resin are flexible copolymers containing at least one 'hard' and one 'soft' monomer, a hydroxyl-containing monomer to provide reactive sites and possibly an acid monomer (see p. 151 for examples). They may be used as solutions in solvent or as non-aqueous dispersions. They produce extremely durable motor car finishes which can be polished to some extent.

In all cases acidity, even in the resin, catalyses the cross-linking process and combination occurs by a condensation reaction between methylol groups in the nitrogen resins and hydroxyl groups in the plasticizing resin:

$$\text{\small\textasciitilde\hspace{-2pt}CH}_2\text{OH} + \text{HO\textasciitilde} \longrightarrow \text{\textasciitilde\hspace{-2pt}CH}_2\text{O\textasciitilde} + \text{H}_2\text{O}$$

The higher cost of MF resins is justified where lack of discoloration is required at baking temperatures above 150 °C (e.g. in white finishes), where water and chemical resistance must be excellent (e.g. in washing machine finishes) and where the best outdoor durability is required (e.g. in car enamels). These improvements, particularly in resistance and durability, are thought to be due to the absence of the water-sensitive carbonyl groups found in UF resins and the increased cross-linking potential. The latter factor accounts for the shorter hardening times of MF resins. A typical stoving schedule for a nitrogen resin finish might be 30 min at 127 °C.

The relatively low molecular weights of the nitrogen and plasticizing resins allow medium solids at spraying viscosities. The solvents used in these finishes consist usually of some alcohol (probably butanol) and aromatic hydrocarbons, such as xylene. Aliphatic mineral spirits may be present in some primers. Storage stability is generally good, since little or no reaction occurs in the can at room temperature unless the acid content of the paint is high, when a slow viscosity rise on storage can occur.

As mentioned in Chapter 12, with legislation requiring the reduction or elimination of solvent in coatings, higher solids and fully water-borne stoving finishes are steadily evolving. Higher solids require resins of lower molecular weight in order that application viscosity can be maintained. Both acrylic and polyester resins are available and these are used with HMMM resin in stoving systems. Where molecular weight is significantly lower than in normal polymers, the resins are termed **oligomers**. A necessity with oligomers is that of retaining adequate functionality (see discussion for alkyds on p. 182); this is not a problem with polyesters, but is more difficult with acrylic resins, where special formulating techniques may be required which are outside the scope of this book.

Because of the water-solubility of HMMM resins, they can be used with water-reducible alkyds, polyesters or acrylics and with acrylic copolymer latices to produce water-based counterparts of the above stoving finishes. While the reduction in solvent is substantial, such coatings are not usually free of small proportions of water-miscible cosolvents, e.g. 2-butoxy ethanol, and ammonia or amines (pp. 119, 120). An example of a water-reducible polyester–MF stoving finish is given in Table 13.5.

Cold-curing finishes

Resin blends with UF and MF resins etherified with butyl or propyl alcohols and catalysed with stronger acids such as *p*-toluene sulphonic acid ($\text{CH}_3 \cdot \text{C}_6\text{H}_4 \cdot \text{SO}_3\text{H}$), will give good wood finishes with hardening times of a few hours at room temperature. Hardening can be accelerated by force-drying at temperatures which do not distort the wood, e.g. 60 °C. On more stable wood-based substrates, such as chipboard or hardboard, drying of clear or pigmented acid-catalysed finishes can be accomplished in a few minutes at over 100 °C or with infrared heating. MF resins give better resistance to water, solvents and chemicals than UF resins do, but cure more slowly at room temperature. The nitrogen resin is plasticized with non-drying or semi-drying alkyds and a solution of acid catalyst is supplied in a separate container. The finish and catalyst are mixed in a convenient ratio, e.g. 10:1 by volume, at the beginning of the day and the catalysed material remains usable at least for the remainder of the working day and usually for much longer.

These finishes compete chiefly with nitrocellulose (NC) lacquers in wood finishes and, although they are harder and more resistant, suffer by comparison for drying performance during the early stages of drying, when dust pick-up can be heavy. NC lacquers can become dust free in 1–2 min, while unadulterated acid-catalysed (AC) wood finishes take 5–10 min. Since any compatible resin may be used in the AC finish, it has been possible to produce 'combination finishes' containing amino resins, alkyds and nitrocellulose, with compromise drying times and film properties. Use of the higher molecular weight polymer necessitates a reduction from the medium solids (35–45%) of full AC finishes, though the user is still better off than with NC lacquers (15–30% solids).

Finishes based on acrylamide-based acrylic nitrogen resins

Unlike UF and MF resins, acrylamide-based acrylic nitrogen resins need not be blended with another component to obtain flexibility; this can be built into the acrylic copolymer by the appropriate choice of monomers (see examples on p. 193). Thus resins with ester linkages in their main polymer chains (e.g. alkyds, polyesters) can be avoided. A cross-linked polymer network can be built up which is resistant to hydrolysis by alkalis in detergents. Coatings can therefore be formulated which are suitable for domestic appliances, being additionally hard but flexible.

This family of resins is also widely used for the formulation of paints for pre-painted coiled metal strip, which may subsequently be fabricated into caravan exterior skins, etc. The maximum amount of methyl methacrylate is included to give good durability, but flexible monomer is needed to allow the painted sheet to be bent and formed.

Best results are obtained if the only acid present is part of the copolymer. Domestic appliance finishes cure in 20 min at 177 °C and coil coatings in 1 min, with the metal reaching a peak temperature of 230–240 °C. Temperatures can be reduced, with some loss in properties, by the addition of a little acid catalyst, which is usually required when the film is hardened with small amounts of MF. Other resins used to modify the film are epoxies, polyesters and alkyds.

Like other acrylic copolymers, these self-cross-linking acrylic nitrogen resins may be produced as latices or in water-soluble form and formulated into water-based paints.

Toxicity

Concentrations of formaldehyde in air are intensely irritating to humans and provide an unacceptable working environment. Additionally, as a result of tests on rats at much higher concentrations, some strongly disputed conclusions have been drawn that formaldehyde might be a human carcinogen. The ACGIH in the USA have therefore set a ceiling short-term exposure limit (STEL) of 0.3 p.p.m. which should not be exceeded at any time during the working day. Steps are being taken by all suppliers to minimize the free formaldehyde contents of all resins and coatings. Good extraction and working practices maintain workplace concentrations below the STEL. The nose and eyes are extremely good detectors of potentially hazardous concentrations of formaldehyde, as mentioned above.

Typical formulations based on nitrogen resins are given in Tables 13.1–13.5.

Table 13.1 Formulations for a clear acid-catalysed wood-finish

	Clear coating	wt%
Resin solutions	Beetle alkyd resin BA509	48.0
	Beetle urea resin BE678	24.5
Solvent	Butanol	27.5
		100.0
	Activator	
	p-Toluene sulphonic acid	20.0
	Distilled water	2.0
	Isopropanol	78.0
		100.0

Activate 100 parts coating:2 parts activator.
Spraying solids: 40%. Air-dry, or force-dry 10 min at 80 °C

Beetle BA509 is a short–medium semi-drying oil alkyd at 50% in xylol–butanol.
Beetle BE678 is an isobutylated UF resin at 66% in isobutanol.
Both resins are made by B.I.P. Ltd.

Table 13.2 Formulations for a green metallic basecoat for motor cars (solvent-borne)

	wt%	
Aluminium paste Sparkle Silver 3334 AR (62% Al)	1.7	} Pigments
Irgalite Green GLN	1.0	
Cellulose acetate butyrate CAB-531 20% in butyl acetate	23.7	} Resin solutions
Uracron CR226 XB	11.4	
Uramex MF822B	1.8	
Xylene	33.5	} Solvents
Butyl acetate	16.4	
Solvesso 100	7.5	
Polyethylene wax Cerafak 106 (10% in xylene)	3.0	
	100.0	

Spray 12–15 μm onto filler, at viscosity
16 sec DIN 4 cup. Flash off 5–7 min,
and spray on clear coat 35–45 μm.
Stove 25 min at 140 °C

Uracron CR 226 XB is an acrylic copolymer (containing hydroxyl groups) at 50% in xylene and *n*-butanol. Uramex MF822B is a butylated MF resin at 72% solids in *n*-butanol. Both resins are made by DSM Resins BV.
Solvesso 100 is an aromatic hydrocarbon solvent blend, BR 160–178 °C supplied by Exxon Chemical.
Sparkle Silver 3334 AR is supplied by Silberline Ltd.

Table 13.3 Formulation of flexible coil coating enamel for appliances (solvent-borne)

	wt %	
Rutile titanium dioxide Ti-pure R-960	25.0	
Eastman PA-1-2CNCp polyester	56.3	} Pigment resins
Cymel 301 melamine resin	3.6	
Eastman C-11 ketone	8.07	
Ethyl 3-ethoxy propionate	1.51	} Solvent
Butoxy ethanol	3.02	
Nacure 1419	1.9	Catalyst
Acrylic flow agent	0.6	Flow control

Application by roller coater to pretreated aluminium at 20–25 μm. Cure for 30 s, 216 °C PMT (peak metal temperature). Application solids: 50.0%, adjusted with solvent blend in formulation.

Eastman PA-1-2CNCp is a cyclohexane dimethanol, neopentyl glycol, 1,6 hexane diol, isophthalic acid, terephthalic acid, 1,4-cyclohexane dicarboxylic acid polyester of acid value 4 and 60% solids in Solvesso 150 solvent. The formulation is available from Eastman Chemical (publication N-33OB).

Eastman C-11 ketone is a ketone solvent blend with BR 200–240 °C supplied by Eastman Chemical.

Cymel 301 is a hexamethoxy methyl melamine resin at 100% solids supplied by Cytec.

Nacure 1419 is an acid cure catalyst made by King Industries.

Solvesso 150 is an aromatic hydrocarbon solvent blend, BR 182–203 °C supplied by Exxon Chemical.

Table 13.4 Formulation of high solids stoving alkyd–MF enamel

The following are ground together to form a millbase:

	wt%	
Rutile titanium dioxide Ti-pure R-900	29.97	Pigment
Amoco resin HS-7 (82.8% in methyl isobutyl ketone)	45.23	Resin
FC-430 (33% in propylene glycol methyl ether acetate)	0.27	Flow control additive
Ethyl 3-ethoxypropionate	4.17	Solvent
to which is then added		
Cymel 303 melamine resin	9.35	Resin
Surfonyl 465	0.08	Surfactant
Exkin 2	1.52	Anti-skin additive
Nacure 5225	0.27	Cure catalyst
Manosec Co 18	0.11	Cobalt drier
Methyl ethyl ketone	9.03	Solvent
	100.00	

Application solids: 76.4%. Application is to phosphated cold-rolled steel to give a metal film thickness of 30 μm. Stove for 30 min at 120 °C.

Amoco HS-7 is a soya fatty acid, trimethylol propane, neopentyl glycol, isophthalic acid alkyd of acid value 6. The formulation is available from Amoco Chemicals (publication GTSR-75B).

Cymel 303 is a hexamethoxy methyl melamine resin at 100% solids supplied by Cytec.

FC-430 is a flow additive supplied by 3M Company.

Nacure 5225 is a blocked dodecylbenzene sulphonic acid cure catalyst made by King Industries.

Surfonyl 465 is a surfactant produced by Air Products.

Exkin 2 is an anti-skin agent produced by Huls America Inc.

Manosec Co 18 is a drier solution containing 18% cobalt and supplied by Rhone Poulenc.

Table 13.5 Formulation of water-borne stoving polyester–MF enamel

Polymer dispersion preparation[a]	wt%	
Amoco resin WB-138 (90% in diethylene		
glycol monobutyl ether)	55.6	Resin
Dimethylethanolamine	1.3	Solubilizing additive
Deionized water	43.1	Diluent
	100.0	
Enamel preparation	wt%	
Rutile titanium dioxide Ti-pure R-900	21.48	Pigment
Polyester resin dispersion (49%)	43.95 ⎫	Resins
Cymel 303 melamine resin	5.37 ⎭	
Dimethylethanolamine	0.57	Solubilizing additive
FC-430 (33% in propylene glycol		
methyl ether acetate)	0.15	Flow control additive
Deionized water	28.48	Diluent
	100.00	

[a] The dispersion is prepared by slowly adding the hot resin solution to the agitated solution of dimethylethanolamine in deionized water, which is then adjusted to pH 7.5–8.0 with further dimethylethanolamine. The product is a dispersion of resin in water, and of milky white appearance.

pH is adjusted to 8.0. Application solids: 49%. Application is to cold-rolled steel to give a final film thickness of 30 μm. Following 5 min flash-off, stove for 20 min at 177 °C.

Amoco WB-138 is a neopentyl glycol, trimethylol propane, isophthalic acid, adipic acid, trimellitic anhydride polyester of acid value 35. The formulation is available from Amoco Chemicals (publication TM-141).

Cymel 303 is a hexamethoxy methyl melamine resin at 100% solids supplied by Cytec.

FC-430 is a flow additive supplied by 3M Company.

Fourteen

Epoxy coatings
paints based on epoxy resins

Under this heading come a wide variety of finishes based on the epoxy resins. These resins involve, either in their preparation or in their cross-linking, the reactions of the epoxide ring, which are described in Chapter 4. A brief revision of these reactions would be worthwhile before proceeding further.

The variety of finishes is so wide that their uses will be described later in detail, while the characteristic 'epoxy' properties will be more easily understood when the resin structures are seen. Let us proceed to the resins.

Epoxy resins

These are usually prepared from epichlorhydrin and a dihydroxy compound (Fig. 14.1). The latter is usually a diphenol – in particular the compound called bisphenol A – though it can be a dihydric alcohol. Two reactions of the phenolic hydroxyl bring about the polymerization:

- condensation with chlorine to eliminate HCl;
- addition to epoxide, opening the ring. It should be noted that this produces one hydroxyl group.

Thus this type of epoxy resin contains a maximum of two epoxide rings *at the ends* of the molecule and a number (which may be zero) of hydroxyl groups along the chain. These groups make the resin a polar one and ensure good adhesion to polar or metallic surfaces. The next thing to note is that the polymer chain contains only carbon–carbon and ether linkages. Both are very stable. We have seen the C–C stability in acrylic polymers, and in the preparation of nitrocellulose we noted that very severe conditions were required to break the ether linkage. Both linkages are much stronger than ester linkages, which are saponified by alkali. Epoxy resins, therefore, have good chemical resistance. Again, the phenolic hydroxyls, which lead to poor colour in phenolic resins, are entirely converted to ether links in epoxy resins, which are therefore of much better colour. Epoxy resins on their own are low to medium molecular weight, relatively brittle materials and are useful film-formers only

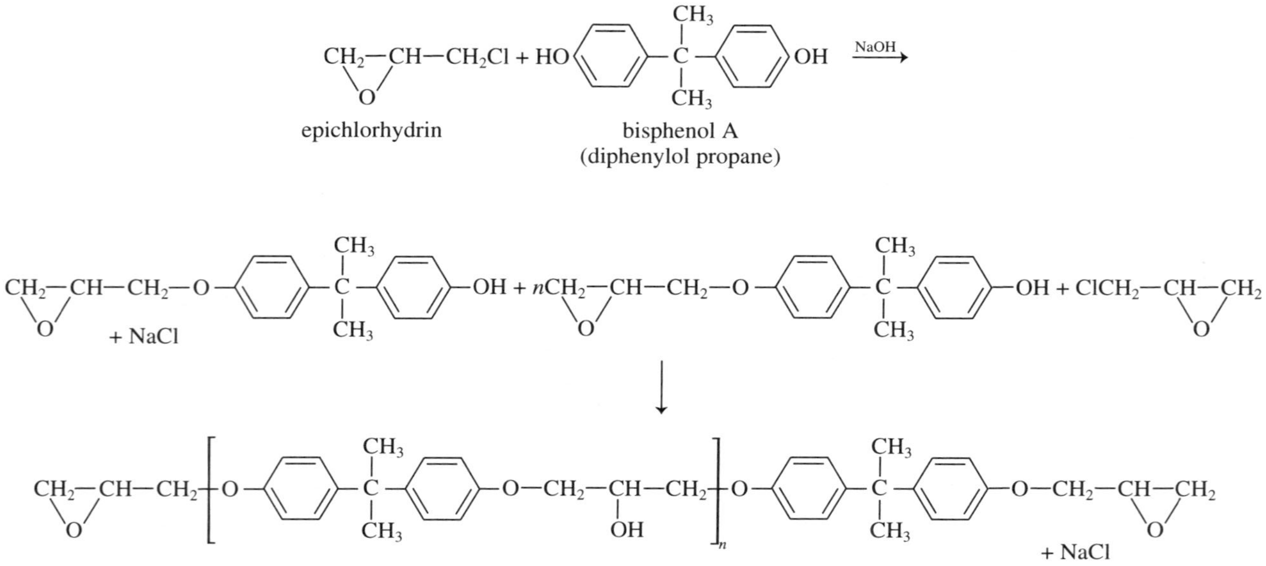

Fig. 14.1 Preparation of epoxy resin from epichlorhydrin and bisphenol A.

when cross-linked with other molecules. Cross-linking takes place through the reactive epoxide rings and hydroxyl groups. These are well separated by three carbon atoms, two oxygen atoms and two benzene rings. Because the cross-links cannot be closely spaced the resins give flexible cross-linked films. However, the aromatic epoxy resins strongly absorb UV light and are degraded by it, leading to chalking if used in topcoats outdoors.

Unmodified epoxy resins vary from the viscous liquid 'diepoxide 0' (molecular weight 340 and with $n = 0$ in the general formula above) to solid polymers of molecular weights up to 8000 ($n = 26$). The resins become solid at molecular weights above 700. While a number of standard grades can be purchased, it is more and more usual for resin producers to take the 'diepoxide 0' resin, and to react this with bisphenol A to form **chain extended** epoxy resins in their own resin plant. This process is also termed **epoxy advancement**. An advantage of this process is that the producer may then introduce other diphenols or hydroxy functional materials to modify the properties of the extended epoxy resin.

Like all polymers, epoxy resins contain a mixture of molecular species. Not all molecules are terminated at both ends by epoxide groups. Low molecular weight diluents, containing perhaps one epoxy ring, are often included to make the 'diepoxide 0' types less viscous. Resins containing these diluents are described as 'modified' epoxies. The solvents for epoxy resins are given in Tables 9.1 and 9.3. It should be noted that the petroleum hydrocarbons are non-solvents.

Three types of alternative epoxy resin are worth a special mention. First is the **epoxy novolac resin**, which contains more than two epoxide rings. Such a resin is made from a phenolic novolac (p. 177) and epichlorhydrin and usually contains three to four epoxy rings per molecule:

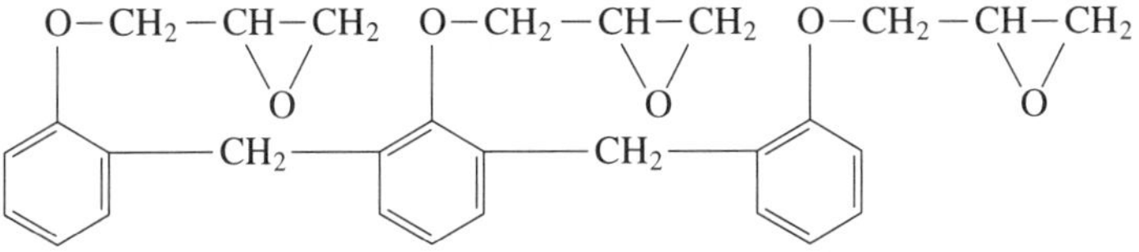

Second is the **cyclo-aliphatic epoxy resin**, in which the epoxide ring is incorporated onto one of the sides of a cyclohexane ring, e.g.

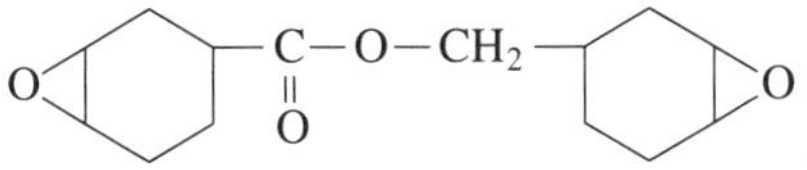

Finally, it is possible to produce an **epoxide functional acrylic resin** by copolymerizing glycidyl methacrylate

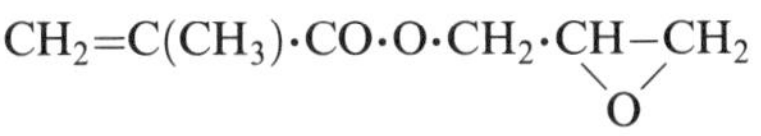

$$CH_2{=}C(CH_3){\cdot}CO{\cdot}O{\cdot}CH_2{\cdot}CH{-}CH_2$$

with other acrylic monomers.

We can regard epoxy resins in two ways: as cross-linking resins and as polyhydric alcohols. Let us consider the second aspect first, since it leads to the possibility of further resins.

Epoxy resins as polyols

Epoxy-ester resins

These can be considered as alkyd equivalents. They are usually made by esterifying the epoxy and hydroxyl groups with oil fatty acids. The epoxy groups are more reactive to carboxyl than are hydroxyl groups and, in the process of reaction, create a hydroxyl group on the carbon atom adjacent to the ester linkage. Thus each ideal bisphenol A epoxy polymer molecule has two sites of epoxide reactivity and $n + 2$ sites of hydroxyl reactivity.

If n and the amount of fatty acid used are both high, the esterified epoxy molecule will have a high functionality of oil side chains. Should the oil be prone to heat bodying (e.g. it is a drying oil), gelation could easily occur as the temperature is raised to complete the esterification. The oil side chains would cross-link the epoxy backbones. For this reason, n is usually kept below 4 (molecular weight 1400), though non-drying or semi-drying oils can be used with resins up to $n = 9$ (molecular weight 2900).

Acid values of 0–10 are usual. Oil length is taken from the fraction of potential ester linkages formed. For example, a resin with average value of $n = 3.7$ can form an average of 7.7 linkages (four from the epoxy groups) per molecule. If only three linkages are made with fatty acids, this is a fraction of 0.4 of the possible total, or 40%. Between 30 and 50% esterification is described as short oil, 50–70% as medium oil and 70–90% as long oil length. Properties vary with oil length very much as they do in alkyds. Resins of a given oil length dissolve in solvents suitable for alkyds of similar oil length. Uses are also similar, but epoxy-ester resins have better adhesion and chemical resistance than alkyds and the stoving versions are harder for the same flexibility. They are more used for primers, however, on account of colour and poorer weathering properties.

Some of the drying fatty acid chains can be partially maleinized to permit emulsification when the carboxyls have been partially neutralized and the resin dispersed in water–solvent mixtures. These water-borne epoxy-esters can be used as binders for **anodic electrodeposition primers** (the electro-deposition process is described on p. 119).

Epoxy-alkyds

Epoxy resins may be used as polyols in alkyd preparation. Since the lowest functionality possible – even with liquid epoxies – is four (two epoxide groups), they are usually used with other polyols such as glycerol and the functionality is previously reduced to three by reaction with fatty acids. Epoxy resins of molecular weight up to 1400 have been used in alkyds. The usual epoxy properties are conferred on the resin, the extent of change being dependent on the amount of epoxy resin incorporated.

Epoxy coatings

These may contain epoxy resins, epoxy-esters, epoxy-alkyds or other possible epoxidized resins. Both stoving and air-drying variants are possible. UV curing finishes are dealt with later in Chapter 16.

Stoving finishes

Short–medium oil epoxy-esters can be used in stoving finishes, with or without nitrogen resins, in place of the corresponding alkyds. They make excellent primers and surfacers, for use on metal goods.

The epoxy resin molecules cross-link with each other and with a variety of other resins.

Cross-linking with resins containing hydroxyl groups

In this category are the phenolic and amino resins. The principal reaction is addition of hydroxyl to the epoxide ring (as in the preparation of epoxy resins), but condensation reactions between the hydroxyl groups occur to a lesser extent.

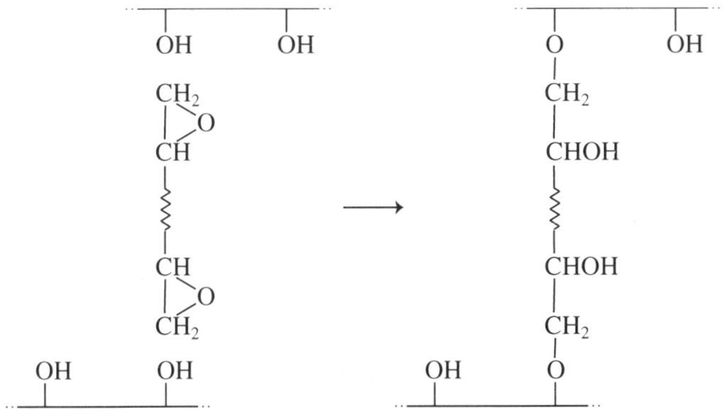

(only reacting groups shown)

Epoxy-phenolic finishes are among the most chemically resistant known. Between 20 and 30% phenolic resin is used with epoxy resin of molecular weight 1400 and the stoving schedule is 20 min at 180–205 °C. Phenolic resins alone give excellent chemical resistance, but cross-linking with epoxy resin improves adhesion and impact resistance. The poor colour of the coatings is due to the phenolic resin. The two resins may be partially reacted together before being used in a paint. Since part of the hardening reaction will then have been carried out before painting, shorter stoving times are possible for the paint user. Also, resins that are incompatible on cold-blending can be made compatible by partial combination in the resin kettle. This broadens and cheapens the range of phenolic resins that can be used. The main uses for these finishes are as can and drum linings.

Epoxy–UF resin combinations are almost as chemically resistant as epoxy-phenolics and have better colour. They make excellent corrosion-resistant primers for coating of coiled metal strip. Films 5 μm thick are stoved for 30–60 s to a peak metal temperature of 240–250 °C.

Cross-linking with resins containing carboxyl groups

Here esterification of the epoxy groups is the means of cross-linking. This method provides another means of obtaining a **thermosetting acrylic finish**.

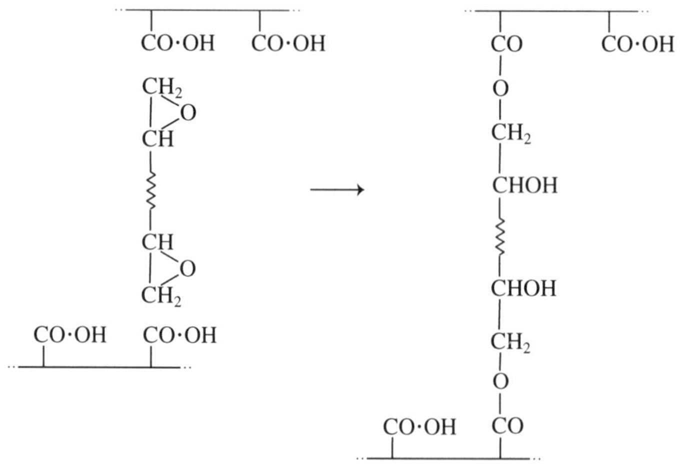

(only reacting groups shown)

The acrylic copolymer is made to contain about 4–12% of acrylic or methacrylic acid and the minimum number of carboxyl groups per molecule for cross-linking (three) is usually considerably exceeded. Stoving temperatures are 170–180 °C, or 150 °C if basic catalyst is added. Such combinations have limited shelf life due to slow reaction in the can. Domestic appliance acrylics are usually as described in Chapter 13, often with minor inclusions of epoxy resin to upgrade resistance properties.

Epoxy resin–MF resin–alkyd finishes

These can be considered as reacting by both routes, since the unreacted acid in the alkyd may take part. Epoxy resin of molecular weight 900 ($n = 2$) is used, since alkyd–epoxy resin compatibility is poor at higher molecular weights. Stoving is at temperatures of 150 °C and above. Improvements over alkyd–melamine finishes occur in adhesion, flexibility, chemical and water resistance and mar and abrasion resistance. Uses are similar, particularly where finishing industrial and domestic equipment is concerned, since the excellent adhesion of the epoxy resin can allow the user to omit a primer.

Powder coatings

Powders based on epoxy resins, or on an epoxy curing reaction, have held dominant positions in the process of powder coating since serious growth began in 1964 and still have such a large share of sales that it is appropriate to introduce the topic in this chapter.

A powder coating is made by mixing molten resin, pigments, cross-linking agent and additives uniformly, extruding the melt, breaking up the solidified extrusion and finally milling to produce a powder with a particle size distribution of approximately 10–100 μm. The pigmented powder is then applied, either by immersing the article in a fluidized bed of powder (powder kept mobile by the passage of air through it) or, more usually, by special electrostatic spray-guns. The film is formed when the particles are sintered in an oven, and cross-linking also occurs.

Powder coatings do not readily give satisfactory films at thicknesses below 35–45 μm, and require high temperatures for stoving and special equipment for handling and application. However, they form good-looking, tough thick coatings and contain no solvent. Wastage is very low, but colour changes in one application unit can lead to speckled finishes, unless scrupulous change-over procedures are followed.

Powder coatings are not necessarily based on epoxy resins; the first powder coatings were based on thermoplastic polymers, and polyolefin, vinyl polymer and polyamide coatings are still used. While those based on epoxy resins are very prominent, there are also polyester, polyurethane and acrylic thermosetting types which we will describe here.

To be suitable for powder coatings, resins should melt for pigmentation at temperatures well below that required for cross-linking, should be brittle and amenable to dry milling when cold, should be stable in the presence of cross-linker when stored at room temperature and finally, at stoving temperatures, should melt and fall rapidly in viscosity before cross-linking. Epoxy resins admirably meet these requirements.

Cross-linkers for epoxy powders can be nitrogen resins or polyamides (p. 212), but two more popular types require special mention.

Dicyandiamide (p. 191) and its derivatives cross-link at 180 °C by reaction of the amino and imino (=NH) groups with the epoxy ring:

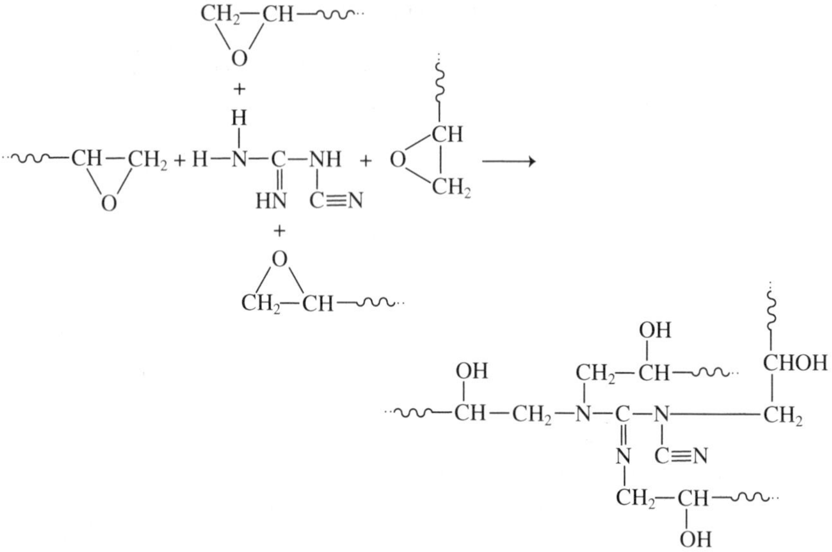

$$CH_2-CH-\wedge\wedge$$
$$\diagdown O \diagup$$

$$+$$

$$H$$

$$\wedge\wedge-CH-CH_2 + H-N-C-NH + O \overset{CH}{\underset{CH_2}{\diagup}}\longrightarrow$$
$$\diagdown O \diagup \qquad HN \quad C\equiv N$$

$$+$$

$$O$$
$$CH_2-CH-\wedge\wedge$$

$$\overset{OH}{\underset{CH_2-CH-\wedge\wedge}{|}} \quad CHOH$$
$$\overset{OH}{\underset{\wedge\wedge-CH-CH_2-N-C-N}{|}} \qquad CH_2$$
$$\underset{N \quad C\equiv N}{\overset{\|}{}}$$
$$CH_2-CH-\wedge\wedge$$
$$\underset{OH}{|}$$

Anhydrides cross-link epoxy resins by reacting first with the hydroxyl groups. Carboxyl groups are produced in the process and these then react with the epoxy ring. Stoving is at 180–200 °C.

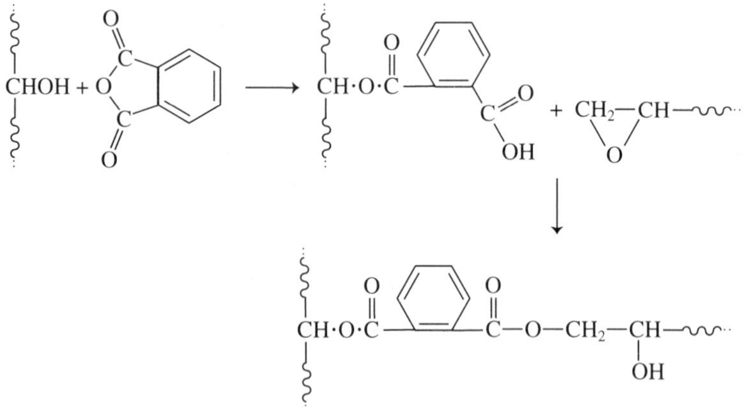

$$CHOH + O \qquad \longrightarrow \qquad CH \cdot O \cdot C \qquad + CH_2-CH-\wedge\wedge$$

$$\downarrow$$

$$CH \cdot O \cdot C \qquad C-O-CH_2-CH-\wedge\wedge$$
$$\underset{OH}{|}$$

Both types of cross-linking reaction can be promoted by the addition of catalysts. Polyesters may first be encountered in powder coatings in the so-called hybrid type where carboxyl-terminated polyester is mixed with epoxy resin; they are hence equally epoxy and polyester powders. These show some improvements on the yellowing and weathering properties of the pure epoxy powders.

A type of polyester powder, not including epoxy resin but curing by an epoxy mechanism, is the type cured with triglycidyl isocyanurate (TGIC).

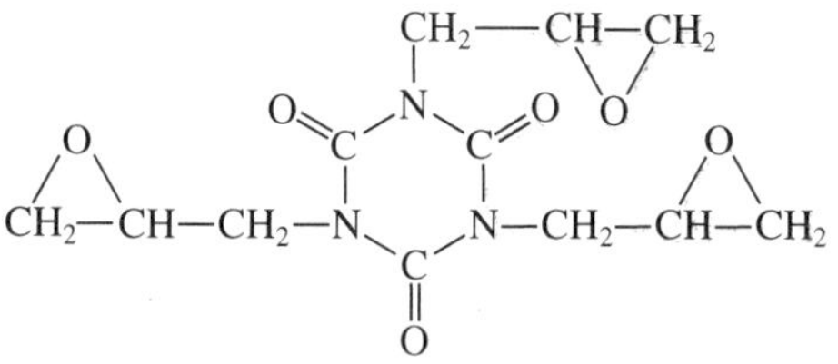

This molecule has three epoxy or glycidyl groups which react on stoving with acid groups on the polyester. These powders have improved exterior durability over those containing epoxy resin. Use of this curing agent is however threatened due to its toxicity.

In order to complete coverage of powder coatings, we must mention here the urethane cross-linked powders. In these cases we have hydroxy functional resins, which may be either polyester or acrylic. These are cross-linked by blocked isocyanate cross-linkers, principally those blocked with ε-caprolactam as will be described in the next chapter.

Finishes curing at normal temperatures

Long oil epoxy-esters, based on drying oil fatty acids, can be used in varnishes, but with poor resistance to chalking out-of-doors, they do not compete with long oil alkyds in decorative paints.

Epoxy resins proper may be cross-linked by reaction with chemicals containing amino groups. The finishes produced may be classified as follows.

Finishes cured by polyamines

The addition reaction between epoxide ring and primary and secondary amino groups that are not directly attached to a benzene ring occurs at temperatures above 15 °C and provides a means of curing the resins at room temperature. To cross-link diepoxide resins, the amine must have a functionality of at least three, i.e. at least three amino hydrogen atoms.

Simple polyamines such as ethylene diamine, $H_2N \cdot CH_2 \cdot CH_2 \cdot NH_2$ (functionality four) and diethylene triamine, $H_2N \cdot CH_2 \cdot CH_2 \cdot NH \cdot CH_2 \cdot CH_2 \cdot NH_2$ (functionality five) can be used, being supplied as an 'activator' solution. This is added to the paint (which contains the epoxy resin) just before it is used. The amines are not catalysts, since they are consumed in the chemical reaction. Epoxy resins of low molecular weight ($c.900$) are preferred. Cross-linking might occur thus:

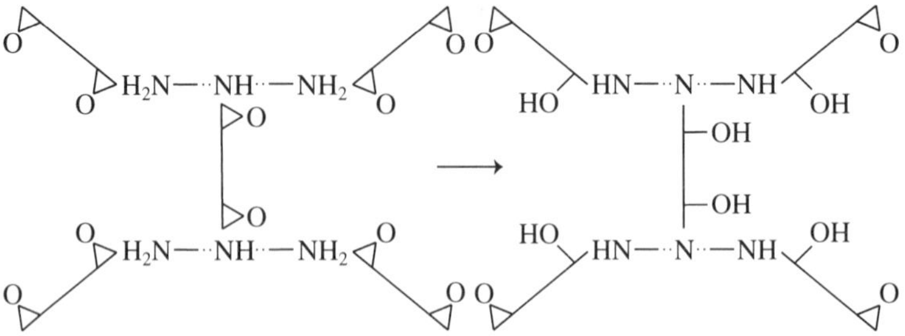

$$NH-CH_2-CH\text{-}\!\!\text{\tiny www}\!\!\text{-}CH-CH_2-NH$$

[chemical structures diagram]

Such paints are prone to surface exudation. The risk of suffering from this fault can be minimized by allowing the paint to stand 12–16 h before it is used. This is quite safe, since pot lives are arranged to be about 48 h at normal temperatures.

In view of this, and the handling hazards associated with simple amines, an alternative, more reliable and safer method is to pre-react some epoxy resin with an excess of amine, so that an adduct containing unreacted amino hydrogen atoms is produced.

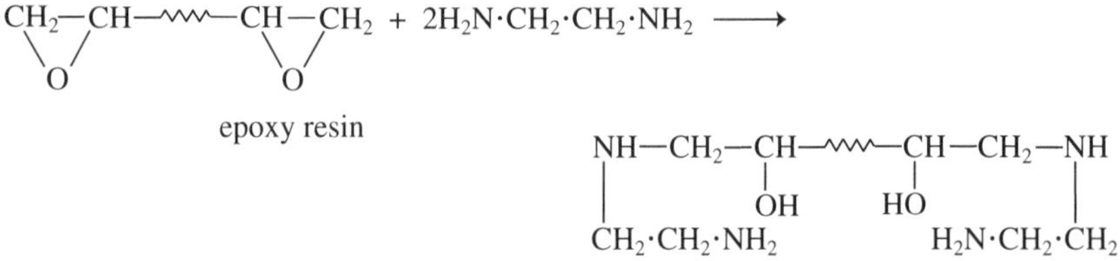

$$CH_2\text{--}CH\text{--}\!\!\text{\tiny www}\!\!\text{--}CH\text{--}CH_2 + 2H_2N\cdot CH_2\cdot CH_2\cdot NH_2 \longrightarrow$$

epoxy resin

$$
\begin{array}{ccc}
NH-CH_2-CH\text{-}\!\!\text{\tiny www}\!\!\text{-}CH-CH_2-NH \\
\quad\quad\quad\quad | \quad\quad\quad\quad | \\
\quad\quad\quad\quad OH \quad\quad HO \\
CH_2\cdot CH_2\cdot NH_2 \quad\quad\quad\quad H_2N\cdot CH_2\cdot CH_2
\end{array}
$$

adduct (functionality 6)

Ester and ketone solvents, which react with amines, must not be present. The adduct is a polyamine and can therefore be used to cross-link epoxy resins. It has advantages in that it is odourless (simple amines are unpleasant to handle), part-reacted (which speeds up drying) and free from simple amines (which cause exudation and are toxic). It can be seen now why it is an advantage to age paints cured by simple amines: amine adducts are formed during the aging.

Finishes cured by polyamide resins

Polyamide resins are prepared from polyamines and dimer fatty acids. Dimer fatty acids are produced from unsaturated drying (or semi-drying) oil fatty acids by Diels–Alder reaction. A possible dimer acid might be

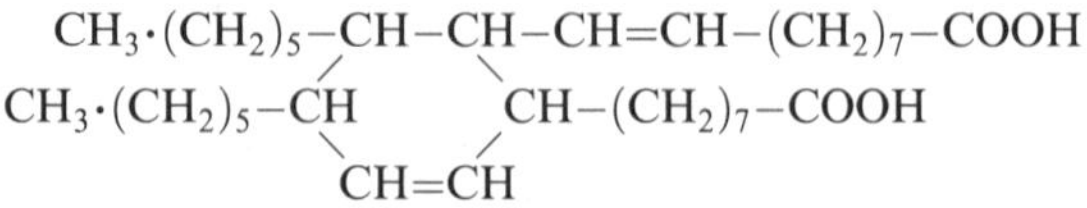

$$CH_3\cdot(CH_2)_5\text{--}CH\text{--}CH\text{--}CH\text{=}CH\text{--}(CH_2)_7\text{--}COOH$$
$$CH_3\cdot(CH_2)_5\text{--}CH \quad\quad CH\text{--}(CH_2)_7\text{--}COOH$$
$$CH\text{=}CH$$

This can be thought of as $HOOC-\!\!\wedge\!\!\wedge\!\!-COOH$ and will react with polyamine to form polyamide resin:

$$n HOOC-\!\!\wedge\!\!\wedge\!\!-COOH + n H_2N- \cdots -NH_2 \longrightarrow$$

$$HOOC-\!\!\wedge\!\!\wedge\!\!-CO \cdot OH- \cdots - \underbrace{NH \cdot CO}_{\substack{\text{amide} \\ \text{linkage}}} -\!\!\wedge\!\!\wedge\!\!-CO \cdot NH- \cdots \text{etc.}$$

$$+ 2n - 1\, H_2O$$

If excess amine is used and some triamine is included, branched polyamides can be produced, which contain unreacted amino groups at the ends of chains. They can be thought of as high molecular weight *polyamines* and as such are clearly suitable for curing epoxy resins.

The polyamide resins increase the flexibility of the film, the amine groups being well spaced in the large molecules, but, because amide groups have now become linkages in the polymer structure, alkali resistance is slightly reduced. Amides are hydrolysed by alkali, though not readily. Cure is slower and pot lives are longer. At air temperatures above 15 °C, the films are surface-dry in a few hours, but full hardness and resistance is only obtained in 7 days. Initial force-drying (at 40–60 °C) will reduce this period. Pigment dispersion can be carried out in the polyamide resin solution; this may give easier dispersion in some cases.

Epoxy–polyamine or polyamide finishes are suitable wherever chemical resistance is required, chalking cannot occur (indoors) or can be accepted and stoving is not possible. The chemical resistance obtained does not equal that of epoxy–phenolic and epoxy–urea finishes, but it is good for a paint drying at atmospheric temperatures. Typical outlets are the maintenance painting of large industrial installations and equipment and the priming of aircraft. With a combination of hardness, flexibility and adhesion, the coatings are also suitable for wood and concrete (e.g. floor coatings). The inclusion of coal tar in the amine pack (in an amount approximately equal to the weight of epoxy resin) upgrades water and corrosion resistance, e.g. for undersea protection of metal (though the colour ranges are limited).

Solventless finishes

It has been pointed out that the simplest epoxy 'resins' are liquids. Undiluted 'diepoxide 0' has a viscosity of 36–64 poises (3.6–6.4 Pa s) at 23 °C, but this viscosity can be lowered further by reactive diluents. The 'modified' resin can have a viscosity as low as 5 poises (0.5 Pa s). The diluents have an epoxide functionality of only one, so they do not contribute to cross-linking, but they are chemically bound into the final cross-linked film. Since these liquid epoxies may be cured by liquid polyamines or polyamides, no further solvent is required for application. Thus the entire liquid paint can be converted to cross-linked solid without heating, so that it may be called 'solventless', or a '100% solids' finish.

The main outlet for such a finish is as an anti-corrosive lining for storage tanks, both on land and in oil tankers. Maximum corrosion resistance is obtained only (1) if the steel is cleaned by grit blasting and (2) if it is covered by about 125 μm of coating. The grit-blasting process leaves a pitted 'profile' on the steel. Ideally, the maximum amplitude of this profile should be determined and a measured thickness of coating equal to three times that amplitude should be applied. In practice it is usually necessary to apply 200–250 μm of coating to protect the high spots adequately. Such thicknesses are only achieved in four or more coats with finishes containing solvent and each coat must be allowed to dry partially before the next is applied. Labour costs are naturally high and the process is long. An oil tanker in dock would at the same time be running up high docking charges. This market is therefore made for a 100% solids epoxy finish, which can be applied at 100–400 μm in one coat, even though the paint itself is very expensive (since solvents are relatively cheap and resins – especially epoxies – are not). Such thick coatings are prevented from sagging by inclusion of mineral thickeners. The viscous finish is applied by airless spraying, where the coating is forced through a nozzle under high pressure. Large areas are coated in a very short time.

In many solventless systems (see Polyesters, Chapter 16), there is an appreciable volume contraction in the change from liquid linear polymer and co-reactants to solid cross-linked polymer. This can put the adhesive bond to the substrate under great strain and can cause loss of adhesion, pulling away from edges and related defects. Solventless epoxy systems are remarkable in that they show negligible contraction.

The solventless epoxy film shows typical epoxy finish properties, but is inevitably less flexible than usual because (1) the films are thicker and (2) the close spacing of the reacting groups leads to a high density of cross-links. The film is, however, very tough and does not shatter easily.

Because the reactants are not diluted in solventless coatings, pot lives are short (0.5–2 h). Because of short pot lives, special twin-feed airless spray equipment is often used. Aromatic polyamines are preferred for curing, so that hardening times of 4–12 h can be obtained even at 0–10 °C. This better low-temperature cure cannot be obtained in outdoor topcoats, where colour is important, since the aromatic polyamines discolour.

As well as the ingredients already mentioned, tertiary amino phenolic catalysts, such as tri(dimethylaminomethyl) phenol,

$$(CH_3)_2N-CH_2 \quad \overset{OH}{\diagup} \quad CH_2-N(CH_3)_2$$

$$\underset{N(CH_3)_2}{\overset{CH_2}{|}}$$

are often included. These produce polymerization of the epoxy resin with itself:

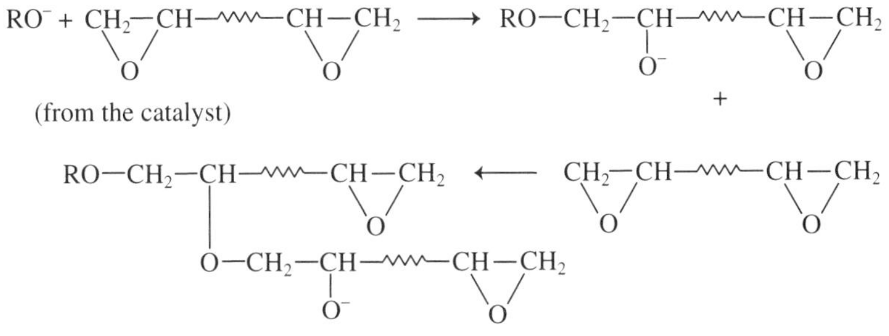

$$RO^- + CH_2-CH-\!\!\!\text{\scriptsize www}\!\!\!-CH-CH_2 \longrightarrow RO-CH_2-CH-\!\!\!\text{\scriptsize www}\!\!\!-CH-CH_2$$

(from the catalyst)

+

$$RO-CH_2-CH-\!\!\!\text{\scriptsize www}\!\!\!-CH-CH_2 \longleftarrow CH_2-CH-\!\!\!\text{\scriptsize www}\!\!\!-CH-CH_2$$

$$O-CH_2-CH-\!\!\!\text{\scriptsize www}\!\!\!-CH-CH_2$$

and so on

About seven epoxy groups are reacted in this way for each nitrogen atom in the catalyst. Phenols are also included and act as accelerators in the curing reaction. Silicone resins may be added as flow agents and dibutyl phthalate has a plasticizing action.

Water-based epoxy coatings

It is possible with epoxies, as with most resin systems, to produce water-based versions. In the simplest system the epoxy resin is emulsified with suitable surfactants. The co-reactive material can be, for example, a UF resin emulsified with it to produce a stoving composition, or a water-soluble polyamine supplied in a second pack for room temperature cure.

   Electrodeposited resins based on epoxy resins can be of both the anodic and cathodic types. The anodic types based on maleinized epoxy esters have already been mentioned in this chapter. **Cathodic electrodeposition resins** may be illustrated by the reaction product of epoxy resin with diethylamine:

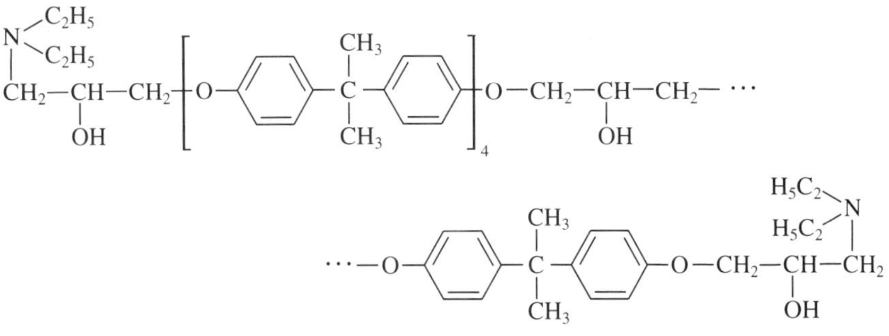

This resin will be partly neutralized with acetic or lactic acid, and co-emulsified with a blocked isocyanate cross-linker (p. 225). This type of resin is now the standard protective primer coating on the majority of the world's motor cars. Pigmentation will include anti-corrosive pigments.

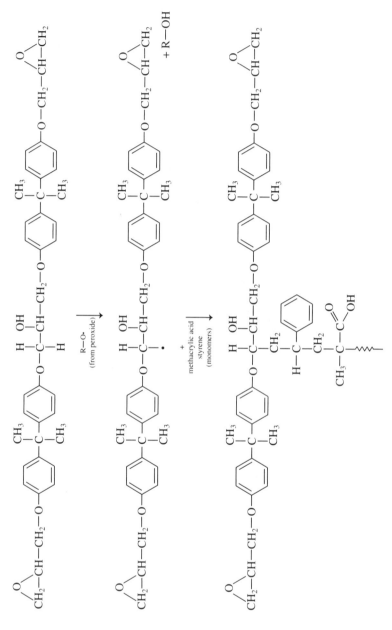

Fig. 14.2 Preparation of an epoxy acrylic graft copolymer.

Water-based epoxy resins not applied by electrodeposition are mostly of the anodic type and maleinized epoxy esters are frequently used. Another class is the **epoxy acrylic graft copolymer**, where the epoxy resin is rendered water-soluble by attaching side chains of acrylic polymer containing a high amount of methacrylic acid. The process of grafting is similar to that of chain transfer (p. 67), where acrylic chains are attached by hydrogen abstraction. This process is illustrated in Fig. 14.2, where C−· represents a radical site, and ⌇ represents an extended acrylic chain containing both methacryclic acid and styrene components.

The equations show how attack by a peroxide creates a site on which a styrene–methacrylic acid copolymer side chain will grow. The acid groups are then partly neutralized with amine to provide solubilization. Not every epoxy molecule is grafted, and the product contains both unreacted epoxy and separate acrylic resin, where the ungrafted epoxy will remain insoluble in water but emulsified by the graft copolymer. The final product is hence an emulsion. These resins cross-link on stoving by reaction of the epoxide and acid groups which are essentially unreacted in the resin-making process, which is carried out at lower temperatures. These resins are used in high volume in beverage can interior spray lacquers.

Some examples of different types of epoxy paints are given in Tables 14.1–14.4.

Table 14.1 Formulation of one-pack paints: epoxy-phenolic clear interior can coating

	wt%	
Epikote 1007	28.8	} Resin
Bakelite 100	7.2	
1-Methoxy propan-2-ol	27.4	} Solvents
Methyl isobutyl ketone	27.4	
Methoxy propyl acetate	9.1	
Phosphoric acid	0.1	Catalyst
	100.0	

Apply by roller coater at 36% solids.
Stove 10 min at 200 °C.
Bakelite 100 is a phenolic resole from Bakelite AG.
Epikote 1007 is a bisphenol A epoxy resin of molecular weight 2900 and is a product of Shell Chemicals.

Table 14.2 Formulation of one-pack paint: cathodic electrodeposition primer

	wt%	
Carbon black Special Schwartz 4	0.25	Black pigment
Basic lead silicate	0.13	Anti-corrosive pigment
China clay ASP 100	1.13 ⎱ Mineral thickeners
Talc Microtalc MP 15-38	1.0 ⎰
Epoxy amine adduct JK512	13.03	Resin
Blocked isocyanate	4.83	Cross-linker
Lead soap (soluble) 6% Pb	0.23	Cure catalyst
Lactic acid	0.65	Neutralizing agent
Demineralized water	78.75	
	100.00	

Apply 20 μm by electrodeposition onto cathode at 15% solids.
Stove 30 min at 180 °C.
Epoxy amine adduct JK512 is the reaction product of Epikote 1001 bisphenol epoxy resin with diamines and with a Cardura E10/diamine adduct. Further details can be obtained from Shell Chemicals.

Table 14.3 Formulation of two-pack solventless finish

Finish			Activator	
	wt%			wt%
Rutile titanium dioxide	6.4	Pigment	Hardener HY830	48.5
Barytes	39.8	Extender	Hardener HY850	51.5
Bentone 27	3.2	Thickener		100.0
Araldite GY250	47.2	Resin		
Dibutyl phthalate	3.4	Plasticizer	Mixing ratio:	
	100.0		finish/activator = 3.3:1 by wt	

Drying times: tack-free 4–5 h, hard 8–14 h at 7–20 °C.
Pot life: 1 h.
Araldite GY250 is an unmodified liquid epoxy resin, viscosity 22.5–27.5 Pa s at 21 °C, made by CIBA Polymers.
Bentone 27 is a treated clay from Rheox Inc.
Hardeners HY830 and HY850 are aromatic amines of 9 and 35 Pa s respectively at 21 °C. HY850 also includes an accelerator. Both are made by CIBA Polymers.

Table 14.4 Formulation of powder coating: polyester epoxy powder for interior use

	wt%	
Rutile titanium dioxide Ti-pure R-900	26.8	Pigment
Eastman polyester PC-17-4N	35.0 ⎫	Resin
EPON 2002 epoxy resin	35.0 ⎭	
Vestagon B-31	0.1	Curing agent/catalyst
Modaflow Powder 111 (Monsanto)	2.05	Flow aid
Benzoin	1.0	UV absorber
Anti-oxidant	0.05	Anti-oxidant

Process by blending and extrusion at temperature not exceeding 95 °C to avoid pre-reaction. Cure for 20 min at 180 °C following electrostatic spray application. Eastman polyester PC-17-4N is a neopentyl glycol–terephthalic acid–trimellitic anhydride polyester of acid value 92 at 100% solids. The formulation is available from Eastman Chemical Ltd (Eastman publication N-281C).
Epoxy resin EPON 2002 is a bisphenol A epoxy resin from Shell Chemicals.
Vestagon B-31 is a curing agent and accelerator for epoxy coatings, and is made by Hüls AG.

Fifteen

Polyurethanes

isocyanate-based coatings

Polyester finishes contain polyester resins which, in turn, contain predominantly ester linkages. Polyurethane finishes need not contain polyurethane resins; however, all polyurethane finishes contain isocyanates or their reaction products. The reader should be familiar with isocyanate reactions before proceeding further with this chapter. As we shall see, the urethane linkage is not necessarily predominant in the dry paint film.

Since there is a variety of isocyanate reactions, the formulator has considerable scope. A number of different approaches have been used, and have led to a number of one-pack and two-pack paints.

One-pack paints

Urethane oil and alkyd finishes

Oils can contain hydroxyl groups in the fatty acid portion (e.g. castor oil), or can be converted to diglycerides and monoglycerides when they are heated in the presence of *poly*hydric alcoh*ols* (**polyols**). This is the 'monoglyceride stage' that we have already described as part of the alkyd-making process (p. 180). By addition of diisocyanate, the hydroxylic oil can be chemically 'bodied', the molecules being linked together by urethane linkages e.g.

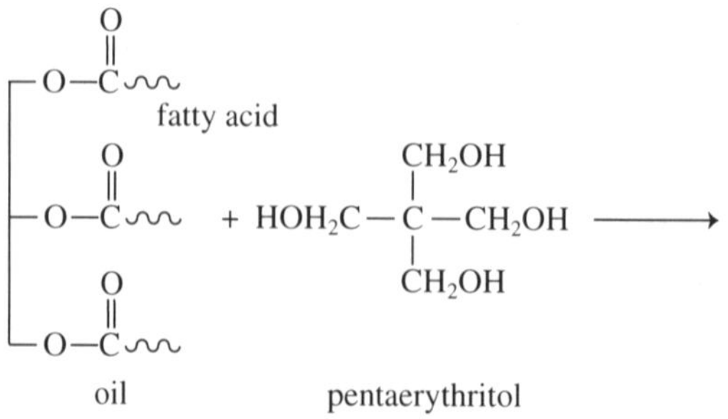

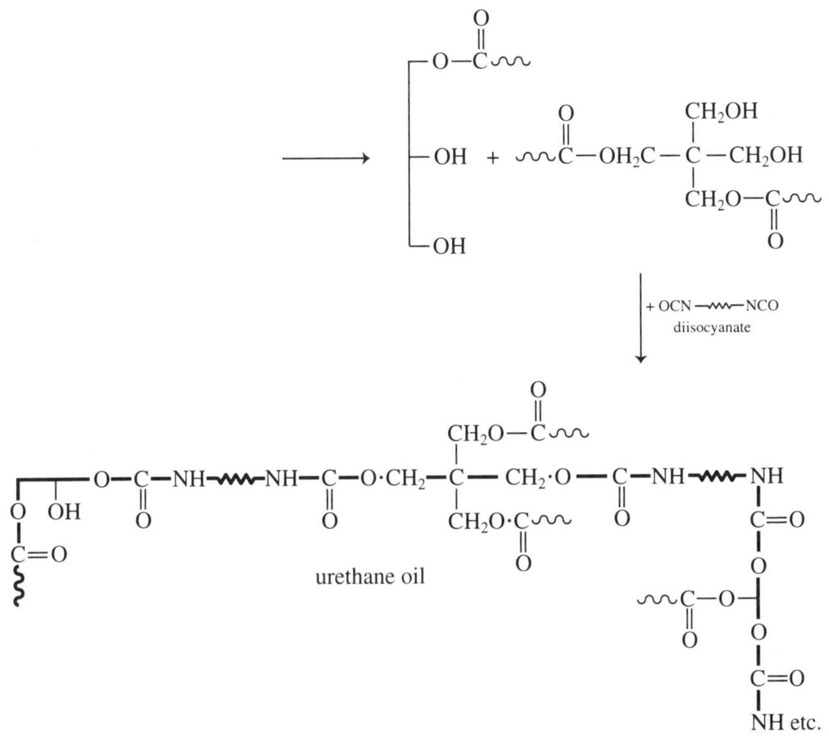

$$
\begin{array}{c}
\text{O}\\
\parallel\\
\text{—O—C}\backsim\backsim
\end{array}
$$

$$
\longrightarrow \quad
\begin{array}{l}
\text{—OH} \quad + \quad \backsim\backsim\text{C—OH}_2\text{C—C—CH}_2\text{OH}\\
\text{—OH}
\end{array}
$$

with
$$
\begin{array}{c}
\text{O}\\
\parallel\\
\quad\quad\quad\quad \text{CH}_2\text{OH}\\
\backsim\backsim\text{C—OH}_2\text{C—C—CH}_2\text{OH}\\
\text{CH}_2\text{O—C}\backsim\backsim\\
\parallel\\
\text{O}
\end{array}
$$

$$
+ \text{OCN}\longrightarrow\backsim\backsim\text{NCO}
$$
diisocyanate

urethane oil

NH etc.

An excess of isocyanate groups is not used, so that the final product, a **urethane oil**, contains no unreacted —NCO. It is therefore stable to moisture, non-toxic and differs from bodied oil in that it contains urethane linkages and can be much higher in molecular weight. The amount of di-isocyanate used and the molecular weight of the product vary with the extent of conversion of oil to mono- and diglyceride and with the relative proportions of these two species. The product has an oil length which is either 'long' or 'medium' (see the definition of the oil length of an alkyd), and the structure resembles the alkyd with urethane links from the diisocyanate replacing the ester links from the difunctional acid.

A **urethane alkyd** is an alkyd in which some of the dibasic acid is replaced by diisocyanate. The ester links are formed first in the usual way, the diisocyanate is added and the remaining hydroxyls are reacted at 80–95 °C to form the urethane linkages (p. 223). It can be seen that the urethane alkyd shares the features of the urethane oil and the oil-modified alkyd.

As with the urethane oils, no unreacted isocyanate groups are permitted in the final alkyd.

Ordinary drying-oil alkyds are restricted in molecular weight by the difficulty of getting esterification to occur when the concentration of carboxyl and hydroxyl groups becomes low near the end of the preparation. Esterification can be continued only by increasing the temperature. This now rises into the region where heat-bodying of the unsaturated fatty acid chains can occur. The coupling of fatty acid chains cross-links the alkyd polyester backbones and can lead to irreversible gelation. Therefore the reaction must be stopped. Since urethane alkyds can be completed at low temperatures where heat-bodying cannot occur, higher molecular weight molecules can be obtained with better control of the final resin-making stages. They dry faster than ordinary alkyds because (1) the molecules are initially bigger and (2) they have a higher fatty acid functionality for oxidative drying.

Urethane alkyds are basically similar to ordinary alkyds and may be drying or non-drying, long oil or medium oil. Short oil urethane alkyds are theoretically possible, but are difficult to make.

Some authors apply the terms 'urethane alkyd'. 'urethane oil' and 'uralkyd' exclusively to urethane oils as defined here. The urethane alkyds of this book are then called 'urethane-modified alkyds'.

The urethane linkage is resistant to alkalis and thus urethane oils and alkyds have better alkali resistance than ordinary alkyds. This is usually accompanied by better water resistance. Urethane oils and alkyds possess a property common to all polyurethane finishes: good abrasion resistance. It is claimed that they are superior to oils and alkyds for the dispersion of difficult organic pigments and carbon black, but in pigmented form they are prone to chalk and lose gloss earlier than normal alkyds and oleoresinous finishes.

The main outlets for urethane oils and alkyds are in varnishes for floors, boats and general use, and in undercoats and in industrial maintenance finishes (where gloss retention is less important than long-lasting film integrity and resistance to water and chemicals). Urethane oils and alkyds may be blended with ordinary long oil alkyds in glossy decorative paints in order to increase their resistance properties as described above.

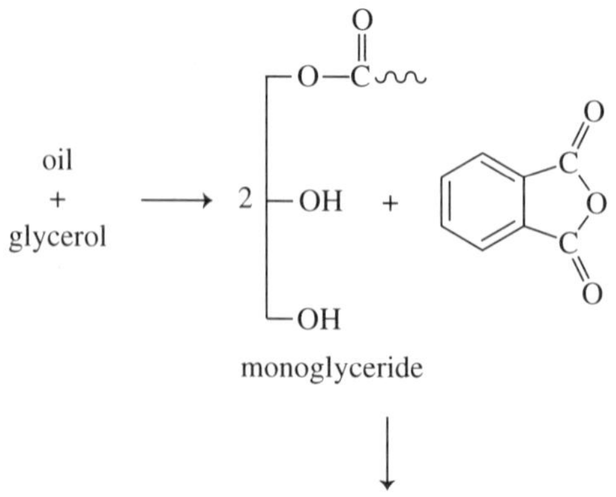

monoglyceride

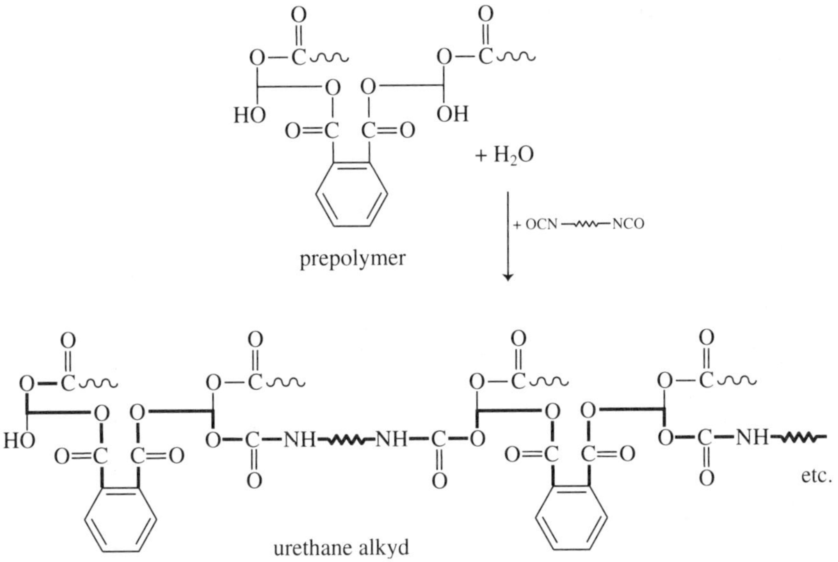

prepolymer

$+ H_2O$

$+ OCN -\!\!\!\wedge\!\!\!\wedge\!\!\!- NCO$

urethane alkyd

etc.

The diisocyanate most commonly used in both urethane oils and alkyds is toluene diisocyanate (p. 224).

Moisture-curing prepolymers

Finishes of this type do not form urethane linkages during the drying process. Like the first group of finishes, they dry by the second mechanism of Chapter 7, by reaction with the atmosphere. This time it is the water vapour in the air, not the oxygen, which reacts chemically with isocyanate in the finish. If a polyisocyanate is present, polymerization can occur:

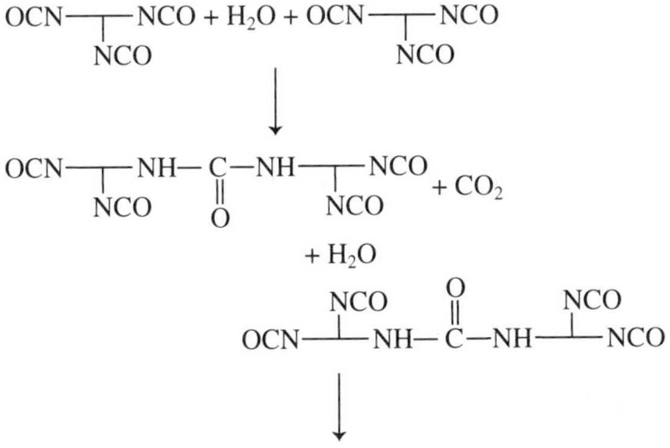

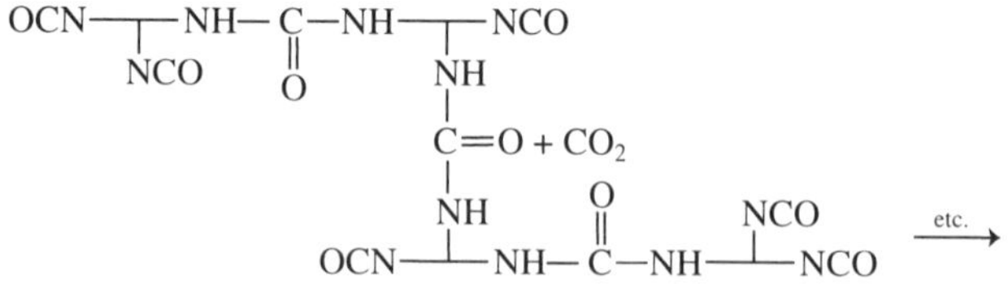

$$\text{OCN}\underset{\text{NCO}}{\overset{}{\rule{0pt}{0pt}}}\text{—NH—}\underset{\overset{\|}{O}}{C}\text{—NH—NCO}$$

(structure with NH, C=O + CO$_2$, NH, and)

$$\text{OCN—NH—}\overset{O}{\underset{\|}{C}}\text{—NH—NCO} \quad \text{etc.} \longrightarrow$$

(substituted urea
linkage)

Drying is faster if the initial polyisocyanate molecules are large. Because of its wide usage in polyurethane foams, the cheapest diisocyanate available is the low molecular weight toluene (tolylene) diisocyanate (or TDI). This material is volatile, toxic and gives rise to marked yellowing in paint films. However, TDI can be converted into a larger molecule termed an **adduct** by reaction with a polyol such as trimethylol propane:

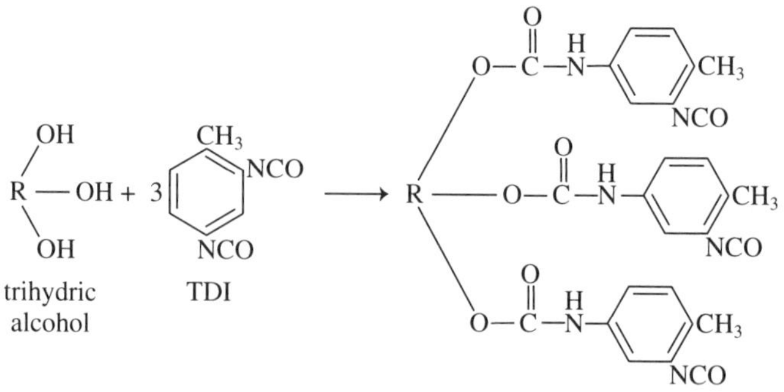

triisocyanate

This reaction occurs because the —NCO group *para-* to the methyl group is several times more reactive than the one *ortho-* to it and will therefore react with the hydroxyl groups preferentially. Some *ortho-* groups will, of course, react and there is an added complication in that commercial TDI contains some of the 2,6-isomer. Thus side reactions leave some unreacted TDI in the final mixture, but methods of reducing this to a very low level have been devised and the resultant 'prepolymer', as it is called, is safe for spraying with appropriate ventilation arrangements (p. 234). Any fairly high molecular weight polyol can be used. Castor oil is such a material. Polyethers (p. 229) are often used.

By the route shown, or any other, a fairly high molecular weight polyisocyanate with a minimum functionality of three (isocyanate groups) is produced. This is the film-former of the finish, reacting with water to give a cross-linked polymer film. The carbon dioxide produced in the process is formed slowly and diffused through the film without causing 'bubbling' or 'popping'. The drying rate varies enormously with the humidity. Force-'drying' by steam is possible.

As the prepolymer reacts with moisture, this must be rigorously excluded from the can, since it will not form a skin on the paint but will eventually gel it right through. Another difficulty is the moisture found on most pigment surfaces. This must be removed, possibly by reaction with monoisocyanate, before the prepolymer can be pigmented. If it is not, gelation occurs. Stability may be further enhanced by addition of moisture scavengers to the coating, such as ethyl orthoformate, $HC(\cdot O \cdot C_2H_5)_3$, or **molecular sieves**. The latter are sodium or potassium aluminium silicates with a finely porous crystal structure. Only small molecules, such as water, can penetrate these fine pores and they are then strongly adsorbed within them. The main outlet for this type of finish – which has the typical polyurethane virtues of toughness and abrasion resistance – is as a floor varnish.

Blocked isocyanate stoving finishes

At sufficiently high temperatures, the reaction products of isocyanates and certain other compounds decompose to reform the isocyanate:

$$R-N=C=O + HO-\langle\bigcirc\rangle \underset{150\,°C}{\overset{\substack{\text{room}\\\text{temperature}}}{\rightleftharpoons}} R-NH-\overset{\displaystyle O}{\overset{\|}{C}}-O-\langle\bigcirc\rangle$$

Hot phenol is an unpleasant by-product to release from a paint and polyisocyanates 'blocked' with phenols have not met with widespread use in coatings. However, the use of polyisocyanates blocked with ε-caprolactam is a popular route to **polyurethane powder coatings**, using typically the aliphatic isocyanate, IPDI, whose structure is given near the end of this chapter. Powder coatings have already been described in Chapter 14.

$$R-N=C=O + HN\underset{\substack{C\\\|\\O}}{\overset{}{\diagup}}(CH_2)_5 \underset{175-180\,°C}{\rightleftharpoons} R-NH-\overset{\displaystyle O}{\overset{\|}{C}}-N\underset{\substack{C\\\|\\O}}{\overset{}{\diagup}}(CH_2)_5$$

The use of blocked isocyanates has proved an important method for cross-linking cathodic electrodeposition coatings (p. 215). The hydroxyl groups on the resin can be cross-linked by a blocked isocyanate such as

$$\underset{\text{NCO}}{\overset{CH_3 \quad NCO}{\langle\bigcirc\rangle}} + 2CH_3 \cdot (CH_2)_3 \cdot \overset{C_2H_5}{\underset{}{CH}} \cdot CH_2OH \qquad \text{2-ethyl hexanol}$$

$$\underset{\substack{\text{175–190 °C}\\\text{[+ dibutyl tin}\\\text{dilaurate catalyst]}}}{\Big\updownarrow}{\scriptstyle <100\,°C}$$

$$\underset{\text{NH·CO·O·C}_8\text{H}_{17}}{\overset{CH_3 \quad NH\cdot CO\cdot O\cdot C_8H_{17}}{\langle\bigcirc\rangle}}$$

Another polyisocyanate used commercially is diphenylmethane-4,4'-di-isocyanate (MDI) and also finds application here. The epoxy and blocked isocyanate are mixed together and the tertiary amino groups of the modified epoxy resin are partially neutralized with acetic acid before the blend is emulsified in water.

These blocked isocyanates are also reactive components in water-borne stoving surfacers used in car plants, where they may be blended with solubilized polyester and sometimes with amino resin (Chapter 13).

Fully reacted urethane solution and dispersions

Just as diisocyanates can be substituted for acids in the alkyd structure to produce a urethane alkyd or oil, so a simpler structure similar to a polyester can be made. Diols such as those described in the next section as being suitable for use in polyesters, and lower molecular weight polyethers, are reacted with diisocyanates to form linear high molecular weight polyurethanes. These are used as clear lacquers and applied as very low solids solutions on flexible substrates such as plastics and wood.

Polyurethane emulsions and dispersions in water may also be made by incorporating polar or hydrophilic groups as described in Chapter 9. They are fully reacted, containing urethane and urea groups, are of high molecular weight, and may contain cosolvent. The chemistry of these is complex and beyond the scope of this book. They are used for the same applications as the solution polymers, and as primers for a range of plastic surfaces.

Two-pack paints

Two-pack paints, now known generally as 2K coatings, are the largest volume urethane coatings produced, of which those under the second heading below are most popular. The term 2K comes from the German for 2-pack (2 komponenten); in Germany the term 1K is used for one-pack finishes.

Activated prepolymer

This type of finish is an extension of the one-pack moisture-curing type. It is virtually the same finish, to which a small quantity of catalyst is added just before use. Tertiary amines, such as *N,N*-dimethylethanolamine $[HO \cdot CH_2 \cdot CH_2 \cdot N(CH_3)_2]$, are most frequently used as catalysts. Even if the moisture content of the air is low, the finish will harden rapidly, but the more reliable drying performance is obtained at the expense of a limited pot life. Unfortunately, the activator is non-polymeric and so cannot be used for pigment dispersion. Pigments are dispersed in prepolymer and must be free from moisture, as in the one-pack composition. The uncatalysed finish is water-sensitive, presenting a shelf-life storage problem as before.

Polyhydroxylic resin–isocyanate finish

In this finish the price of 'pot life' is paid to gain the following advantages:

* faster room-temperature drying than that of urethane oils and alkyds, by a mechanism independent of the atmosphere;
* easier pigment dispersion in a resin free from isocyanate;
* shelf-life stability of the paint, by isolation of the moisture-reactive component in an activator pack;
* extreme formulating versatility, because of the wide variety of ingredients available for both packs.

The method is to place in the 'paint' pack a solution of a linear or branched polymer containing hydroxyl groups, plus suitable additives. The activator pack contains a solution of a moderate molecular weight polyisocyanate, such as that described earlier in this chapter in the section on moisture-curing prepolymers. Since the hydroxyl spacing in the resin can be made large, it is not necessary to have widely spaced isocyanate groups to get flexibility in the cross-linked film.

Three types of polyisocyanate are available for use in these paints: the polyol adduct already encountered, the polyisocyanurate and the biuret.

Polyisocyanurates are produced when isocyanates react with themselves at high temperatures.

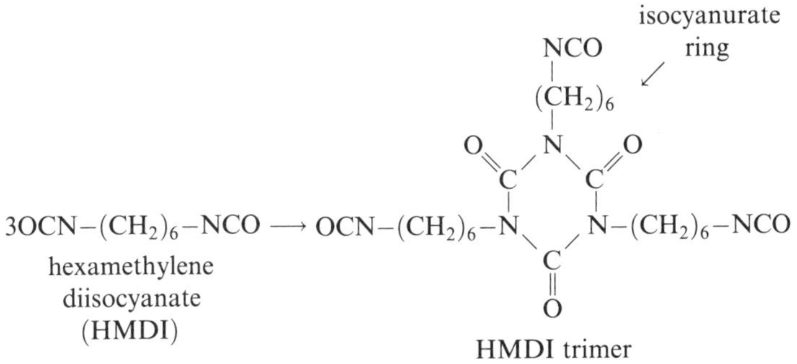

$$3OCN-(CH_2)_6-NCO \longrightarrow$$

hexamethylene
diisocyanate
(HMDI)

HMDI trimer

In the equation above a trimer is shown, but more complicated structures will also be formed, due to reaction of some of the three unreacted trimer $-NCO$ groups with further HMDI. Dimer may also be present.

Another typical compound used is that formed by reacting three molecules of aliphatic diisocyanate, hexamethylene diisocyanate, with one of water, to produce a triisocyanate containing urea linkages known as a

biuret (together with more complex molecules not shown in the following idealized equation):

$$3OCN-(CH_2)_6-NCO + H_2O \longrightarrow \begin{array}{c} HN-(CH_2)_6-NCO \\ | \\ C=O \\ | \\ N-(CH_2)_6-NCO + CO_2 \\ | \\ C=O \\ | \\ HN-(CH_2)_6-NCO \end{array}$$

The resin, as has been stated, may be any resin containing hydroxyl groups. This includes castor oil, alkyds, nitrogen resins, epoxy resins, cellulose derivatives and so on, provided always that, if a mixture is used, the ingredients are compatible with one another. However, the most popular resins used are saturated polyesters, acrylic resins and, to a lesser extent, polyethers.

Although expensive, saturated polyester resins have excellent colour retention and water resistance. They perform better than polyethers, since ether linkages are water-sensitive (diethyl ether, $C_2H_5 \cdot O \cdot C_2H_2$, is partly soluble in water) and in polyethers they are repeated throughout the polymer chain without the separation by long, water insensitive hydrocarbon chains which could offset their effect.

Saturated polyester resins have been defined in Chapter 12 and their use with nitrogen resins was described in Chapter 13. The resins used in polyurethane finishes might contain a selection from the ingredients list on p. 181 and, amongst others:

1,3-butane diol	$HO \cdot CH_2 \cdot CH_2 \cdot CHOH \cdot CH_3$
1,4-butane diol	$HO \cdot CH_2 \cdot CH_2 \cdot CH_2 \cdot CH_2 \cdot OH$
1,6-hexane diol	$HO \cdot CH_2 \cdot CH_2 \cdot CH_2 \cdot CH_2 \cdot CH_2 \cdot CH_2 \cdot OH$
1,2,6-hexane triol	$HO \cdot CH_2 \cdot CHOH \cdot CH_2 \cdot CH_2 \cdot CH_2 \cdot CH_2 \cdot OH$

Again hydroxyl groups are to be found at the ends of chains. The number present is greater than that likely to be found in an ordinary saturated polyester, because the excess of hydroxyl groups over carboxyl in the resin ingredients is much higher than that usually allowed. For this reason also, some unreacted hydroxyls from triols are likely to be found in the middle of polymer chains:

$$\cdots -O \cdot CO\text{-}\!\!\wedge\!\!\wedge\text{-}CO \cdot O \underset{\underset{CH_2OH}{|}}{\rule{0pt}{0pt}} O \cdot CO\text{-}\!\!\wedge\!\!\wedge\text{-}CO \cdot O-OH$$

(only ester links and hydroxyl groups shown)

The formulation of hydroxyl-containing acrylic copolymers has already been discussed on p. 151 and their use with nitrogen resins on p. 195. Hard, extremely flexible and durable coatings are formulated by using these acrylic resins with aliphatic polyisocyanates to avoid yellowing (see Yellowing below). Typical compounds used are the isocyanurate trimers of hexamethylene diisocyanate or isophorone diisocyanate (pp. 227, 233). These are now widely used on account of their lower polarity and hence viscosity.

Polyethers are made by reacting ethylene oxide or propylene oxide with a polyol in the presence of an acid (BF_3) or basic (NaOH) catalyst, e.g.

$$
\begin{array}{c}
CH_2OH \\
| \\
CHOH \\
| \\
(CH_2)_3 \\
| \\
CH_2OH
\end{array}
\;+\; 3n \;
\underset{O}{\overset{}{CH_2{-}CH{-}CH_3}}
\;\xrightarrow[\substack{120\text{--}200\,^{\circ}C \\ \text{catalyst}}]{\text{pressure}}\;
\begin{array}{c}
\qquad\qquad CH_3 \\
\qquad\qquad | \\
CH_2{\cdot}(O{\cdot}CH{\cdot}CH_2)_n{\cdot}OH \\
| \qquad\qquad CH_3 \\
| \qquad\qquad | \\
CH{\cdot}(O{\cdot}CH{\cdot}CH_2)_n{\cdot}OH \\
| \\
(CH_2)_3 \quad CH_3 \\
| \qquad\quad | \\
CH_2{\cdot}(O{\cdot}CH{\cdot}CH_2)_n{\cdot}OH
\end{array}
$$

| 1,2,6-hexane triol | propylene oxide | polyether |

It can be seen that the polymer chains contain multiple alkali-resistant ether linkages and end in hydroxyl groups. Thus the number of hydroxyl groups in the polymer is the number of hydroxyl groups in the polyol starting ingredient. Molecular weights vary from 400 to 4000.

The design of the resin containing hydroxyl groups is important to film properties, since the spacing of hydroxyl groups along the chain will influence the flexibility. Close spacing will mean frequent cross-links and poor flexibility. Pronounced branching in the resin also tends to the same result. Dense cross-linking is usually accompanied by increased hardness and improved resistance to water, solvents and chemicals.

2K coatings as described above, using saturated polyesters and hydroxyl acrylic resins, have significant use for painting of aircraft and the refinishing of motor cars, where their rapid cure at room temperature is advantageous. They are also found in the factory finishing of furniture, where again usage is high. They are expensive in comparison with acrylic lacquers, so cost can be minimized by formulating the acrylic resins of low hydroxyl functionality, in order that smaller amounts of the costly isocyanate are needed. Inevitably, if this is done, low cross-link densities give a chemical resistance which is inadequate where requirements are exacting.

From this emphasis on hydroxyl groups, it might be thought that the hardening process consists solely of urethane formation between resin and polyisocyanate. This is obviously not so, since water vapour is usually present and the isocyanate reactions with water must proceed in competition with the reaction with hydroxyl. This will lead to consumption of isocyanate groups and linking of isocyanate molecules by urea linkages. With a difunctional isocyanate this tends to space the resin chains farther apart, giving increased flexibility:

$$
\begin{array}{ccc}
\text{OH} & \text{OH} \\
& + & + 2\,\text{OCN}\!-\!\!\!\sim\!\!\!\sim\!\!\!-\!\text{NCO} \\
\text{OH} & \text{OH}
\end{array}
$$

No moisture /   \ + H$_2$O

$$
\begin{array}{cc}
\text{O} & \text{O} \\
| & | \\
\text{C}{=}\text{O} & \text{C}{=}\text{O} \\
| & | \\
\text{NH} & \text{NH} \\
\{ & \{ \\
\text{NH} & \text{NH} \\
| & | \\
\text{C}{=}\text{O} & \text{C}{=}\text{O} \\
| & | \\
\text{O} & \text{O}
\end{array}
$$

$$
\begin{array}{cc}
\text{OH} & \text{O} \\
& | \\
& \text{C}{=}\text{O} \\
& | \\
& \text{NH} \\
& \{ \\
& \text{NH} \\
& | \\
& \text{C}{=}\text{O} + \text{CO}_2 \\
& | \\
& \text{NH} \\
& \{ \\
& \text{NH} \\
& | \\
& \text{C}{=}\text{O} \\
\text{OH} & \text{O}
\end{array}
$$

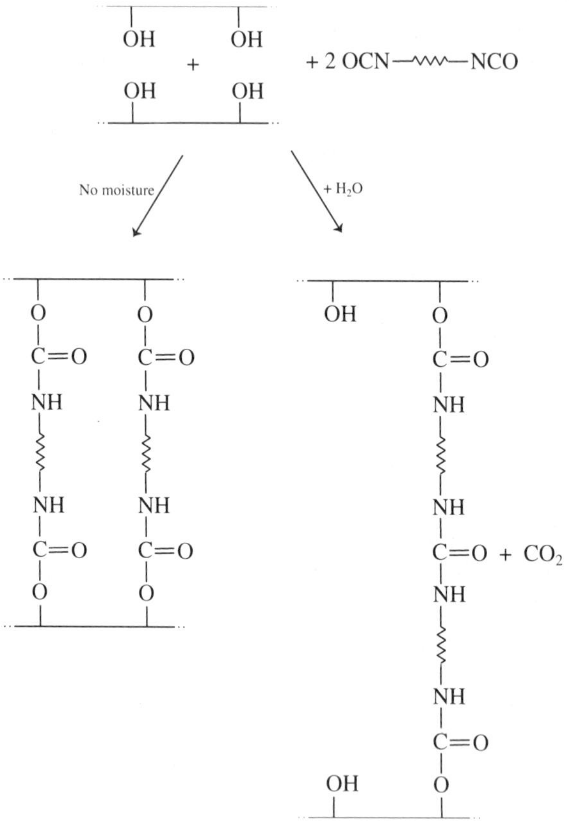

However, if the functionality is greater than two, branching occurs and increased overall isocyanate functionality results:

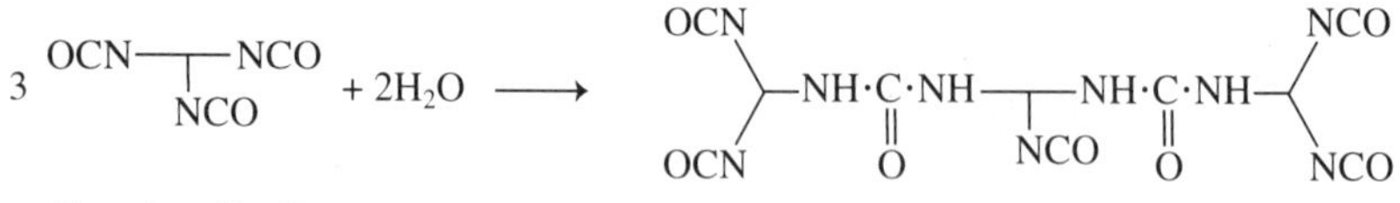

$$
3\ \ \text{OCN}\!-\!\!\underset{\text{NCO}}{\text{NCO}} + 2\text{H}_2\text{O} \longrightarrow \underset{\text{OCN}}{\overset{\text{OCN}}{>}}\!\!-\text{NH}\cdot\underset{\text{O}}{\overset{\parallel}{\text{C}}}\cdot\text{NH}\!-\!\!\underset{\text{NCO}}{|}\!\!-\text{NH}\cdot\underset{\text{O}}{\overset{\parallel}{\text{C}}}\cdot\text{NH}\!-\!\!\underset{\text{NCO}}{\overset{\text{NCO}}{<}}
$$

(functionality 3)   (functionality 5)

It is therefore not practical to match the number of available resin hydroxyl groups in the paint pack with the same number of isocyanate groups in the activator pack and it is usual to have a ratio of $-\text{NCO}/-\text{OH}$ of 1.1:1 to 1.3:1, allowing for reaction with moisture. Increased flexibility is gained by reducing the ratio; greater hardness and chemical resistance result from increasing it, together with faster curing.

In addition to the uses already mentioned, polyol–isocyanate two-pack finishes have wide applications for floors and boats; as corrosion-resistant metal finishes and for coating plastics, rubber, concrete and masonry. Exterior durability is good if ingredients are correctly chosen and paints are well formulated.

Reactive polyurethane dispersions

Having described the reactivity of the isocyanate group to water, it may come as a surprise that it is possible to prepare aqueous dispersions, where isocyanate reactivity remains. While these coatings do have a fairly short 'pot life', the groups are sufficiently reduced in their contact with water, by reason of being in dispersed form, to remain unreacted for a sufficiently long time in practice. Also because water evaporation and particle coalescence are relatively fast, reaction with water can be controlled and limited during application and drying.

A low viscosity, specially modified polyisocyanate is mixed with a previously prepared polyol dispersion when the isocyanate disperses into the polyol dispersion. This dispersion remains stable over a few hours, long enough to allow application. The chemistry of these systems is complex, and not fully disclosed, but can be based on both aliphatic and aromatic isocyanates, which are usually solvent-free. This is a new 2K technology now finding application in all markets using solvent-borne 2K urethane coatings.

Vapour curing

The reactions of isocyanates in both prepolymer and polyhydroxylic resin–isocyanate finishes are catalysed by various catalysts, including tertiary amines. If the catalyst is added to the coating, fast cure times at room temperatures and relatively short pot lives result. However, if the amine is withheld from the coating and introduced as a vapour, either by vaporization into the first stage of a curing chamber, or by injection into the atomizing air of a spray-gun or the directing air of an air-assisted electrostatic spray gun, then the coating is cured rapidly, catalysed by permeation of the amine into it. The pot life of the polyhydroxylic resin–isocyanate two-pack coating is considerably lengthened, while the shelf-life of the prepolymer coating is long if moisture can be excluded successfully.

N,N-dimethylethanolamine is one of the preferred amines, because its combination of good catalytic activity, adequate volatility and relatively low toxicity is an effective compromise. Amines are not equally effective catalysts for all types of isocyanate and care must be taken about selecting aliphatic isocyanates in particular (p. 233) to achieve best cure rates.

Vapour curing has been shown to work successfully in small industrial painting installations and has potential benefits on metal castings, which absorb heat intended for curing, and on some plastic substrates (e.g.

GRP), where surface defects can cause blow holes in the paint coating during stoving.

Solventless finishes

Since liquid polyols (e.g. castor oil, polyethers, low molecular weight polyesters) and liquid isocyanates (e.g. diphenylmethane diisocyanate) are available, it is theoretically possible to produce 100% solids finishes. However, carbon dioxide has great difficulty in diffusing through thick films of such viscous coatings and bubbles of the gas form holes and craters. The coatings will only be successful if moisture from all sources can be reacted or absorbed harmlessly. The use of molecular sieves for this purpose has been recommended. Coatings with pot lives of 30–60 min are applied by manual spreading with various implements on concrete floors and road surfaces. Twin-feed spray-guns may also be used.

General matters

One or two matters common to polyurethane finishes of all types must now be covered briefly.

Solvents

Except for urethane oil and alkyd finishes, all polyurethanes require solvents that will not react with isocyanates; alcohols and ether-alcohols are obviously excluded. The water content of the solvents should be negligible. Manufacturers often label their solvents 'urethane grade', implying low water content. Solvents readily miscible with water, e.g. acetone, should be treated with suspicion as they can take up water from the atmosphere. Obviously the exact choice of solvent will vary considerably with the finish composition.

Additives

The usual thickeners and flow agents can be used. The usual driers and anti-oxidants are used in urethane oils and alkyds. Catalysts which speed up the curing reactions (and reduce pot life) are tertiary amines and metal salts, particularly tin salts. The latter should be checked for toxicity before inclusion in paint.

General properties

In spite of the wide variety of types of polyurethane finish, certain characteristics are common to the family as a whole. These are toughness and abrasion resistance, combined with flexibility, good chemical resistance and good adhesion.

An outstanding feature of paints that dry by reaction of isocyanate is that hardening will occur even at 0 °C. The finishes can therefore be used in conditions unsuitable for epoxies and acid-catalysed nitrogen resin finishes.

Yellowing

In Chapter 6 it was shown that colour is associated with certain chemical groups and that many of these contain nitrogen. Thus any film-former containing nitrogen atoms should be suspect for colour retention, until it has been shown to be otherwise. It is not just a matter of whether chromophoric groups are present in the freshly formed polymer, but more whether they are likely to be produced as decomposition products on exposure to air, light, moisture and warmth. Polyurethanes based on aromatic isocyanates are most prone to yellowing (just as epoxies based on aromatic amines discolour; Chapter 14). The effect of aromatic rings on colour is covered in Chapter 6. Prepolymers made from aliphatic isocyanates do not yellow, but are generally less reactive than their aromatic counterparts. This cure rate problem is overcome by converting aliphatic diisocyanates into biuret or isocyanurate trimers (pp. 228 and 227).

Other diisocyanates of interest, of which three are aliphatic, are

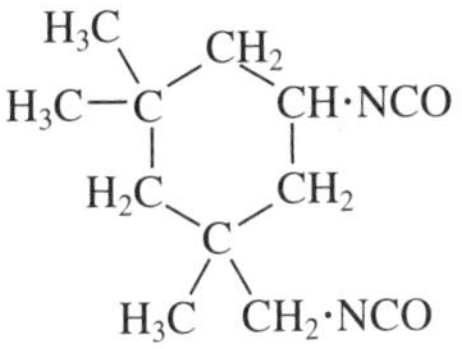

isophorone diisocyanate
(IPDI)

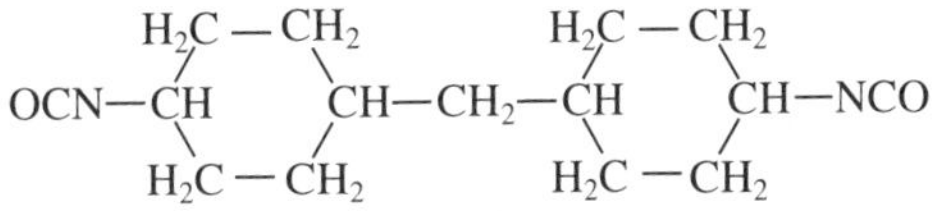

dicyclohexyl methane-4,4'-diisocyanate
(H_{12}MDI)

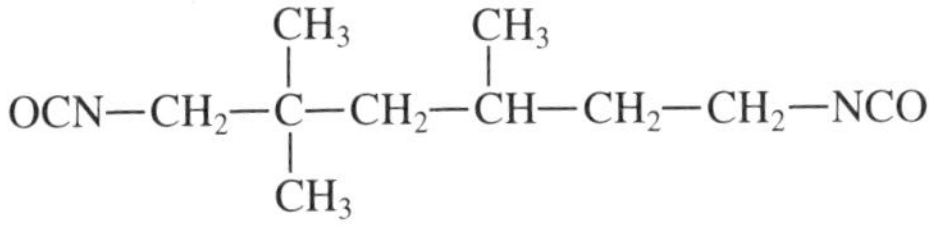

2,2,4-trimethyl-1,6-hexane diisocyanate
(TMDI)

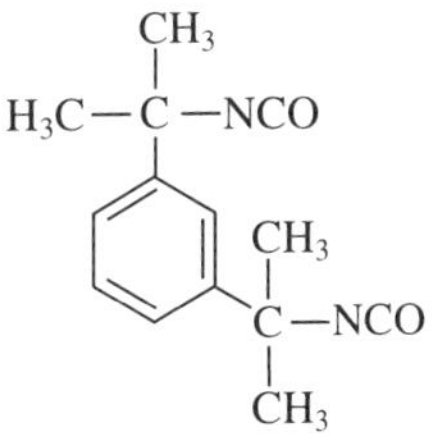

m-tetramethylxylylene diisocyanate
(TMXDI)

Formulations based on non-yellowing aliphatic isocyanates with good durability outdoors are now well established, but cost significantly more than corresponding products based on aromatic isocyanates.

Toxicity

Volatile isocyanates, such as TDI and hexamethylene diisocyanate, are
severe irritants to the eyes, nose and throat. They are also respiratory
sensitizers, causing asthmatic symptoms in sensitized individuals. Higher
molecular weight polyisocyanates for paints therefore should contain very
low levels of these isocyanates, so that the isocyanate content of the air in
the workspace remains below the OEL or TLV (p. 135). Even if they contain
only low levels of volatile isocyanates, precautions must be taken to prevent
inhalation of spray droplets containing unreacted isocyanates (either
excellent extraction of overspray in the spray booth, or the wearing of
fresh air hoods by those in the vicinity). The uncured liquid finishes can
also be skin irritants.

Finishes based on urethane oils and alkyds and blocked isocyanates do not
present the same hazards, since care is taken in their manufacture to ensure
complete reaction and the absence of free isocyanate groups.

Some examples of polyurethane formulae are given in Tables 15.1–15.3.
(An example using blocked isocyanate was given in the cathodic electrocoat
formulation in Table 14.2.)

Table 15.1 Formulation of oil-modified polyurethane white decorative gloss finish

	wt%	
Rutile titanium dioxide	28.7	} Pigment
Zinc oxide	1.5	
Polyurethane 2388	65.5	Resin solution
Cargill N lecithin	0.3	Pigment dispersant
Anti-oxidant	0.1	To improve stability in the can
Cobalt octoate (6% Co)	0.1	} Driers
Calcium octoate (5% Ca)	0.3	
Zirconium complex (6% Zr)	0.3	
Mineral spirits	3.2	Solvent
	100.0	

Ready for brushing at 63% solids.
Polyurethane 2388 is a urethane oil–alkyd based on safflower oil at 50% solids in mineral spirits
(white spirit) and is made by Cargill Inc., USA.
Cargill N lecithin is soya lecithin.

Table 15.2 Formulation of moisture-cure prepolymer floor finish

	wt%	
Polyurethane 3651	66.4	Film-former
Xylol	33.2	Solvent
Dimethylethanolamine[a]	0.4	Catalyst[a]
	100.0	

Ready for application to wood or concrete floors at 40% solids. Polyurethane 3651 is a moisture-curing prepolymer at 60% solids in xylol and 'Cellosolve' acetate and is made by Cargill Inc., USA.

[a]Catalyst may be added just before use to speed up cure at low humidities (when a pot life of 8–24 h is obtained) or it may be omitted altogether.

Polyurethanes

Table 15.3 Formulation for force-dry red automotive repair topcoat and room-temperature dry, high solids automotive clearcoat

Force-dry red automotive repair topcoat (2K)			Room-temperature dry high solids automotive clearcoat (2K)	
	wt%			wt%
Sicomin Red L3330S	5.4	Pigment		
Sicoecht Scarlet P5RH	0.6			
Uracron CY499E	74.0	Resin	Uracron CY474E	75.0
Xylene	5.0		Solvesso 100	7.5
Butyl acetate	5.6	Solvent	Methoxy propyl acetate	7.5
Methoxy propyl acetate	4.4		Butyl acetate	7.5
Butyl diethylene glycol acetate	0.8		Butyl glycol acetate	0.7
Levelling/anticrater agent Urad DD25 (50% in xylene)	0.2		Levelling/anticrater agent Byk 331	0.2
Floating/flooding additive Disperbyk 160	1.0	Additive	Tinuvin 292	0.5
Silicone additive Baysilone OL17 (1% in xylene)	2.4		Tinuvin 1130	0.5
Dibutyl tin dilaurate (1% in xylene)	0.6	Catalyst	Dibutyl tin dilaurate (1% in xylene)	0.6
	100.0			100.0
Desmodur N3390	28.0	Isocyanate	Desmodur N3390	24.0
Thin to spray viscosity 20 s DIN 4 cup Force-dry 30 min at 80 °C			Thin with xylene, butyl acetate, methoxy propyl acetate 60:20:20 to 18 s DIN 4 cup. Application solids 53%. Pot life 8 h. Dust-free time 30 min, tack-free time 130 min.	

Uracron CY 499E is a hydroxy acrylic resin at 75% solids in butyl acetate.
Uracron CY 474E is a hydroxy acrylic resin at 70% solids in butyl acetate. Both are supplied by DSM Resins.
Desmodur N3390 is HMDI isocyanurate trimer supplied by Bayer AG.
Solvesso 100 is an aromatic solvent blend, BR 160–178 °C, supplied by Exxon Chemicals.
Byk 331 is a levelling aid supplied by Byk-Chemie GmbH.
Tinuvin 292 is a hindered amine light stabilizer (HALS). Tinuvin 1130 is a benztriazole UV stabilizer. Both are supplied by CIBA Specialty Chemicals.

Sixteen

Radiation curing and unsaturated polyesters
finishes curing through unsaturation

In Chapter 12 we mentioned that the departure from the 100% solids of the old oil paints to the lower solids of modern finishes was a consequence of the demand for paints with better performance. The paint industry's present need to provide these improved properties at higher solids has also been mentioned. We have already seen in Chapter 14 how the epoxy–polyamine system meets these requirements. The finishes under discussion in this chapter achieve very high solids and freedom from solvent by another approach.

A 100% solids paint is achieved when the film-forming ingredients are fluid enough for application and are essentially involatile. Since the polyester used with polyester–glass fibre mouldings fulfils the first part of these conditions and goes some way towards satisfying the second part, it is clearly a potential paint material. The development of the plastics material into a practical polyester finishing system for wood was completed first in Germany in the 1950s. The result was a smooth, hard, glass-like finish with excellent resistance to most forms of damage that occur in domestic usage. Another significant application has been in spray fillers for car refinishing, but their exploitation in other paint markets has been limited.

The development since the 1960s of extremely rapid methods of curing these coatings with ultraviolet and electron beam radiation has led to an extension of the chemistry to compositions based on resins and monomers containing acrylic unsaturation, and to the cationic UV curing of epoxy compositions. All of these coatings will also be discussed in this chapter.

Ingredients of unsaturated polyester coatings

Though discussing unsaturated polyesters, this section introduces special features of all the coatings covered in this chapter that cure by a free radical mechanism.

The major ingredient of the unsaturated polyester coating is the **unsaturated polyester resin**, and Chapters 12 and 15 gave an idea of what is meant by a polyester resin and, in particular, a saturated polyester resin. Although a drying-oil alkyd is, in a sense, an unsaturated polyester resin, the term has come to be applied solely to polyester resins based on components which introduce unsaturation *directly into the polyester backbone*. This unsaturation must be capable of direct chain growth copolymerization with vinyl monomers. To give a linear polymer, any of the dibasic acids or dihydric alcohols mentioned in Chapters 12 and 15 may be used, but the resin should include some unsaturated components. These are usually, though not necessarily, acids, e.g.

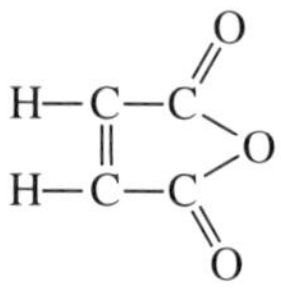

maleic anhydride

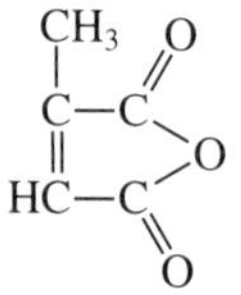

citraconic anhydride

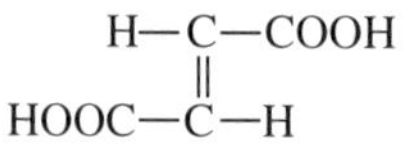

$$H-C-COOH$$
$$\parallel$$
$$HOOC-C-H$$

fumaric acid

$$CH_2$$
$$\parallel$$
$$C-COOH$$
$$\mid$$
$$CH_2 \cdot COOH$$

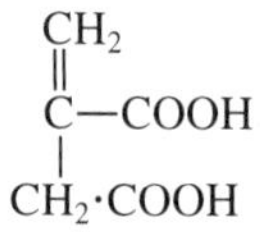

itaconic acid

Part of a typical unsaturated polyester chain might be

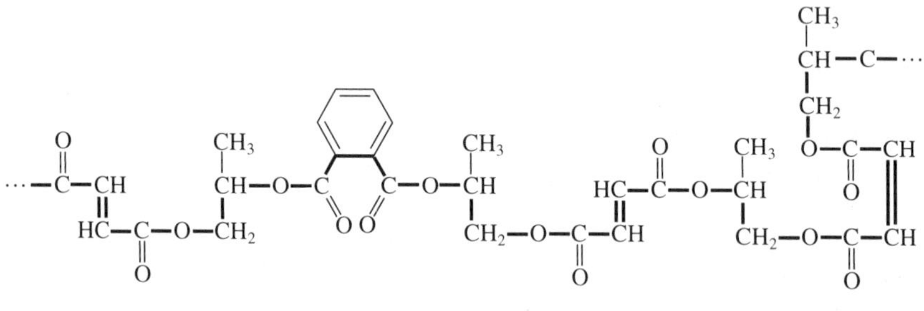

This would be made from 1,2-propylene glycol, phthalic anhydride and (probably) maleic anhydride (since the cis-maleic form of the rigid double bond

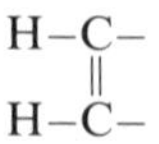

$$H-C-$$
$$\parallel$$
$$H-C-$$

largely changes to the trans-fumaric form,

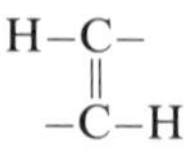

$$H-C-$$
$$\parallel$$
$$-C-H$$

at the temperature of the resin cook). Typical molecular proportions might vary from 22:10:10 to 22:5:15 (a slight excess of hydroxyl in each case). Normal alkyd methods of preparation are used, with, of course, inert gas over the liquid to prevent oxidation of the unsaturation. Since there is no monoglyceride stage, all the ingredients are usually present from the start of the reaction. If isophthalic acid is used, it is reacted with the diol before the maleic anhydride is added. If this is not done, unreacted insoluble iso-phthalic acid will remain at the end of the resin preparation.

The principal difference between this resin and the unsaturated oligomers used in radiation curing is that the latter are of lower molecular weight and the unsaturation is typically situated at the ends of the molecules, as will be described.

The second ingredient of all these coatings is the **copolymerizable monomer**. Since the resin component, especially when that component is polyester resin, is highly viscous, the monomer must be extremely fluid and a solvent for the resin, so that the combined solution is fluid enough for application. When the resin is cooling, but still fluid, the monomer is introduced to produce the resin solution, which is the basis of the finish. In spite of the reactivity of the two components, no reaction can begin unless free radicals are produced in the solution. These radicals can only be produced by the following.

1. Ultraviolet light. This is excluded in storage by the use of light-proof containers.
2. Heat. Normal storage temperatures are quite safe.
3. The presence of an unstable compound. Such a compound may be present in trace quantities, possibly due to (4).
4. Oxygen attack on double bonds or peroxide formation at reactive carbon atoms. Oxygen is never completely excluded from cans although as we shall see later, its presence is less likely to cause polymerization here than with drying oils.

Since there is some chance of polymerization and gelling due to (3) and (4), both the monomer itself and the polyester resin solution need to contain small quantities of inhibitor (Chapter 5). Quantities of less than 0.05% inhibitor in the resin solution will give stability for several months, provided conditions (1) and (2) above are avoided. Further quantities may be added to the paint itself to delay curing. A total of up to about 0.2% might be included.

Any relatively involatile monomer with the vinyl ($CH_2=CH-$), vinylidene ($CH_2=C<$) or allyl ($CH_2=CH \cdot CH_2-$) grouping present is theoretically suitable. In practice, many are not sufficiently reactive with the fumarate double bond. Of the remainder, styrene is very reactive and cheap. It is usually the chief (or only) monomer, but may have a comonomer such as vinyl toluene, methyl methacrylate, ethylene glycol dimethacrylate, triallyl cyanurate or diallyl phthalate.

$$CH_2 \cdot O \cdot CO \cdot \underset{\underset{CH_3}{|}}{C}=CH_2$$

$$CH_2 \cdot O \cdot CO \cdot \underset{\underset{CH_3}{|}}{C}=CH_2$$

ethylene glycol dimethacrylate

$$
\begin{array}{c}
\overset{O}{\overset{||}{C}}-O \cdot CH_2 \cdot CH=CH_2 \\
\underset{O}{\underset{||}{C}}-O \cdot CH_2 \cdot CH=CH_2
\end{array}
$$

diallyl phthalate

$$
O \cdot CH_2 \cdot CH=CH_2
$$

triallyl cyanurate

$$CH_2=CH \cdot CH_2 \cdot O \qquad O \cdot CH_2 \cdot CH=CH_2$$

The solution, having been made stable in the can, must now be made reactive on the coated surface. In the case of unsaturated polyesters, this is done by introducing factor (3) in relatively large quantities. This is an *initiator* (p. 65). In thermally cured, as opposed to radiation cured finishes, the substance chosen to decompose and produce free radicals is an **organic peroxide**. A solution of this material in unreactive solvent forms the activator pack of the two-pack finish. The amount of solvent is chosen to give a suitable finish/activator mixing ratio, e.g. 10:1 by volume. The ketone, diacyl and hydroperoxide types are most frequently used.

Commercial ketone peroxides are not pure and contain substantial amounts of hydroperoxide. These peroxides have already been introduced in Chapter 5.

If the finish is to be stoved, a wide range of peroxides are suitable, depending on the temperature to be used and the pot life required. Lauroyl, 2,4-dichlorobenzoyl and benzoyl peroxides (acyl peroxides), methyl ethyl ketone (MEK) peroxide and cyclohexanone peroxide (ketone peroxides) and cumene hydroperoxide are examples.

However, if the finish is to be cured at room temperature, there is no peroxide which can, on the one hand, remain relatively stable in the activator pack and, on the other, produce free radicals fast enough to give rapid hardening when activator and finish are mixed. Peroxide decomposition is accelerated by the addition of a metal salt from the drier range. The mechanism of the catalysis is as given in Chapter 12 for hydroperoxides. The combination of cobalt and either of the above ketone peroxides is widely used. Another combination is benzoyl peroxide and a tertiary amine, but this has the drawback that the films produced are prone to yellowing.

Since the **accelerator** acts only on the peroxide, it is safely incorporated in the finish pack, though it is often added to pigmented compositions just before use, to avoid slow deactivation during storage, by adsorption on the particle surfaces. On no account should it be mixed with the peroxide directly, since the combination is explosive. Many pure peroxides explode if subjected to sudden shock or heat. They are therefore supplied in a more stable form as pastes, or as solutions in a high boiling liquid (e.g. the plasticizer dibutyl phthalate).

Whichever method of activation is used, decomposition of peroxide leads to free radical attack, probably on the styrene:

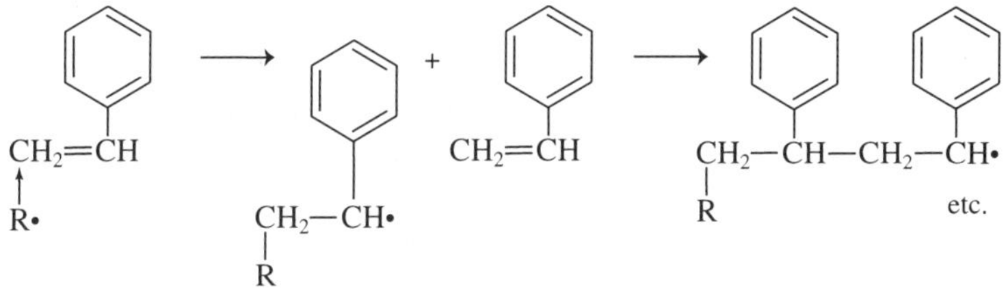

This will be recognized as the mechanism for chain polymerization which was described in Chapter 5 for the formation of polyethylene. However here, after quite a short polystyrene chain is formed, the free radical end encounters polyester unsaturation and copolymerization occurs:

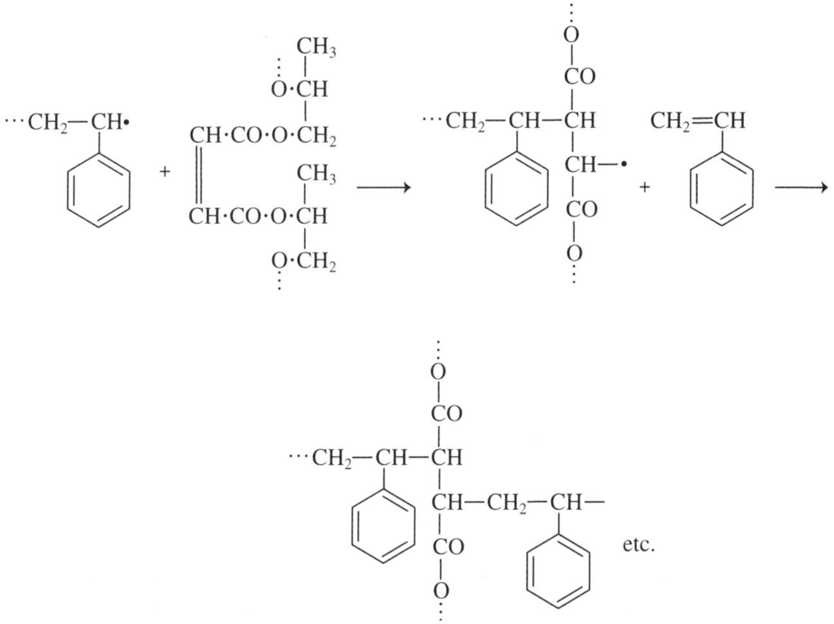

A simplified picture of the resulting cross-linked polymer is given in Fig. 16.1.

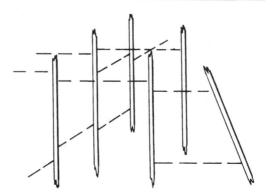

Fig. 16.1 The cross-linking of an unsaturated polyester finish. The broken rods symbolize portions of the polyester chains and the dashes represent styrene molecules.

Air inhibition

We have just described the process occurring in the body of the coating. However, at the surface of the paint a different story unfolds:

As explained in Chapter 12, polyperoxides are stable and styrene polyperoxides will not initiate the cross-linking copolymerization. They decompose eventually to aldehydes. Unreacted styrene (BP 146 °C) evaporates from the surface, leaving only the sticky polyester resin and styrene polyperoxides behind. The surface will never harden.

There are two solutions to this problem. One is to include a very small proportion of paraffin wax, which is soluble in styrene, but becomes insoluble as the finish cures. It separates to the surface of the film, forming a continuous layer which (1) prevents oxygen from reacting with the styrene free radicals and (2) prevents evaporation of styrene. The wax gives a matt surface which is acceptable for lower gloss finishes. If a higher gloss finish is required, it must be removed by mechanical sanding, followed by mechanical polishing to a hard glass-like surface.

The other solution is chemical. Groups which dry by an oxidative mechanism are built into the polyester resin, so that oxygen causes surface hardening, even though styrene polyperoxide is formed. Unsaturated fatty acid groups are not used, but instead the shorter **allyloxy group**: $CH_2=CH-CH_2-O-$. This behaves similarly to non-conjugated unsaturation because it contains a methylene group with a double bond on one side and an oxygen atom on the other. The oxygen atom is like a double bond in that both are strongly electron-attracting. The methylene group is very prone to oxidation by oxygen (p. 172), when hydroperoxide forms and oxidative drying proceeds as in Chapter 12. In addition, the allyl $(-CH=CH-CH_2-)$ double bond will polymerize slowly with the other ingredients by chain growth polymerization. The allyloxy group can be built into the polyester chain by using, as part of the original glycol, an allyl ether of a polyol with three or more hydroxyl groups, e.g.

$$CH_2 \cdot O \cdot CH_2 \cdot CH=CH_2 \qquad\qquad CH_2 \cdot O \cdot CH_2 \cdot CH=CH_2$$
$$|\qquad\qquad\qquad\qquad\qquad\qquad\quad |$$
$$CHOH \qquad\qquad\qquad CH_3 \cdot CH_2 \cdot C \cdot CH_2OH$$
$$|\qquad\qquad\qquad\qquad\qquad\qquad\quad |$$
$$CH_2OH \qquad\qquad\qquad\qquad CH_2OH$$

glycerol monoallyl ether   trimethylol propane monoallyl ether

The disadvantage of this solution lies in the long surface hardening time. However, cure is complete in a few minutes if infrared heating and waxy additives are used and promoters, such as ethyl acetoacetate, $CH_3 \cdot CO \cdot CH_2 \cdot CO \cdot O \cdot C_2H_5$, are included to enhance the action of the cobalt. The allyloxy compounds also raise the cost of the finish. Nevertheless, this solution to air inhibition is generally preferred for matt and semi-matt clear and pigmented finishes.

Unsaturated polyester finishes

We have covered the main reactive ingredients, so that the final thermally cured finish may be summarized thus:

Finish pack	*Activator pack*
Pigment (if required)	Peroxide
Polyester resin	Solvent (to adjust mixing ratio)
Monomer	
Inhibitor	
Accelerator (if required)	

As wood finishes, thick films (125–250 μm) can be applied to verticals in one coat without sagging, by the inclusion of the usual thickening additives. The resin solution is easily pigmented, but care must be taken to avoid overheating (leading to gelation) in the milling process.

For refinish filler, similar high build (250 μm) can be achieved by applying 60–70 μm coats in successive passes. Otherwise, primer surfacer coats of 50 μm may be applied in one double or two single coats.

Pot life and methods of activation

If slow hardening is acceptable, or if heat can be used to cure the finish, the simple two-pack arrangement described above will give pot lives long enough to allow the user to finish a large number of articles or a large area with one batch of activated finish. But polyester finishes are capable of really rapid hardening at room temperature.

Both car refinishers and mass production furniture makers require rapid hardening. Touch-dry times of 15–30 min are not uncommon, and mechanical sanding and, in the case of furniture, polishing can be done after 2–4 h. To achieve this hardening rate, the peroxide and accelerator levels must be adjusted so that the pot life of the above mixture is only *10–20 min*. The finish must be activated in small separate batches and spray-guns must be rinsed out scrupulously after each batch has been used, to prevent blockages due to gelled finish. To avoid the drawbacks of this 'pot-mix' system, for woodfinishing several alternative activation methods have been devised, of which one is the contact process.

Contact process

In this process the peroxide is added to a simple, ultra-low solids, non-reactive lacquer. A thin coat of lacquer is applied. This is followed by polyester finish containing accelerator. Free radicals again form at the inter-coat boundary and curing will proceed, even through a second coat of polyester applied several minutes after the first. This process completely avoids the pot life problem – a big gain – but it has several disadvantages. Delayed solvent evaporation from the lacquer (or basecoat) can cause the entire finish to 'sink' after a few weeks' aging, showing up wood grain pattern, for example, in what was previously a smooth surface. In furniture finishing, special peroxide-resistant dyes must be used in the stains applied under the basecoat. If an appreciable area is 'missed' by the basecoat, either curing will not occur in that area or the area will be comparatively soft. Finally, the basecoat dry film must be thin to minimize sinkage. Therefore basecoat solids must be low and a lot of wasted solvent must be paid for by the customer.

Nevertheless, this is the most widely preferred of all methods. Nowadays, when so much furniture is made from pre-finished flat boards, the basecoat can be applied accurately in thin coats by roller coater and thus can contain higher solids. The finish is then applied by a curtain coater, which is positioned immediately after the roller coater on the production line.

General properties of polyester finishes

Polyester finishes are very hard, tough, resistant to solvents and resistant to moderately hot objects (e.g. a teapot or a lighted cigarette). These are extremely useful properties for a furniture finish, clear or pigmented.

Polyester films are not flexible, in the ordinary sense, unless soft. Sufficient flexibility to withstand the expansion and contraction of wood substrates as the humidity varies is obtained, with hardness, by spacing out the cross-links. This can be done by decreasing the proportion of unsaturated acid in the polyester resin, or by using long-chain, flexible saturated acids or alcohols.

The ordinary maleic–phthalic–propylene glycol wood finish compositions have poor resistance to alkali, since the ester linkages are easily saponified, but more resistant polymers can be made at higher cost from isophthalic acid and from the reaction product of propylene oxide and bisphenol A, shown below.

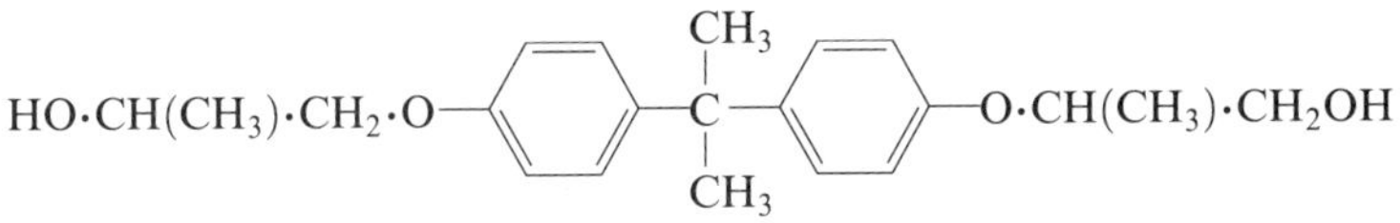

$$\text{HO·CH(CH}_3)\text{·CH}_2\text{·O} \overset{\displaystyle \text{CH}_3}{\underset{\displaystyle \text{CH}_3}{\text{C}}} \text{O·CH(CH}_3)\text{·CH}_2\text{OH}$$

Radiation curing

On p. 239 ultraviolet light was listed as one cause of free radical formation. Polyester clear coatings stored in untinted glass bottles will polymerize in days or weeks. This process can be accelerated considerably and used as a method of curing the coating. This can be done by including in the coating a substance which is thermally stable, but which is readily decomposed to free radicals in UV light. Examples of such substances, which are called photo-initiators, are given below:

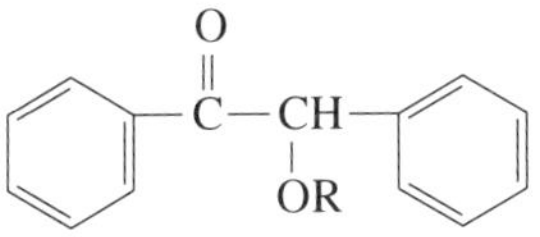

benzoin ethers 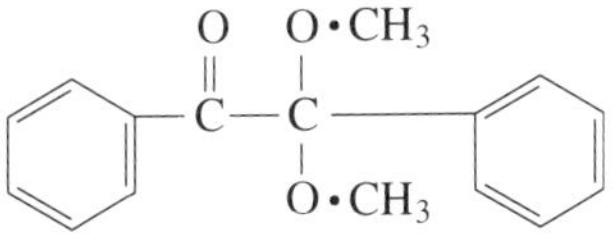   benzil dimethyl ketal

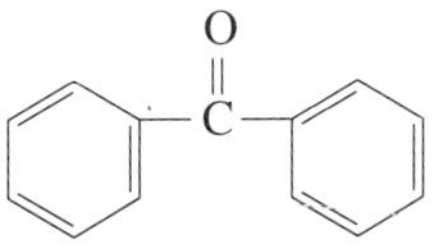

benzophenone
(used with tertiary amines)

Powerful UV lamps, available in tube form, are now rated up to $100 \, \text{W cm}^{-1}$ of lamp length, and these can be mounted over a conveyor on which flat, coated articles (e.g. wood-based panels, metal sheet, paper or card) are carried. Hardening of the coating can be achieved by a few seconds of exposure to such lamps, because of the very efficient conversion of UV energy to free radicals. Yet the coating, being thermally stable, is a one-pack product.

The unsaturated polyester coatings of the type earlier described are nowadays mostly replaced with other resin types of lower viscosity. Instead of an unsaturated polyester resin, lower molecular weight polymers may be used, often called acrylic **oligomers**. These oligomers are for example the reaction products of acrylic acid and end groups in epoxy resins (**epoxy acrylates**) and in saturated polyesters (**polyester acrylates**). The linking reactions are shown diagrammatically:

$$\boxed{\text{EPOXY}} \!-\! \underset{\displaystyle \overset{O}{\triangle}}{CH-CH_2} \; + \; \underset{\text{acrylic acid}}{HOOC\cdot CH=CH_2} \; \longrightarrow$$

$$\boxed{\text{EPOXY}} \!-\! \overset{\displaystyle OH}{CH} \!-\! CH_2OOC\cdot CH=CH_2$$

$$\boxed{\text{POLYESTER}} \!-\! OH \; + \; \underset{\text{acrylic acid}}{HOOC\cdot CH=CH_2} \; \longrightarrow \quad \boxed{\text{POLYESTER}} \!-\! OOC\cdot CH=CH_2$$

where two or sometimes more acrylate groups may be attached to the epoxy or polyester.

Another more varied class termed **urethane acrylates** are obtained by linking the hydroxy monomer, such as hydroxyethyl acrylate, and a polyol, normally oligomeric, with a diisocyanate. Again this is shown diagrammatically:

$$\boxed{\text{OLIGOMERIC POLYOL}} \!-\! OH \; + \; OCN\!-\!R\!-\!NCO \; + \; \underset{\text{hydroxyethyl acrylate}}{HOCH_2CH_2OOC\cdot CH=CH_2}$$

$$\downarrow$$

$$\boxed{\text{OLIGOMERIC POLYOL}} \!-\! O\overset{\displaystyle O}{\overset{\|}{C}}\!-\!\overset{\displaystyle H}{\overset{|}{N}}\!-\!R\!-\!\overset{\displaystyle H}{\overset{|}{N}}\!-\!\overset{\displaystyle O}{\overset{\|}{C}}OCH_2CH_2OOC\cdot CH=CH_2$$

and typically, two or more acrylate groups are attached to the oligomer.

The oligomeric polyol may be a low MW flexible polyester or a polyether (pp. 228–9). Urethane acrylates can possess a range of properties dependent on the contributions of the softer polyol centre block and the outer urethane hard segments.

All of these oligomer types are used dissolved in acrylic monomers. Suitable ones have low volatility, low odour and are not severe skin irritants. Generally these are di- or triacrylate esters of typical polyols, e.g.

$$\begin{array}{l} CH_2\cdot CH_2\cdot CH_2\cdot O\cdot CO\cdot CH=CH_2 \\ \;\;| \\ CH_2\cdot CH_2\cdot CH_2\cdot O\cdot CO\cdot CH=CH_2 \end{array}$$
1,6-hexane diol diacrylate

$$\begin{array}{l} \quad\quad\quad CH_2\cdot O\cdot CO\cdot CH=CH_2 \\ \quad\quad\quad\;\;| \\ CH_3\cdot CH_2\cdot C-CH_2\cdot O\cdot CO\cdot CH=CH_2 \\ \quad\quad\quad\;\;| \\ \quad\quad\quad CH_2\cdot O\cdot CO\cdot CH=CH_2 \end{array}$$
trimethylol propane triacrylate

Acrylates of polyols which have been reacted with ethylene or propylene oxides are now frequently used. This includes acrylates of the reaction product of propylene oxide and bisphenol A, shown earlier (p. 245).

Methacrylate oligomers and monomers as well as allyl and vinyl monomers can also be used, but rates of cure decline in the order

$$\text{acrylates} > \text{methacrylates} \gg \text{allyl} > \text{vinyl}$$

Acrylate coatings are generally harder than polyester coatings and can be tougher or more flexible in thin films. Ingredients can be chosen to be substantially involatile at room temperature, while styrene is similar in volatility to xylene. Acrylates and methacrylates are, however, just as prone to air inhibition as polyester–styrene systems.

In free radical curing of UV coatings, air inhibition is usually overcome by creating so many free radicals in such a short time that the oxygen is 'swamped'; although polyperoxides are formed, many successful carbon–carbon chain propagations also can occur. This kind of 'overkill' in the chemistry of the coating can be expensive and equipment has been developed to cure coatings with UV under a blanket of nitrogen. The complexities and costs of such equipment have not met with general favour. However, the use of an inert gas blanket is accepted as an integral part of radiation curing initiated by exposure to electron beams.

In **electron beam curing**, flat paint films are passed under an electron source, so that energetic electrons impact with the coating over its full width. Coating compositions are very similar to those used with UV sources (the order of reactivity of the chemical types is the same), but they do not contain free radical initiators of any type. Instead, the bombardment of high energy electrons creates free radical sites at random in the coating and polymerization is so rapid that cure takes place in a fraction of a second, provided that oxygen is excluded by a nitrogen blanket. While electron accelerators are very expensive, one significant success of this technique has been in the high volume printing of cartons for beverages.

Most pigments absorb UV light quite strongly and for this reason, in paint applications, UV curing has been largely confined to clear coatings for heat-sensitive substrates. Other successful applications have been in lithographic and silk screen inks, since inks, though heavily pigmented, are applied in very thin layers (2–3 μm). Electron beam curing is, however, equally effective with clear and pigmented coatings.

UV curing of epoxies

The epoxy ring

$$\overset{\displaystyle O}{\overset{\displaystyle /\diagdown}{-CH-CH_2}}$$

looks somewhat like a carbon–carbon double bond $-C=C-$, so it may not be surprising to learn that it is capable of chain growth polymerization by a chain reaction. Free radicals are not, however, involved. Normal free radical initiators (e.g. organic peroxides, Chapter 5) do not bring about the polymerization. Polymerization is brought about by Lewis acids (e.g. BF_3), but it is so rapid and vigorous that they cannot be used as paint activators. Instead, we require complex salts with large, unstable organic cations and large, generally inorganic anions, e.g.

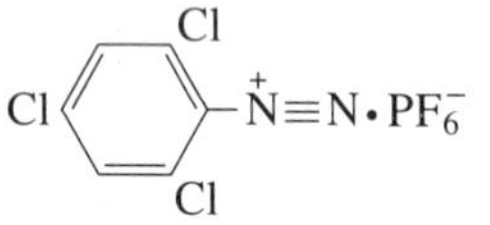

2,4,6-trichlorobenzene diazonium   di-(*p-t*-butylphenyl) iodonium
hexafluorophosphate   hexafluoroantimonate

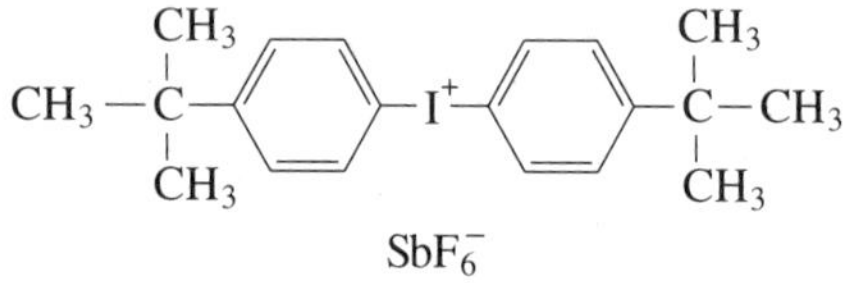

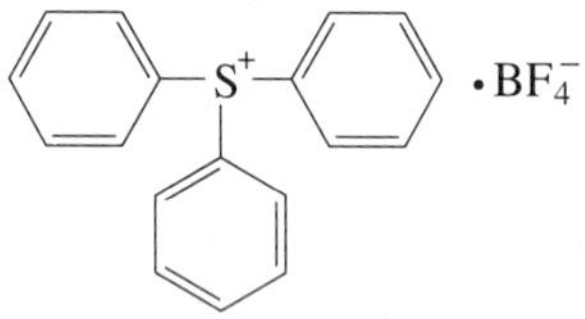

triphenyl sulphonium tetrafluoroborate

All of the above 'onium' salts are relatively stable when included in epoxy coatings but, on exposure to UV light, they are decomposed to form reactive cations. The full mechanism of cationic polymerization processes is complex; a complication in using these processes is that bases, water and smaller anions, e.g. Cl^-, which readily attack positive sites in organic compounds, inhibit polymerization. We may illustrate the cationic polymerization of epoxies by assuming that initiation occurs via a proton:

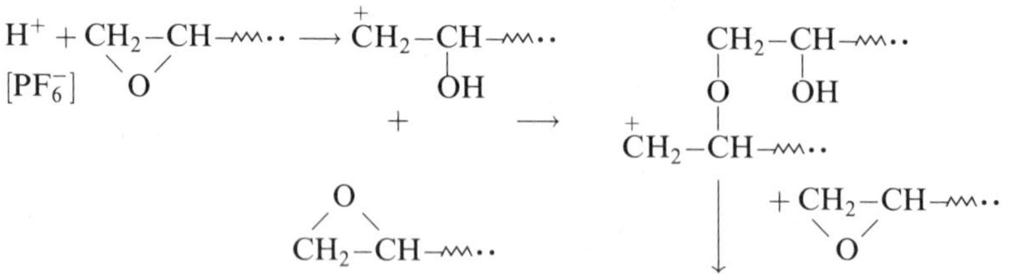

and so on

After decomposition of the photo-initiators by UV light, polymerization is very rapid and films may be substantially cured in less than a second. Further polymerization can occur at room or higher temperatures after UV illumination has ceased.

The epoxy novolac, cyclo-aliphatic epoxy and epoxide-functional acrylic resins mentioned in Chapter 14 may be used to form coatings with good adhesion to metal and are particularly suitable for the exterior surfaces of cans. Cyclo-aliphatic epoxides give fastest cure, novolacs have the highest epoxy functionality and dihydric alcohol diepoxides impart flexibility. Viscosity is reduced by monoepoxides and small proportions of alcohols (which terminate in chain growth). The polymerization is not inhibited by oxygen, but is inhibited by moisture. The coatings need not contain any significant amount of volatile material, i.e. they are almost 100% solids.

Examples of unsaturated polyester and UV-curing acrylic coatings are given in Tables 16.1 and 16.2.

Table 16.1 Formulation of clear wood finish (with wax)

Varnish		wt%	
Thickener	Fine particle silica	1.5	} Disperse on roller mill
Resin solution	{ Polyester resin (maleic–phthalic–propylene glycol) at 67% solids in styrene	35.8	
	{ Polyester resin solution (as above)	40.7	
Monomer	Styrene	15.2	
Additives	{ Wax solution in styrene (2% wax)	5.0	
	{ Cobalt octoate (6% Co)	1.2	
	{ 2% Hydroquinone in acetone	0.6	
		100.0	

Activator		
Peroxide	50% Methyl ethyl ketone peroxide in dibutyl phthalate	25
Solvent	Ethyl acetate	75
		100

Mixing ratio (for normal spray application, pot-mix or dual feed):

Varnish/activator, 10:1 by volume

Pot life: 10 min
Fully hardened: 4 h } 15–20 °C

Alternatively, the finish may be applied over the following basecoat:

Basecoat		
Resin	{ HX30–50 (or RS $\frac{1}{2}$ s) nitrocellulose (dry weight)	4.5
	{ Cyclohexanone resin[a]	4.5
Peroxide	Cyclohexanone peroxide powder	6.0
Solvents	{ Methyl isobutyl ketone	17.0
	{ Acetone	50.0
	{ Ethyl acetate	16.0
	{ Isopropyl alcohol	2.0
		100.0

Spraying solids: 15%

The polyester finish should be thinned 12:1 by volume (varnish:thinner) with a 20% solution of dibutyl phthalate in methanol

[a]Cyclohexanone resin is made by polymerizing the ketone, with or without aldehyde, by heating under pressure with alkali.

Table 16.2 Formulation of low gloss UV-curing varnish for wood or paper

		wt%
Matting aid	Fine particle silica Syloid 166	9.0
Oligomer solution	Ebecryl 608	34.0
Monomers	{ OTA 480	26.0
	{ Tripropylene glycol diacrylate	26.0
Photo-initiators	{ Benzophenone	3.0
	{ Irgacure 651	2.0
		100.0

Apply by roller coater at 100% solids.
Cure under one UV lamp ($80\,W\,cm^{-1}$ of length) at conveyor speed 20–$40\,m\,min^{-1}$.
The following materials in the formula are made by U.C.B. s.a.:
Ebecryl 608 is an epoxy-acrylate at 75% in OTA 480.
OTA 480 is a polyol triacrylate.
Irgacure 651 is a photo-initiator (benzyl dimethyl ketone acetal) supplied by CIBA Specialty Chemicals.

Seventeen

Chemical treatment of substrates

In Part Two of this book so far we have considered the general principles by which all paints are formulated, the non-resinous ingredients that are used and, finally, six groups of paints classified by their drying mechanisms.

All these coatings are applied to a substrate to produce a paint system, e.g. substrate, primer, undercoat, topcoat. Whether the overall result is satisfactory will depend on how these layers adhere together, how they perform individually and where the weakest link is.

Sometimes the weakest link is at the substrate–primer interface (e.g. poor adhesion), or in the substrate itself (e.g. wood suffering from wet rot). For this reason it is necessary now to consider the substrate itself and its particular chemistry.

Before we can paint the substrate to produce a successful system, it may be necessary to treat the substrate in some way to upgrade its performance in that system. Such treatments are designed either (1) to improve the substrate or (2) to prevent its degradation.

There are numerous reasons why it may be necessary to improve the substrate. It may be poorly produced, e.g. too rough to finish. It may have a natural weakness, e.g. chipboard is too porous. It may have been damaged or contaminated, e.g. dented or greasy metal. It may have deteriorated during storage, e.g. rusty steel.

Many of these problems can be dealt with by physically treating the substrate to produce a clean, level, sound surface suitable for painting. Such treatments include the planing and sanding of wood, the degreasing and sand-blasting of steel, and the filling and sealing of damaged or porous surfaces. Because they are essentially physical in nature, these treatments will not be discussed further in this chapter. We shall confine our attention to those cases in which a chemical treatment of the substrate is necessary, either for improvement of it, or more often to prevent its degradation.

In particular, we shall look at the improvement of plastic surfaces for better paint adhesion, the protection of wood against biodegradation and the protection of metals against corrosion.

Improving plastics for paint adhesion

Different plastics present different painting problems but, on the whole, once clean they do not often present problems of paint adhesion. Exceptions are the polyolefin plastics, e.g. polyethene and polypropylene. Because these polymers lack polar groups, they have low energy surfaces which are difficult to wet and are not readily penetrated by solvents. In addition to this, the surface layer of the plastic, a few nanometres thick, is often different to the bulk, being low in molecular weight and weak. Paint adhering to this layer will pull the layer away and peel off, thus appearing to have poor adhesion.

To obtain good results on these plastics, it is necessary to modify the surface of the plastic, to cross-link the weak layer into the plastic below and to oxidize it, thus creating polar groups and a higher energy surface. The preferred treatments for doing this are corona discharge (for plastic films) and flame treatment (for moulded articles).

In **corona discharge** the plastic film passes at $1-2\,\mathrm{m\,s}^{-1}$ over an insulated metal roller, which is electrically earthed and $1-2\,\mathrm{mm}$ under a bar electrode. Through the electrode passes an AC current oscillating at high frequency (about $20\,000\,\mathrm{Hz}$) and at a high voltage (about $20\,000\,\mathrm{V}$). From this electrode there is a continuous discharge towards the plastic film and earthed roller. Sparks can be seen, but additionally a *plasma* is produced. A plasma is a gas (in this case, air) in which the gas molecules are broken down into individual atoms, free radicals, ions, electrons and photons (in this case, by the high energy discharge).

These highly reactive particles impact with the plastic surface, oxidizing it and disrupting C–H and C–C bonds to a depth of $5-50\,\mathrm{nm}$. Subsequent recombinations of the C· free radicals can cross-link the weak surface layer into the bulk polymer and the new surface will contain polar groups, such as $C{=}O$, $C{-}OH$, $COOH$, $C{-}O{-}NO$ and $C{-}O{-}NO_2$. These changes are reflected in a rise in the critical surface tension of the plastic from e.g. 31 to $47\,\mathrm{mN\,m}^{-1}$. This surface is more readily wetted by paints. Adhesion is improved because surface layers are no longer weak and because of strong attractions between polar groups in paint and plastic (p. 95).

In **flame treatment** the oxidizing portion of a gas flame ($6-10\,\mathrm{mm}$ from the tip of the blue inner cone) is brought into contact with the surface of the plastic for a brief period ($0.02-0.1\,\mathrm{s}$). The flame temperature is high ($1100-2800\,^{\circ}\mathrm{C}$) and the flame is a plasma. Thus the plastic is modified by oxidation by the plasma and by recombination of free radicals created in the polymer with free radicals in the plasma. Changes are observed to a depth of $4-9\,\mathrm{nm}$ and wetting and adhesion are improved. This is a popular technique for improving the adhesion of printing inks to the surfaces of moulded polyolefin containers.

The above techniques are generally preferred, but any means of surface oxidation can be effective, such as treatment with oxidizing acids (strong sulphuric, chromic) or halogens, or treatment with photo-initiators (p. 245) followed by irradiation with UV light.

Protection of wood against biodegradation

All surfaces are continually exposed to bacteria and the spores of fungi. The latter are particularly destructive to wood. The fungi which cause the most damage to wood are the members of the Basidiomycetes family, which includes both brown and white rots. Brown rot is so called because it attacks only the cellulosic material in wood, leaving behind the lignin which is brown in colour. White rots, on the other hand, destroy both cellulose and lignin. Commonly encountered members of this family include *Serpula lacrymans* (a brown rot) commonly known as 'dry rot' and the wet rots *Coniophora puteana* (brown rot) and *Phellinus contiguous* (a white rot). Two other families are also important in the degradation of wood: these are the ascomycetes and the deuteromycetes (collectively known as soft rots). Apart from those fungi which cause structural degradation, many other fungi cause the surface disfigurement known as bluestain or sapstain. This drastically reduces the value of timber.

Perfectly sound-looking wood is almost sure to have spores on its surface and there may even have been penetration into the cells below. Even if the painter is very observant, he or she may well miss signs of degradation already present in the wood surface and covering organisms with paint will not prevent trouble later. This is especially so if design deficiencies in the wood structure aid penetration of water, encouraging the moisture content of the wood to rise above 20%, the level necessary for the fungi which cause rot to become active. They will then grow and multiply under the coating, feeding on the wood beneath and causing rot damage, or erupting through the coating to form pinhole damage and cause staining.

However, fungi can be prevented from attacking the wood by treating it chemically and keeping its moisture content to a minimum. Chemical treatment may be solvent-borne or water-borne. Active ingredients include pentachlorophenol, tributyltin oxide and copper and zinc naphthenates (solvent-borne) and mixtures of copper sulphate, sodium or potassium dichromate, and hydrated arsenic pentoxide (water-borne). These act in a variety of ways to destroy or inhibit the growth of fungi.

It is equally important to get these preservatives deep into the wood. For best results special equipment is required and the job is best done in the factory after seasoned pieces have completed all stages of machining. The most effective techniques use vacuum and/or pressure to displace air from the cells and force preservative into them. If remedial action has to be taken outside a factory, pressure injection, dipping, brushing and spraying may be used (in order of decreasing penetration).

Special low-viscosity preservative-containing primers can also add markedly to the life of an exterior paint system by destroying organisms on or in the surface layers of the wood and also by penetrating open joints to seal end-grain, thus helping to keep the moisture content of the wood as low as possible.

Protection of metals against corrosion

Corrosion is essentially the conversion of metal to a hydrated form of oxide. In the presence of oxygen and water, the overall corrosion reaction of steel is

$$4Fe + 3O_2 + 2H_2O \longrightarrow 2Fe_2O_3 \cdot H_2O$$

On pages 17 and 18, electrolysis at inert electrodes was discussed. If the electrodes are not inert, then at the anode metal dissolution occurs and metal cations are formed, e.g.

$$Zn - 2e \longrightarrow Zn^{2+}$$

The surface of steel is never uniform and, if it is in contact with a thin layer of aqueous electrolyte, then small electrical imbalances from site to site will lead to the formation of an electrolytic cell. At anodic sites, the anodic reaction

$$4Fe - 8e \longrightarrow 4Fe^{2+} \text{ (oxidation)}$$

occurs. The cations formed migrate through the electrolyte towards more negative, cathodic sites. At these, the *cathodic reaction*

$$4H_2O + 2O_2 + 8e \longrightarrow 8OH^- \text{ (reduction)}$$

is occurring and the anions produced migrate from thence towards the anodic sites. Within the metal, electrons flow from anodic to cathodic sites and within the electrolyte, migrating ions meet to form soluble ferrous hydroxide:

$$4Fe^{2+} + 8OH^- \longrightarrow 4Fe(OH)_2$$

When sufficient oxygen is present, this is oxidized to insoluble hydrated ferric oxide:

$$4Fe(OH)_2 + O_2 \longrightarrow 2Fe_2O_3 \cdot H_2O + 2H_2O$$

Thus it can be seen that corrosion (dissolution of the metal) is an electro-chemical process. Rusting does, however, require an additional oxidation step. If the final corrosion product can be formed as a strongly adherent, insoluble and impermeable layer on the metal surface, corrosion will decline, but if it is loose or permeable, the corrosion process will continue. In real life, the above processes are modified by the presence of other chemical species, even in trace amounts, and on steel the layer is commonly loose and permeable.

In fact, corrosion can be inhibited by one of the following techniques:

• keeping the surface completely dry, so that no conducting electrolyte can form on it;
• starving cathodic areas of oxygen;

- forming a film impermeable to electrons on cathodic sites;
- Forming a film impermeable to metal cations on anodic sites.

Coating metal with paint might be assumed to be taking preventive action by all four mechanisms, but in fact no paint is *completely* impermeable to water or oxygen, so that after an initial delay, barrier properties are lost and corrosion eventually begins. It is therefore necessary either to include inhibitive ingredients within the paint (anti-corrosive pigments) or to apply an inorganic chemical pretreatment before painting, or to do both. The objectives of using inorganic chemical pretreatments, often called **conversion coatings**, are twofold:

- to *passivate* the surface by forming upon it a relatively stable, strongly adherent, corrosion-inhibiting layer;
- to provide a surface to which paint coatings readily adhere.

The conversion coatings are thin and easily damaged. Subsequent layers of paint protect the films from damage and, if they contain inhibiting pigments, repair faults and damage where they occur. The overall result of paint plus pretreatment is often prevention of corrosion for years.

The use of inorganic chemical pretreatments on metal will now be illustrated by considering the chromating of aluminium and the phosphating of steel.

Chromating of aluminium

Aluminium alloys are widely used in aircraft construction, extrusions for building purposes, skins of caravans (aluminium coil) and beverage cans. In each case they are protected by organic surface coatings on interior and exterior surfaces. Prior to coating, the metal is pretreated to improve corrosion protection, normally with a chromate treatment. The most widely used of the well established chromating treatments are the amorphous chrome phosphate process.

Amorphous chromate treatments contain hexavalent chromium (chromates, CrO_4^{2-}, and dichromates, $Cr_2O_7^{2-}$) and fluoride (F^-) and may contain an accelerator to speed up the reaction. They are strongly acidic (pH 1.5–2) and operate at temperatures between 20 and 40 °C. These treatments are preferred for aircraft components and coil-coated aluminium and produce a golden yellow-coloured coating. Coating weights vary between 100 and 600 $mg\,m^{-2}$.

Amorphous chrome phosphate treatments contain hexavalent chromium fluoride and phosphate and operate at about pH 1 and 40–50 °C. They are preferred for can coatings (especially can end stock) and give colourless to green films with coat weights of 50–500 $mg\,m^{-2}$.

Not only are there a wide variety of commercial chromate formulations in use, but the conversion coatings formed also differ in composition and experts disagree about the detailed chemistry. There is, however, general agreement that in amorphous chromate treatments the following reactions are involved:

- Acid attack upon the metal:

$$2Al + 6H^+ \longrightarrow 2Al^{3+} + 3H_2$$

- Reduction of hexavalent chromium to trivalent chromium:

$$3H_2 + 2Cr_2O_7^{2-} \longrightarrow 2Cr(OH)_3 + 2CrO_4^{2-}$$

- Formation of hydrated aluminium oxide:

$$2Al^{3+} + 4H_2O \longrightarrow Al_2O_3 \cdot 3H_2O + 6H^+$$

- Formation of aluminium chromate:

$$2Al^{3+} + 3CrO_4^{2-} \longrightarrow Al_2(CrO_4)_3$$

- Formation of complex chromic chromates:

$$2Cr(OH)_3 + CrO_4^{2-} \longrightarrow Cr(OH)_3 \cdot Cr(OH)CrO_4 + 2OH^-$$

The role of the fluoride is to assist dissolution of the aluminium, even when in oxide form.

The amorphous chrome phosphate process forms coatings consisting largely of aluminium and chromium phosphates (see Phosphating below). It is preferred by the can industry because only trivalent chromium is found in the formed coating. In view of growing concerns about the toxicity of hexavalent chromium, this is an essential feature for containers for foodstuffs.

Concerns about toxicity and effluent disposal have led to other approaches for the pretreatment of aluminium. **No-rinse pretreatments** contain hexa- and trivalent chromium with organic polymer or silica and are applied at controlled film weights with no subsequent removal of excess by rinsing. They therefore present no effluent disposal problem. **Chrome-free pretreatments** (essentially special phosphating treatments) avoid chromium in the workplace altogether and have been widely adopted for two-piece beer and beverage cans.

Phosphating of steel

The main phosphate pretreatments for steel are iron phosphate and zinc phosphate types. Zinc phosphates are preferred for best corrosion resistance out-of-doors.

Iron phosphate coatings are amorphous coatings and very thin (0.1–1 g m^{-2}). The working solution contains primary sodium or ammonium phosphates, together with other ingredients which may include an oxidizing accelerator, and surfactants to combine degreasing with the chemical treatment. The pretreatments operate between pH 3 and 5.5 at temperatures between 50 and 80 °C. The process begins with attack on the steel, forming primary ferrous phosphate:

$$8Fe + 16NaH_2PO_4 + 8H_2O + 4O_2 \longrightarrow 8Fe(H_2PO_4)_2 + 16NaOH$$

About a half of this is oxidized to ferric phosphate, which is insoluble and deposits:

$$4Fe(H_2PO_4)_2 + 4NaOH + O_2 \longrightarrow 4FePO_4 + 4NaH_2PO_4 + 6H_2O$$

The other half of the ferrous phosphate is converted to ferric hydroxide as the pH rises:

$$4Fe(H_2PO_4)_2 + 12NaOH + O_2 \longrightarrow 4Fe(OH)_3\downarrow + 4NaH_2PO_4 + 4Na_2HPO_4 + 2H_2O$$

and on drying:

$$4Fe(OH)_3 \longrightarrow 2Fe_2O_3 + 6H_2O$$

The final coating contains about 40% ferric oxide and the rest is ferric phosphate.

Zinc phosphate coatings are crystalline and somewhat thicker ($1-5\,g\,m^{-2}$). The bath or spray contains primary zinc phosphate, $Zn(H_2PO_4)_2$, phosphoric acid and oxidizing accelerators. The pretreatments operate between pH 1.5 and 3.3 at temperatures between 25 and 90 °C. The process begins with acid attack on the steel, oxidation (assisted by the accelerator) and precipitation of some ferric phosphate as above. However, the zinc phosphate is in a finely balanced equilibrium with the other species:

$$4Zn^{2+} + 3H_2PO_4^- \longrightarrow ZnHPO_4 + Zn_3(PO_4)_2\downarrow + 5H^+$$

The removal of even small amounts of hydrogen ion by acid attack on the steel:

$$Fe + 2H^+ \longrightarrow Fe^{2+} + H_2$$

upsets the equilibrium locally. As a result, the reaction moves to the right to create more hydrogen ion, producing as it does so sparingly soluble secondary zinc phosphate and insoluble tertiary zinc phosphate. The latter deposits at the sites where reaction has taken place, forming a protective crystalline layer, mainly hopeite, $Zn_3(PO_4)_2 \cdot 4H_2O$, though some phosphophyllite, $Zn_2Fe(PO_4)_2$, is also formed in the layers nearer the substrate.

Both iron and zinc phosphate coatings contain small amounts of porosity, probably caused by the escape of hydrogen in the earlier stages of phosphating. The free pore area can be reduced by rinsing in hot dilute (0.01–0.05%) solutions of chromic acid. Insoluble iron chromates are formed in the pores, passivating these areas and increasing the corrosion resistance almost fourfold. Thus the protective properties of phosphating and chromating are combined. However, in many cases the phosphate pretreatment is designed to give good results without the chromate rinse, either for reasons of economy, or because of concerns about the toxicity of hexavalent chromium and the disposal of the effluent. Areas such as car manufacturers now look towards nickel and chrome free treatments.

Iron phosphating is widely used on panel radiators, refrigerators and other items not subject to severe exterior exposure. Zinc phosphating is used to protect motor car bodies and other articles which require maximum corrosion protection. Since phosphating works well on zinc, it is used on electro-zinc for washing machines and on hot-dipped galvanized steel coil-coated for the cladding and roofing of buildings. The use of zinc-coated steels in automobile bodies is also increasing rapidly, so that phosphate pretreatments have to be designed to treat both zinc and steel surfaces.

Eighteen

Health, safety and environment

This chapter brings together issues which have already been touched on earlier in this book. These are concerned with the health and safety of the individual and the protection of the environment from all aspects of paint preparation and usage. Some source books are listed in the Appendix.

Health and safety

Personal health and safety

In the UK the *Control of Substances Hazardous to Health Regulations (COSHH)*, updated in 1994, require all persons at work to know the safety precautions to take, so as not to endanger themselves or others through exposure to substances hazardous to health. This applies to the manufacture, testing and use of paint.

A wide range of materials are used in paint making and a variety of chemical reactions occur in both the preparation and the use of paint. It is essential to think of all ingredients as chemicals which have hazards associated. For example, short term exposure to acids and alkalis can cause corrosive burns, a variety of materials can cause skin and eye irritation, and dusts and vapours can give rise to lung irritation. Inhaling higher levels of solvent vapour can cause short term intoxicating effects and extended skin contact can result in disorders such as dermatitis. Longer term exposure to a number of chemicals can result in cumulative harmful effects or sensitization.

Many of these dangers can be prevented by sensible general precautions, and some with extra care because of specific dangers. Good general practice is to avoid unnecessary exposure to chemicals. Contact can be made with skin and eyes during handling, with the lungs by breathing dust or vapour, and by swallowing after transfer from the hands. These can be prevented by control of the work environment and by use of protective clothing. Basic precautions are always to work in conditions of good ventilation, to use some form of skin and eye protection and to wash before eating or drinking. The strictest precautions are those required against cancer-causing chemicals, though these are not likely to be present in paint supplied for application.

Precautions during the application of paint depend on the methods of application and curing used. Minimizing contact with solvent vapour requires air extraction; spray operations will be carried out in spray booths where contaminated air and excess spray droplets are treated. For some materials such as isocyanates, spray operators may be required to wear air-fed hoods. Factory environments are now regulated for all emissions including vapours, dusts and liquid and solid wastes, and these must be monitored and controlled.

Substances already noted as hazardous are lead and chromate pigments (Chapter 8), solvents (Chapter 9), lead driers (Chapter 12), formaldehyde (Chapter 13) and isocyanates (Chapter 15). Though hazards were not specifically mentioned, users also need to be aware of the toxicity of fungicides and algicides (Chapter 10), the general unpleasantness of unsaturated monomers (particularly the potential of acrylates as skin irritants), the dangers of peroxides (Chapters 13 and 16) and finally the hazards of low molecular weight amines and epoxides (Chapter 14). Although not a chemical hazard, but requiring care, is exposure to UV radiation (Chapter 16), which has the potential to cause severe eye and skin damage. For many substances there are maximum permitted exposure levels in the workplace, defined by Occupational Exposure Limits (OELs) or TLVs (p. 135). For some materials, e.g. lead and isocyanates, regular health checks are a legal requirement; those with allergic conditions may be restricted from handling isocyanates (both in paint use and in resin making) and anhydrides (used in resin making), since these are respiratory sensitizers.

Prevention of fire and explosion

Sources of ignition must be avoided when handling flammable materials, and care must be taken that only approved (non-sparking, low surface temperature) electrical equipment is used. The **Lower Explosive Limit** (LEL) for a solvent is the minimum concentration of that solvent in an air–vapour mixture through which flame will spread from an ignition source. These limits for solvents are published and ventilation must be arranged to keep solvent concentrations well below the LELs (usually solvent levels are, in any case, kept well below LELs for personal safety reasons). In addition, the dusts from many organic and inorganic powders can form explosive mixtures with air. When nitrocellulose is being handled, addition methods must be carefully controlled and non-sparking metal tools should also be used. Finally, precautions against static electricity must always be taken when larger quantities of bulk solvent are being handled, by using earthing straps on containers. Conductive shoes or shoe earth-straps should also be worn by factory workers.

Written information on hazards

In the UK, regulations including the *Chemical Hazard Information and Packaging Regulations (CHIP 2)* require that all hazardous substances, whether paint or paint ingredients, should be properly labelled when they are supplied, or consigned by road. Additional regulations cover international transport by sea, air, road and rail. Before supply and transport, substances must be assessed and, depending on the outcome, containers may have to be labelled as 'Flammable', 'Highly Flammable' or 'Extremely Flammable', and 'Irritant', 'Harmful', 'Corrosive', 'Oxidizing', 'Toxic' or 'Very Toxic' labels may be necessary. Labels identifying danger to the environment and marine pollutants may also be required. Paints must be supplied with instructions for safe use. As well as mentioning safety precautions discussed in the preceding sections, it may be necessary to highlight the need for segregated storage of paints and solvent thinners; in any case, storage in the workplace is bad practice. Recommendations for disposal of waste paint and used containers are also necessary. Regulations concerning labelling for the users of paint require the use of both '*risk phrases*', relating to particular hazards associated with the material, and '*safety phrases*' indicating safety precautions required. As an example, the label on a container of decorative gloss paint, prepared using a long oil alkyd and thinned with white spirit, must include the health and safety information: 'Flammable, Keep away from sources of ignition, No smoking, Contains white spirit, work only in areas of good ventilation.'

Finally, **Material Safety Data Sheets** with property and composition information are required for users. Mandatory headings include identification of the material and company supplying; composition, or information on the ingredients; hazard identification; first aid measures; fire fighting measures; accidental release measures; handling and storage; personal protection; chemical and physical properties; toxicological and ecological information; disposal considerations and transport information.

Protecting the environment by good formulation

Reduction in the **volatile organic content** (VOC) of coatings, contributed by the solvents present, is now required because solvents, principally hydrocarbons, contribute through a reaction chain to an increase in ground level pollution (**photochemical ozone**). The hole in the ozone layer in the stratosphere concerns chlorofluorocarbons (CFC) and is a very different issue. Also of concern to formulators is the removal of lead, chromium and cadmium, which are long term environmental pollutants in paint residues.

Compliant coatings are those coatings which meet legal requirements for low VOC, or requirements arising from voluntary industry agreements. Compliant coatings include powder coatings (Chapter 14), unsaturated polyester and radiation curing finishes (Chapter 16) and aqueous emulsions (Chapter 11), which all contain little or no solvent. In addition, alkyd, polyester and acrylic coatings can be made at high solids with reduced solvent, or in water-borne solution, or emulsified in water using techniques already described. Reformulation of solvent-borne coatings to make them water-borne requires awareness of the special properties of water (Chapter 9).

While VOC laws and agreements define maximum solvent levels, voluntary eco-labelling schemes, of which the first was the *German Blue Angel scheme* (so called because the container symbol resembles a blue angel), provide incentives to formulate to higher 'green' standards. Included in these schemes are targets for a maximum of 1% of aromatic solvents (xylene and toluene) in decorative paints.

Life Cycle Analysis (LCA) is a phrase describing the cradle to grave evaluation of a product's environmental impact. It requires consideration of raw material acquisition and the production, use and disposal of paint (and the painted product). The results of this type of evaluation assist the development of products and processes, improve compliance with regulations, reduce cost associated with waste, and supply information to those responsible for eco-labelling. The *Environmental Management System* (EMS), also referred to as ISO 14000, is a developing scheme enabling measurement, monitoring and evaluation of environmental performance, and LCA is a part of this scheme.

Appendix A

suggestions for further reading

These publications cover the main paint ingredients and some special features referred to in Part Two of this book. The expensive reference books (marked *) can be consulted through technical libraries.

General paint technology

The Chemistry of Physics of Coatings, edited by A. R. Marrion (Royal Society of Chemistry, London 1994). An introductory text.

Federation Series on Coatings Technology, Federation of Societies for Coatings Technology, Blue Bell, PA, USA. A series of nearly 30 booklets covering most important coatings topics.

* *Organic Coatings: Science and Technology*, Vols 1 and 2, Z. W. Wicks, F. N. Jones and P. S. Pappas (Wiley-Interscience, New York, 1992–4). These two volumes comprehensively cover current coatings technology, particularly polymer chemistry aspects in Volume 1.

* *Surface Coatings*, Vols 1 and 2, Oil and Colour Chemists' Association, Australia (Chapman and Hall, London, 3rd edition 1992 and 2nd edition 1984 respectively).

* *Paint and Surface Coatings: Theory and Practice*, edited by R. Lambourne (Ellis Horwood, Chichester, 1987).

Paint formulation

Principles of Paint Formulation, R. Woodbridge (Blackie, London, 1991).

Pigments

Introduction to Pigments, J. H. Braun in *Federation Series on Coatings Technology*.

* *Pigment Handbook Vol. 1: Properties and Economics*, edited by P. A. Lewis, 2nd edition (Wiley-Interscience, New York, 1988).

* *Industrial Organic Pigments*, W. Herbst and K. Hunger (VCH, 1993).

Assessment of pigment properties

* *Paint Testing Manual*, edited by G. G. Sward, 14th edition (American Society for Testing and Materials, 1995). The most comprehensive volume on paint testing available.

Pigment dispersion equipment

* *Paint Flow and Pigment Dispersion*, T. C. Patton, 2nd edition (Wiley, New York, 1979), Chapters 17–26. A comprehensive book, which contains practical information on various mills, as well as a full theoretical coverage.
* *Dispersion of Powders in Liquids*, G. D. Parfitt, 3rd edition. (Applied Science, London, 1981).

Solvents

Solvents, W. H. Ellis in *Federation Series on Coatings Technology*.
* *Industrial Solvents Handbook*, E. W. Flick, 4th edition (Noyes, New Jersey, 1991).

Additives

* *Handbook of Coatings Additives*, Vols 1 and 2, edited by L. J. Calbo (Marcel Dekker, 1987 and 1992).

Application of paint

Application of Paints and Coatings, S. B. Levinson in *Federation Series on Coatings Technology*.
* *Industrial Paint Application*, W. H. Tatton and E. W. Drew, 2nd edition (Newnes-Butterworth, London, 1971).
* *Industrial Paint Finishing Techniques and Processes*, G. F. Tank (Ellis Horwood, London, 1991).

Paint testing

* *Paint Testing Manual*, edited by G. G. Sward, 14th edition (American Society for Testing and Materials, 1995).
* *Paint and Varnishes, Vol. 1: General Test Methods, Vol. 2: Raw Materials* (ISO Standards Handbooks, 1994).

Paint defects

* *Hess's Paint Film Defects*, edited by H. R. Hamburg and W. M. Morgans, 3rd edition (Chapman & Hall, London, 1979). A comprehensive reference book on all defects found in paint films, their causes and cure.

Toxicity

* *Patty's Industrial Hygiene and Toxicology*, Vol. 2 in six parts, edited by G. D. and F. E. Clayton, 4th edition (Wiley-Interscience, New York, 1993–4).
* *Dangerous Properties of Industrial Materials*, N. I. Sax, 9th edition in three volumes (Van Nostrand Reinhold, New York, 1994).
Hazards in the Chemical Laboratory, edited by S. G. Luxon, 5th edition (Royal Society of Chemistry, London, 1992).

Appendix B

functional groups of organic chemistry in systematic nomenclature

Family	Simple example	Systematic name[a]	Trivial name[b]
Alkane (paraffin)	H_3C-CH_3	Eth*ane*	
Alkene (olefin)	$H_2C=CH_2$	Eth*ene*	Ethylene
Alkyne (acetylene)	$HC\equiv CH$	Eth*yne*	Acetylene
Arene (aromatic)	(benzene ring with H's)	Benzene	
Alcohol	H_3C-OH	Methan*ol*	Methyl alcohol
Ether	$H_3C-O-CH_3$	Meth*oxy*methane	Dimethyl ether
Aldehyde	$H_3C-\overset{O}{\overset{\|}{C}}\cdot H$	Ethan*al*	Acetaldehyde
Ketone	$H_3C-\overset{O}{\overset{\|}{C}}-CH_3$	Propan*one*	Acetone
Carboxylic acid	$H_3C-\overset{O}{\overset{\|}{C}}\cdot OH$	Ethan*oic acid*	Acetic acid
Ester	$H_3C-\overset{O}{\overset{\|}{C}}\cdot O-CH_3$	Methyl ethan*oate*	Methyl acetate
Carboxylic acid chloride	$H_3C-\overset{O}{\overset{\|}{C}}\cdot Cl$	Ethan*oyl chloride*	Acetyl chloride
Carboxylic acid anhydride	$H_3C-\overset{O}{\overset{\|}{C}}\cdot O\cdot\overset{O}{\overset{\|}{C}}-CH_3$	Ethan*oic anhydride*	Acetic anhydride
Epoxide	$H_2C\overset{O}{\diagup\diagdown}CH_2$	*Epoxy*ethane	Ethylene oxide
Peroxide	(dibenzoyl peroxide structure)	Dibenzoyl *peroxide*	Benzoyl peroxide
Amine	H_3C-NH_2	Methyl*amine*	
Amide	$H_3C-\overset{O}{\overset{\|}{C}}\cdot NH_2$	Ethan*amide*	Acetamide
Nitrile (cyanide)	H_3C-CN	Ethan*enitrile*	Methyl cyanide
Nitro	H_3C-NO_2	*Nitro*methane	
Nitrate	$H_3C-O\cdot NO_2$	Methyl *nitrate*	
Halide	H_3C-Cl	*Chloro*methane	
Thiol (mercaptan)	H_3C-SH	Methane*thiol*	Methyl mercaptan

[a] Normal group prefix/suffix shown in italics [b] Where different to systematic name

Index

The main reference is given in bold type where appropriate. In order to make the fullest use of this index where chemical substances are concerned, reference may be made to the lists inside the covers for alternative names.

Abietic acid 176
Absolute zero 7
Absorption of light 83, **87–88**
Accelerator 147, 173–174, 241
Acetic acid 19, **40–42**, 49, 52–53
  anhydride **40–42**, 52, 453
Acetone **50**, 52, 123
Acetyl chloride **41–42**, 52, 453
Acetylene *see* Ethyne
Acid 17–22
  anhydride **42**, 52, 53, 210–211
  carboxylic 40–42
  -catalysed finishes **196–197**, 199
  chloride **42–43**, 52, 53
  fatty **41**, 44–46
  Lewis 248
  value (acid number) 21, 119, **181**
Acidity 21
Acrilan 74
Acrolein 172
Acrylamide 151, 193, 197
Acrylic acid 40, **151**, 161, 207, 246
  coatings (unsaturated) 245–247
  finishes, thermosetting 195–196,
    207–208, 228–229, 245–247
  lacquers 153–155, 156
  latices **161**, 196
  nitrogen resins 192–193
  polymers 151
  resins, thermosetting **192–193**, 195
Activator 147
Active hydrogen **48**, 59

Acyl 42
Addition polymerisation *see* Chain
  growth polymerisation
  reactions 188, 211
Additive 93, **136–149**, 232
Adhesion **95–96**, 203, 233, 251–253
ADIB 65
Adipic acid 181
Aircraft finishes 213, 229
Air-inhibition **242–243**, 247
Airless spraying 214
Alcohols 38–40
Aldehydes 49–51
Aldol 50
Algicide 148
Aliphatic 30
Alizarin 85
Alkali 19
Alkanes 30–36
Alkenes 36–37
Alkyd finishes 183–186
Alkyds 142, **179–183**, 194, 228
  emulsified 183, 186
  urethane 221
  vinyl-modified 182
Alkyl 34
Allyl compounds 239, 247
Allylic position 169
Allyloxy compounds **172**, 243
Alkynes 37
Aluminium alkoxide derivatives 174
  chloride 22

Aluminium *cont'd*
  metal 255–256
  oxide **22**, 47
  pigment 107, 110, 162
  silicates 138
Amide group 52
  linkage 53
Amides 52–53
Amine-epoxy resin adduct 212
Amines **51–52**, 119, 196, 212–213, 226,
    231, 240
Amino groups 51
  resins 162, **187**
Ammonia **19, 28**, 49, 51–52, 119, 196
Ammonium acetate 52
  carbamate 53
  carbonate 26
  chloride 21, 28
  ethanoate 52
  hydroxide 19, 20, 28
  ion 28
  nitrate 28
  nitrite 27
Amorphous solids 63
Amphoteric 22
Amyl acetate 42
Angle of incidence 78
  of reflection 78
Anion **17**, 18
Anode **17**, 18, 119
Anthracene 84
Anti-corrosive finishes 213
Anti-fouling additives 148–149
Anti-oxidants 147, 170
Anti-skinning agents 120, **147**
Application 93–95
Aromatic compounds 54–58
Arsenic pentoxide 253
Atom 3
Atomic nucleus 3, 12, 14
  number 3
  weight 68
Attractions, inter-molecular 95
  inter-particle 107, 136–139
Attrition 112
Autodeposition process 162
Autoignition temperature 134
Automotive finishes *see* Motor car finishes

Autophoretic process 162
Autoxidation 168
Auxochromes 84
Azelaic acid 108
Azo group 65, 84
Azomethine group 84
Azoxy group 84

Bactericide 148
Barium hydrogencarbonate 27
  peroxide 25
  sulphate 25
Base 17–22
Basecoat, active 244
Basicity 21
Bentones 139
Bentonite 138
Benzene 54–56
  sulphonic acid 54–55
Benzil dimethyl ketal 245
Benzoic acid **58**, 181
Benzoin ethers 245
Benzophenone 245
Benzoyl peroxide 65, 240
Benzyl alcohol 58
Bicarbonate *see* Hydrogencarbonate
Binder 92
Biocide 148
Biodegradation 253
Bisphenol A **203–204**, 245
Bitumen 35
Biuret 228
Bleeding 104–105, 110
Blushing 144
Boat varnish 222, 230
Boiling point **9**, 22, 122–124, **132–134**
  range 35, 122–124, 133
Bond, chemical 11–14
  coordinate 12
  covalent **12**, 25
  double **36**, 64, 168–173, 175, 239, 248
  hydrogen **22**, 61, 125–128, 133, 143
  ionic 17
  length 55
  single 31
  triple 37
Boric oxide 21
Boron trifluoride 229, 248

Bromine 35
Bromo group 84
Brownian movement 11
Butane 34–35
1,3-Butane diol 228
1,4-Butane diol 228
1,3-Butadiene 36
Butanedioic acid 41
  anhydride 41
Butanoic acid 41
Butanol **39**, 123, 133, 135, 189–191, 193
Butenedioic acid 41
But-2-enal 51
But-1-ene 36
But-2-ene 36
2-Butoxy ethanol **121**, 124, 196, 200
Butyl acetate (ethanoate) 42, 123
  acrylate 65
sec-Butanol 39, 50, 123
tert-Butanol 39, 123
Butyl benzyl phthalate 58, **154**, 156
  butanoate 43
tert-Butyl perbenzoate 66
Butylene 36

Calcium carbide **27**, 191
  carbonate 16, **26–27**, 91
  chloride 14, 16, 26, 28
  cyanamide 191
  hydrogencarbonate 27
  hydroxide 27
  oxide 27, 147, 191
  salts 173–174
Can coatings 165, 207–208, 215–217
ε-Caprolactam 225
Car finishes *see* Motor car finishes
Carbamic acid 60
Carbamide *see* Urea
Carbides 26
Carbohydrate 152
Carbon 14, 23, **25–26**, 191
  black 106, 110, 222
  dioxide 15–16, 23, **25–26**, 53, 223, 232
  monoxide 23, **26**, 27
Carbonate 16, 20, **26**
Carbonic acid 16, 20, 26
Carbonyl group **49**, 84
Carboxyl group 40

Catalyst **24**, 147, 177
Cathode **17**, 18, 119
Cation **17**, 18
Cellobiose 152
Cellulose **151–153**, 253
  acetate 152
  acetate butyrate 152, 199
  colloids 143, 160, 161, 162
  ethers 143
  hydroxyethyl 162
  nitrate **152–153**, 156, 197, 249, 260
  polymers 151–153
  sodium carboxymethyl 143
Cerium salts 174
Chain growth polymerization 64, **69–70**, 151, 159, 182, 238, 241, 243, 248
  reaction 67
  transfer 67
  transfer agent 67
Chalking **97**, 205, 211, 213, 222
Charge transfer 85
Chelates, metal **142**, 162
CHIP 2 261
Chloride 20
Chlorinated rubber 148, 150
Chlorine 24, 35–36, 37, 48
Chloro group 84
Chloroalkanes 35–36, 121, 124
Chloroethane 43
Chlorohydrin 48
2-Chloropropanoic acid 40
Chroman ring 178
Chromates 20, 108, 255–256, 257
Chromating 255–256
Chromic acid **19**, 164, 253, 257
Chromophore 84
Cissing 144–145
Citraconic anhydride 238
Citric acid 19
Clearcoat 92
Coalescing solvent **157**, 161
Cobalt salts 173–174, 240, 243
Coefficient of viscosity 8
Co-grinding of pigments 116
Coil coatings 163, **165**, 166, 195, 197, 200
Colloidal state **11**, 111, 140
Colloids, protective **159**, 160, 161, 162

Colour 77–88
  brightness 86
  chemistry 84–85
  Index 111
  matching 111, **116–117**
  mixtures 85–87
  pastel 104
  primary 86
  secondary 86
  solution 116
Combination finish 197
Complementary colours 86
  pigments 87
Compliant coatings 261
Compound 6
Condensation 10
  polymerization *see* Step growth
      polymerization
  reaction 69–70, 180, 187, 188–189,
      192, 195, 207
Conjugation 45–46
Consistency 94
Contact process 244
Continuous phase 155
Contraction in cross-linking 215, 245
Conversion coatings 255–258
Coordinate bond 12
Coordination driers 174
Copolymer block **69**, 140
  graft 160, **164**, 182, 217
  random 68
Copper 23, 28
Copper II naphthenate 253
  chloride 85
  nitrate 28
  oxide 23
  sulphate 85, 253
Corona discharge 252
Corrosion 254–255
COSHH 259
Cost of solvents 135
Coumarone 177
  -indene resins 177
Covalent bond 12
Cracking of paraffins 36
Critical surface tension **95**, 252
Crystallinity in polymers 72–73
Crystals 61–63

Cumene hydroperoxide **66**, 240
Cuprous oxide 148
  thiocyanate 148
Curtain-coating 93, 244
Cyanamide 191
Cyclohexane 54
Cyclohexanone 123
  peroxide 240
  resin 249
Cyclopentane 34

Dammar 154
Defects, application 94
Diacyl peroxide 66
Diakon 151
Dialkyl peroxide 66
Diallyl phthalate 239–240
Diamond 26, 61, 62
Diazonium salt 248
Dibasic acid 41
Dibutyl phthalate **154**, 161, 215, 241
Di-*t*-butyl peroxide 66
Dichlorobenzene 131
2,4-Dichlorobenzoyl peroxide 240
Dichloroethene 37
1,2-Dichloroethylene 37
Dichromates 253, 255
Dicyandiamide 191, 210
Dielectric constant 18
Diels–Alder reaction 175
Diepoxide 0 **205**, 213
Diethyl ether **47**, 124, 133
Diethylamine 215
Diethylene glycol 121, **124**, 181
  monomethyl ether 124
  triamine 212
Diffraction 81–83
Dihydric alcohol 39
Diisocyanate, aliphatic 227, 229, 233
Dilatancy 138
Diluents 118, 134
  for epoxy resins 205
Dimer fatty acids 212
Dimethyl formamide 74
Dimethylamine 51
N,N-dimethylethanolamine **226**, 231
Dimethylol urea 188
Dipentene **38**, 120, 122

Diphenylmethane-4,4′-diisocyanate **59**, 226, 232
Dipping 93, 94
Diradical 172
Dispersed phase 155
Dispersion of pigments 111–116
  of polymer 160
Disproportionation 67
Dissociation 18–20
  degree of 19
Distemper, oil-bound 159
Dodecyl alcohol 145
Domestic equipment finishes 196, 200, 209
Double decomposition 15–16
Driers 147, 167–168, **173–174**
Drum finishes 207–208
Drying, lacquer 98, 151
  methods, comparison of 100–102
  of conjugated oils 167–168, 171–172
  of non-conjugated oils 167–168, 168–171
  of paint 98–102
  oxidative 167–174
Durability 96, 176, 179, 183

Electrodeposition 93, **119**
  paints 206, 215, 218
Electrolysis 17–18
Electrolyte 17
Electron **3**, 12–13, 14, 17–18, 22, 24–25
  beam curing 247
  valency **12–14**, 17, 22, 25
Electronegativity **14**, 17, 22, 25
Element 3, 6
α-Eleostearic acid **45**, 172
Emulsification 146, 183
Emulsion **11**, 155–157, 158–159
  aqueous 119, **158–160**
  paint 155–164
    application 158
    aqueous 160–164
    film-formation 157–158
    glossy 161
    industrial 162, 164
    oil-based 159
    pigmentation 157–158, 161
    solid 162

polymerization 159–160
Enamel 92
Ene reaction 175
Environmental management scheme 262
Epichlorhydrin 183–185
Epoxides 48–49
  of dihydric alcohols 203, 249
Epoxy-acrylate 246
Epoxy-acrylic finishes 205, 208, 217
Epoxy advancement 205
Epoxy-alkyds 206–207
Epoxy-amino resin finishes 208
Epoxy-ester resins **185–186**, 191
Epoxyethane *see* Ethylene oxide
Epoxy finishes 207–219
  non-stoving 211–215
  powder coatings 209–211, 219
  solventless 213–215, 218
  stoving 207–211, 217–219
  UV curing 247–249
  water-based 215–217
Epoxy–MF–alkyd finishes 209
Epoxy novolac resin **205**, 248
Epoxy-polyamide finishes, 212–213
Epoxy polyamine finishes 211–212, 218
  coal tar-modified 213
Epoxy-phenolic finishes 207–208
Epoxy properties 203–205
Epoxy-resins 198, **203–206**, 246
  cyclo-aliphatic **205**, 248, 249
  modified **205**
Equations, chemical 15–16
Equivalent weights 21
Ester gum 154
  linkage 43
Esters 42–44
Ethanal 40, **50–51**
Ethanamide 52
Ethane **30**, 37, 55
Ethane-1,2-diol *see* Ethylene glycol
Ethanoic acid *see* Acetic acid
Ethanoyl chloride **41–42**, 52, 53
Ethanol **42–43**, 47, 50, 123, 135
Ethanolamine 49
Ethene *see* Ethylene
Ethenyl group 84
Ether linkage **47**, 152, 171, 209
Ether-alcohols 49, 121, 124

Etherification 47, **189**, 192
Ethers 47, 121, 124
Ethoxyethane 47
Ethyl acetate 42, 52, 123, 135
  acetoacetate 243
  acrylate 151, 161, 193
  bromide 51
  cellulose 152
  chloride 43
  ethanoate 42, 52, 123, 135
  hexoic acid 173
  hydrogensulphate 413
  nitrate 43
  orthoformate 225
  sulphuric acid 40, 43
Ethylamine 51
Ethylene **36**, 40, 48, 55, 64
  copolymer latices 161
  diamine 212
  dichloride 37
  glycol 37, **39**, 48, 119, 123, 179, 181
    dimethacrylate 239–240
    monoacetate 469
    monobutyl ether 121, **124**
    monomethacrylate 151
  oxide **48–49**, 51, 119, 229
2-Ethyl hexyl acrylate 161
Ethyne 27, 37
Evaporation **9**, 94
  rate 132–134
Extender **93**, 106, 146

Ferric (Iron III) chloride 22, 24
  hydroxide 257
  oxide 13, 15, 22, 254, 257
  phosphate 256–257
Ferrous (Iron II) chloride 24
  hydroxide 254
  phosphate 256–257
  sulphide 15
Filler 92
Film properties 95–98
Film-former **92**, 93, 101
Finish 93
  for car repainting 229, 236
  for chipboard 197
  for concrete 213, 232
  for hardboard 197

  for plastics 150, 231
  for rubber 231
Flame treatment 252
Flammability 134
Flash point **122–124**, 134, **260**
Flexibility **96–97**, 228, 230
Flocculation **107**, 116, 137–140
Floor finish 213, 222, 225, 235
Flow 93
  agents 145
  cup 129–130
  time 130
Fluoride 232, 233
Force-drying 213
Formaldehyde **50**, 177–178, 187–193, 198
Formalin 188
Formic acid 41
Formula, acid-catalysed woodfinish 198
  acrylic metallic car basecoat 189
  alkyd gloss paint (decorative) 185
  can coating, interior 217
  car lacquer 156
  cathodic electrodeposition primer 218
  chemical 6
  coil coating plastisol 163
  emulsified alkyd decorative paint 186
  emulsion-based undercoat 163
  emulsion paint 163
  high solids stoving enamel 201
  NC woodfinish 156
  oil-modified polyurethane paint 234
  oleoresinous varnish 185
  paint (sand mill dispersion) 115
  polyester coil coating enamel 200
  polyester woodfinish 249
  powder coating 219
  solventless epoxy finish 218
  urethane two-pack clearcoat 236
  urethane two-pack topcoat 236
  UV-curing varnish for paper 250
  water reducible AD gloss paint 184
  water-borne stoving finish 202
  wood primer 185
Free radicals **25**, 64–65, 67, 147, 159, 168–174, 217, 237, 239, 241, 242, 245, 247, 248, 252
  combination of 67

Freezing point 10
Frequency 77
Fumaric acid 41, 238
Functionality of alkyds 182
  of drying oils 172, 173
  of glycerol 181
  of monomers 70–71
Fungi 253
Fungicide 148, 159, 161
Furniture finishes 150, 156, 197, 229, 245

Gas oil 35
Gases 10
  elementary 15
Gel, irreversible 99
Gel strength, measurement 143
Giant molecule 61
Glass temperature **63**, 73, 76
Gloss 79
  loss of 97
  reduction of 146
Glucose 152
Glycerin nitrate 44
Glycerine (glycerol) **39**, 44, 123, 180–181
Glycerol monoallyl ether 243
Glycidyl methacrylate 205
Glycol 39
Glycol ethers 49
Grafting reactions 141, 160, 217
Graphite 26
Grease 35
Grey 88

Hardness **96**, 228, 230, 245
  of water, temporary 27
Hazardous substances 260
Health and safety 259–261
Hemiacetal 50
Heterolytic dissociation 25
Hexamethoxymethyl melamine
  (HMMM) resins **192**, 195, 196
Hexamethylene diisocyanate 59, 227
1,6-Hexane diol 228
1,6-Hexane diol diacrylate 246
1,2,6-Hexane triol 228, 229
Hiding power **105**, 138
High solids paint 166, 183, 196
  speed disperser 112–113

Hindered amine light scatterer (HALS)
  147
Homolytic dissociation 25
Homopolymer 68
Hopeite 257
Hot spray 155
House paints 99, 160–162, 183,
  185–186
Hue **83**, 104, 110, 116
  brilliance of 110
Hydrocarbons, aliphatic **30–36**, 120,
  122, 134
Hydrochloric acid 15, **18–22**, 26, 43, 48,
  50, 59
Hydrogen 13, 17, **23–24**, 28, 37
  bond **21–22**, 57, 125–128, 133, 143
  iodide 47
  ion 18–21
  peroxide 24–25
  sulphide 20
Hydrogenated castor oil **46**, 139
Hydrogencarbonate 16, 20
Hydrogensulphate 19
Hydrogensulphite 19
  compound 50
Hydrolysis 43
Hydroperoxides 65–66, 167–173, 240,
  243
Hydroquinone 67
Hydroxide 16
  ion 18–21
Hydroxyethyl acrylate 246
  methacrylate 151
Hydroxyethyl cellulose 161
Hydroxyl group **38**, 84

Incompatibility 82
Indene 177
Indicator 21
Induction period 167
Industrial equipment finishes 150
Infrared radiations **77**, 197
Inhibitor **67–68**, 147, 239, 243
Initiator **65**, 66, 159, 240
Interfacial tension 144–145
Iodo group 84
Iodonium salt 248
Ions **17–21**, 24

Iron 13, 15
  II salts *see* Ferrous
  III salts *see* Ferric
  oxides (pigments) 110
  salts 174
Isoamyl acetate 42
Isobutane 33
Isobutanol **39**, 123, 191
Isobutyl propanoate 43
Isocyanate group 58–60
Isocyanate:hydroxyl ratio 230
Isocyanates **58–60**, 220–234
  blocked 225–226
Isocyanurates 227
Isomer **33**, 37, 54, 56–57
Isomers, geometrical 37
Isophorone diisocyanate 233
Isopentane 33–34
Isophthalic acid **58**, 183, 195, 238, 245
Isoprene 37
Isopropanol 39, **123**, 133
Itaconic acid 238

Jelly paint 137

Kerosene 35
Ketone peroxide 66
Ketones 49–51, 135
Ketoxime **52**, 185

Lacquer 92, **150–155**
  film-formers 153–155
Lactic acid 41, 119, 215
Lakes 110
Latent heat
  of evaporation 9
  of fusion 7
Latex (latices) 158–160
  core-shell 141–142, 160
Laughing gas 14
Lauroyl peroxide 240
Lead 13
  carbonate 26
  chromate 108
  iodide 85
  monoxide 23, 28
  nitrate 26, 29
  red 108

salts 108, 173
sulphate 25
sulphide 25
white 108
Life cycle analysis 262
Ligand 85
Light 77–83
Lightfastness **104**, 110
Lime 27
Limonene 38, 120
Linoleic acid 44
Linolenic acid 44
Liquids 7–9
Litharge 23
Litmus 17, 21
Livering 108
Lower explosive limit 260
Lucite 151

Magnesium 23, 27
  hydrogencarbonate 27
  nitride 27
  oxide 23
Maintenance finishes 213, 22
Maleic acid 41
  anhydride 119, **238**
Manganese dioxide 24
  salts 173–174, 179
Marine finishes 148–149, 222, 230
Material Safety Data Sheets 261
Melamine 191–192
  formaldehyde (MF) resins **191–192**,
    194–197
Melting point 7, 10
Metal finishes 150, 153, 178, 194–196,
    197–198, 207, 208
Metal primer 176, 207
Metallic elements **6**, 14
Meta-substitution 56
Methacrylic acid 141, **151**, 161, 193,
    207
Methanal *see* Formaldehyde
Methane **30–31**, 34
Methanoic acid 41
Methanol 39–40, 123
Methoxy group 84
Methyl ethyl ketone 50, 123, 131
  peroxide 66, 240

Methyl iodide 47
    isobutyl ketone 50, 123
    methacrylate **65**, 151, 161, 239
    propyl ether 47
Methylamine 51
    hydrochloride 51
Methylene chloride 36
Methylol group 188
    ureas 188
Mica pigment 107
Micelle 159
Microgels 141–142, 160
Micro-organisms 148
Mill 112
    ball 113
    bead 113–114
    high speed disperser 112–113, 115
    roller 113, 115
    triple-roll 115
Mixing schemes 117
Mixture 6
Moisture scavengers 147
Molecular sieves **225**, 232
    weight **68**, 70, 99, 222
Molecule 6
Monoglyceride **180**, 220–222
Monomer **64–69**, 239–240, 247
Monoterpene 38
Montmorillonite 138
Motor car finishes 150, 156, 195, 199, 229
    primers 208, 215
    refinish paints 150, 229, 236, 244
    surfacers 208, 244

Naphthenic acids 173
Natural gas 35
    resins 154
Neopentane 33–34
Neopentyl glycol **181**, 195
Neutralization 20–21
Neutron 3
Nitrate 16, 20, **29**
Nitric acid 19, 28, 36, 43, 57
    oxide 28
Nitrides 27
Nitrites 20, 27–28
Nitro group 84
Nitrocellulose 130, **152–153**, 156, 197, 260

Nitroethane 43
Nitrogen 12, 14, **26–29**, 180, 191, 247
    monoxide 28
    peroxide 28
    resins **187–193**, 228
        effects of alcohol 189–191
        paints based on 194–198
Nitroglycerine 44
Nitroparaffins 36, 121, 124, 135
Nitrophenols 57
Nitroso group 84
Nitrous acid 19, **28**
    oxide 14, **28**
Noble gases 13
Nonanol 190
Non-aqueous dispersions (NAD) 158, **164–165**, 195
Non-metals **6**, 14
Novolac, phenolic 177
Nucleus of atom **3**, 12, 14
Nylon 76

Occupational Exposure Limits **135**, 198, 234, 260
Octadecane 34
Octanoic acid 173
Oil absorption 106
    blown 175–176
    bodied 175–176
    boiled 175
    castor **45–46**, 220, 224, 228, 232
    coconut 45
    conjugated **171–172**, 173, 175
    dehydrated castor 46, 182
    drying 44–45
    essential 38
    fatty 44
    heat-bodied 175–176
    hydrogenated castor **46**, 139
    length of alkyds 182
        of epoxy ester 206
        of oleoresinous varnish 179
        of urethane alkyds 221
        of urethane oils 221
    linseed **45**, 106–107, 167, 176, 182
    lubrication 35
    maleinized 119, 206
    non-conjugated 168–171

Oil *cont'd*
  non-drying 44–45
  paints 99, 176–179
  safflower 45
  semi-drying 45
  soya bean **45**, 182, 183
  stand 175–176
  tall **45–46**, 183
  tung **45**, 172
  urethane 220
Oils 44–46
Oleic acid 44
Oleoresinous finishes **176–177**, 185
Olefins 36
Oligomer 196
  acrylic (unsaturated) 246
  polyol 246
Organosols 75, 158, **165**
Orthophosphate 20
Orthophosphoric acid 19, 20
Ortho-substitution 56
Oxidation **23–24**, 252, 254, 257
Oxides 23
Oxidizing agent **23**, 25, 28
Oxygen 12–15, 17–18, 23–25, 167, 173,
    239, 242, 247

Paint 91–93
  literature 109, 111
Paper finishes 150
Paraffin oil 35
  wax 35, 242
Paraffins 31
Paraformaldehyde (paraform) 188
Para-substitution 56
Particle shape 107
  size **105–107**, 109
Pascal 9
Pascal-second 9
Passivation 255
Patching 97–98
Pauling 14
Pentachlorphenol 253
Pentadecane 34
Pentaerythritol **39**, 181, 220
Pentane 30–34
Peracid ester 66
Periodic table **3–6**, 12

Peroxide linkage 65, 171, 240, 241
  radicals 65, 172
Peroxides, organic 65–66
Personal protection 259–260
Perspex 65, 151
Petrol 36
  light 36
Petroleum 35, 36
  resins 176
  spirit 120, 153
pH scale **21**, 143, 161, 162, 255, 256
Phenol 53, 164, 202
Phenolic condensate 177
  resins 177, **177–179**, 207–208, 217
Phenols **56**, 67, 214
Phenylamine 58
  hydrochloride 59
Phenyl group 56
  isocyanate 59
Phosgene **53**, 59
Phosphates **20**, 255–258
Phosphophyllite 257
Phosphoric acid 19, 20
Phosphorus oxychloride 41
  pentachloride 41
Photoinitiators 245, 248
Phthalic anhydride **58**, 179–183, 195,
    200, 210, 238
Pigment 91, 103
  agglomerates 106
  anti-corrosive **108**, 110, 255
  chemistry 108
  dispersion 111–116
  functions 103
  literature 109, 111
  particle shape 107
  properties 103–108
  selection 111, 116
  surface 106, 138
  surface area 106
  volume concentration 97, 138
Pigments, azo 111
  black 110
  dioxazine 111
  inorganic 109–110
  mica 107, 110
  natural 109
  organic 110

perinone 111
perylene 111
phthalocyanine 111
polymeric 160
quinacridone 111
synthetic 109
white 105, 108, 110
Pine oil 38, **120**, **122**
α-Pinene 38, 120
Pinolenic acid 45–47
Plasma 252
Plastic flow 138
Plasticisers 97, **154**, 161, 163, 166
Plastics, surface treatment 252–253
Plastisol 75, 162, **166**
Platinum 23, 28
Plexiglas 151
Poise 9
Poiseuille 9
Polarity 22
Polishing 97
  mechanical 243
Polyacrylonitrile 73
Polyamide resin 139, **212–213**, 214
Polyamines, aromatic 214
Polybasic acid 41
Polybutyl methacrylate 65, 151
Polycoumarone 177
Polyester acrylate 246
  finishes (unsaturated) **243–244**, 249
  properties 245
  resins 69, 180
    durability 194–195
    saturated 179–180, 194, 195, 200,
      228, 229
    unsaturated 237–242
Polyethers 119, 145, **229**, 232
Polyethylene (polyethene) 65, 74, 252
  glycol 119, 140, 182
Polyhydric alcohol **39**, 220
Poly-isobutyl methacrylate 151
Polyisocyanurates 227
Polylauryl methacrylate 151
Polymer **64–76**, 92
  amorphous 72–73
  branched 71
  chain growth 64–69
  cross-linked **71–72**, 96, 98–100

crystalline **72–76**, 151, 152
  dispersion 160
  linear **70–71**, 97
  molecular weight **68**, 70, 95, 96, 99,
    222
  step growth **69–70**, 76
  thermoplastic 72
  thermosetting 72
Polymerization, cationic 248
  chain growth 64–69
  step growth 69–70
Polymethyl acrylate 160
Polymethyl methacrylate 65, 68, **151**,
  154–155
Polyol 39, **220**
Polyperoxide 242
Polypropylene 262
Polystyrene 75–76, 131
Polytetrafluoroethene 95
Polythene *see* Polyethylene
Polyurethane dispersions 226
  reactive dispersions 231
Polyurethane finishes
  one-pack **220–226**, 234, 235
  powder coatings 225
  properties 232–233
  two-pack **226–231**, 236
Polyvinyl acetate 65, **75**, 160
  alcohol 143
  chloride 65, **74**, 162, 165, 166
Polyvinylidene fluoride 165
Popping 144
Potassium carbonate 26
  chlorate, 24
  chloride 24
  dichromate 253
  hydrogencarbonate 27
  hydroxide 19–20
  nitrate 29
  nitrite 29
  permanganate 37, 40
  sulphate 20
Pot-life **100**, 231, 244
Powder coatings **209–211**, 219
Prepolymers 223–225
Pressure of a gas 10
Pretreatments 255–258
  amorphous chromate 255–256

Pretreatments *cont'd*
  chrome-free 256
  chrome phosphate 255–256
  iron phosphate 256–257, 258
  no rinse 256
  zinc phosphate 257–258
Primary alcohol 39
  carbon 35
Primer 92
  alkyd based 185
  for plastics 226
  oil-based 176
  preservative 253
  -surfacer 92, 226
Propagation of polymerization 67
Propane 30–32
Propanal 50
Propanoic acid 40–41
Propanol 39, 47, **123**, 179
Propanone *see* Acetone
Propenoic acid *see* Acrylic acid
Propylene (Propene) 36, 48
  chlorohydrin 48
  glycol **181**, 238
    monomethyl ether 121, 124
  oxide 48, **229**, 245
Proton **3**, 14, 17
Pseudoplasticity 138
Pulling over 97

Quaternary
  ammonium hydroxide 51
    ions 51, 139
    salt 51
  carbon 35
  ethyl ammonium bromide 51

Radiation curing 245–249
Radiations, electromagnetic 77–78
Reactions, chemical 15–16
Reducing agent 23
Reduction 23–24
Reflection 78–79
Reflow of lacquers 97
Refraction 80–81
Refractive index **80**, 82, 83, 105
Renovation 97–98
Repair 97–98

Resin 64
  natural 64, 154
  water-soluble **109–110**, 176, 179, 182, 186
Resistance to abrasion 209, 225, 232–233
  to alkalis 222
  to chemicals 150, 207–209, 222
  to solvents **97–98**, 229
  to water 107, 222, 229
Resole, phenolic 178
Retarder 147
Ricinoleic acid 44, 45
Ring-opening 48
Risk phrases 261
Road surfacing 232
Rosin 148, **176**, 178
Rust 13, **254**

Safety phrases 261
Salt, common 6, **15**, 61
Salts 19–21
Sand mill 113
Sanding, mechanical 243
Saponification 43
Satin sheen 146
Scavengers, moisture 147, **225**
Scission reactions 171
Sealer 92
Sebacic acid 181
Secondary alcohol **36**, 37, 50
  carbon 35
Second-order transition temperature 63
Shear 111–112
  rate of **8–9**, 136–137
  stress **8**, 136
Shelf-life 100
Shrivelling 174
Silica 106, **138**, 146
Silicates **138–139**, 225
Silicone oils 145, 195
  resins **195**, 215
Silver 48
Sinkage 244
Skinning of paint 99
Soap formation 43
Soaps **41**, 43, 145, 159, 173
Sodium acetate 43
  aluminate 22

borate 21
carbonate 26
chloride 6, **15**
hydrogencarbonate 27
hydrogensulphite 21, **50**
hydroxide 15, **19**, 26, 43, 48, 204, 229
nitrate 26
phenoxide 56
stearate 41
Softening temperature **63**, 73–76
Solids 7
content **94**, 100
Solubility of crystalline solids 10
of polymers 97, **127**, 129
parameter **122–128**, 131
Hansen 128–129
sphere 128–129
Solubilization 159
Solute 10
Solution, saturated 10
Solutions 10
Solvency 121–129
Solvent properties 121–135
Solventless finish 213–215, 218, 232
Solvents 10–11, 94, **118–135**, 139–142
chemical types of 118–121
for polyurethanes 232
Specific gravity of pigments 108
Spectrum, optical 83
solvent 126
Spraying 94
hazards, precautions 260
twin-feed 214, 232
viscosity 129
Stability of suspensions 111–112
thermal, of pigments 109
Stabilization, ionic 158
steric 158
States of matter 6–10
Step growth polymerization 69–70, 76, 165, 179, 188
Stopper 92
Strengths of acids and bases 19
Stress **8**, 137
Strontium hydrogencarbonate 27
chromate 108
Styrene **57**, 122, 161, 182, 193, 239, 241–242

-butadiene polymers 161
copolymers 76
polyperoxides 242
Substitution of benzene 56–58
Substrate 92, **251–258**
Succinic acid 41, 181
Sulphate 15, 20
ion 18
Sulphite 20
Sulphonium salt 248
Sulphur 12, 15, 23
dioxide 23
trioxide 18, 23
Sulphuric acid **18–19**, 23, 24, 40, 43, 46, 47, 54, 253
Sulphurous acid 20
Surface area **106**, 138
optical 78
tension **8**, 95, 120, 144–145
Surfacer 92, 226
Surfactant **145–146**, 158–160
Suspension 10
colloidal **11**, 103
Symbol 3, 6

Tank linings 214
TDI trimer 224
Terephthalic acid 181
Termination of polymerization 67
Terpene resins 177
Terpenes **36**, 120, 122
α-Terpineol 38
Tertiary alcohol 38
carbon 35
Terylene 76
Tetrachloromethane 39
Tetrahydric alcohol 37
Textile finishes 150
T_g *see* Glass temperature
Thermoplastic 72
Thermosetting 72
Thickener 139–142
selection 143–144
Thickeners
associative **140**, 143–144
HEUR 140
mineral **138–139**, 143
resinous **139–143**, 158

Thinner 93
Thio group 84
Thixotropy 94, **137**, 143
Threshold limit values **135**, 198, 234, 260
Tin salts 148, 149, 225, 232
Tinters 117
Tinting strength 103–104
Titanium
  chelates 142
  dioxide 80, **105**, 108, 110, 160
Titration 21
Toluene 57, **122**
Toluene diisocyanate **59**, 223, 224, 225
p-Toluene sulphonic acid 196
Topcoat 103
Toughness 96, 225, 232
Toxicity of anhydrides 260
  of formaldehyde 198
  of isocyanates 234, 260
  of lead 173, 260
  of polyurethanes 234
  of solvents 135, 259
Transition elements 85
Triallyl cyanurate 239–240
Tributyl tin oxide 253
Trichloromethane 36
Tri(dimethylaminomethyl) phenol 214
Triethanolamine 49
Triglyceride 44
Trihydric alcohol 39
Trimellitic anhydride 181
Trimethylamine 51
Trimethylol propane **181**, 195, 224
  monoallyl ether 243
  triacrylate 246
Turpentine 38, 120, 122
Twin-feed spray guns 232
Two-pack paints **100**, 197, 199, 211–215, 218, 226–231, 236, 243–244
Tyndall effect 11

Ultra-violet absorbers 146
  curing 237, 245–250
  hazards 260
  lamps 245
  light **77**, 104, 110, 153, 165, 166, 168, 183, 237, 239, 260
Undercoats 92

Unsaturation 36
Uralkyd 220
Urea **53**, 61, 62, 187–189
  formaldehyde resins 187–191
    water-soluble 188
  linkage **224**, 226, 227
Urethane 60
  acrylate 246
  alkyds **220–223**, 232, 234
  oils **220–223**, 232

Valencies of carbon 30–31
Valency 11–14
  electrons **12**, 13–14, 17
  variable **12–14**, 24
Vapour curing 100, **231**
Varnish **93**, 222, 225
Vaseline 35
Vehicle 92
Versatic acids 161
Vinyl
  acetate 65, 160
    copolymers 161
  chloride 65
    copolymers 75, 161
    latices 161
  group 65
  monomer 239
  toluene **57**, 122, 239
  Versatate 161
Vinylidene chloride copolymer latices 161
Vinylidene group 65
  monomer 239
Viscometers 129–130
Viscosity **8**, 94, 129, 130–132, 136–142
  brushing 183
  coefficient 8
  of emulsion 131–132
  measurement **129–130**, 143
  non-Newtonian **137–142**, 143
  of solutions 130–131
  of solvents 122–124, 131
VOC 261

Water 6, 13, **17–21**, 22, 80, 118–120, 122, 129, 135, 152, 158–160, 253

-based paints **119–120**, 140, 141–142, 160–164, 182, 183–184, 186, 192, 196, 202, 215–217, 218, 226, 231, 262
-solubility of resins **119–120**, 184, 215–217
Wavelength **77–78**, 81, 83, 106
Wax 146
    paraffin **35**, 242
    polyethylene 146
    polypropylene 146
Weathering 97
Wetting **95–96**, 106, 145
White spirit 120, 122
Whitewash 91
Whiting 91
Wire enamels 178
Wood finishes 150, 156, 197, 199, 222, 229–230, 235, 237, 243–245

Wood primers 176, 185
Wood rots 253

Xylene 57, **120**, 122, 125, 128–129, 135

Yellowing **179**, 183, 233

Zinc 24
    chromate 108
    naphthenate 253
    oxide 108
    phosphate 257–258
    sulphate 24
Zirconium 173–174
    chelates 142
    complexes 185, 234

Trivial name	Systematic name
Acetaldehyde	Ethanal
Acetamide	Ethanamide
Acetic acid	Ethanoic acid
Acetic anhydride	Ethanoic anhydride
Acetoacetic acid	3-Oxo-butanoic acid
Acetone	Propanone
Acetonitrile	Ethanenitrile
Acetophenone	Phenylethanone
Acetyl chloride	Ethanoyl chloride
Acetylene	Ethyne
Acrolein	Propenal
Acrylamide	Propenamide
Acrylic acid	Propenoic acid
Acrylonitrile	Propenenitrile
Adipic acid	Hexanedioic acid
Amyl acetate	1-Pentyl ethanoate
Amyl alcohol	Pentan-1-ol
tert-Amyl alcohol	3-Methylbutan-2-ol
Aniline	Phenylamine
Azelaic acid	Nonane-1,7-dioic acid
Benzoic acid	Benzenecarboxylic acid
Benzoyl peroxide	Dibenzoyl peroxide
Benzyl alcohol	Phenyl-methanol
Bisphenol A	2,2-bis(4-Hydroxyphenyl)propane
Butadiene (1,3-Butadiene)	Buta-1,3-diene
Butyl alcohol (n-Butanol)	Butan-1-ol
sec-Butyl alcohol (sec-Butanol)	1-Methylpropan-1-ol or butan-2-ol
tert-Butyl alcohol (tert-Butanol)	2-Methylpropan-2-ol
Butylamine	1-Aminobutane
Butylene	But-1-ene
Butyraldehyde	Butanal
Butyric acid	Butanoic acid
Caprolactone	1,6-Hexanoide
Carbon tetrachloride	Tetrachloromethane
Catechol	Benzene-1,2-diol
Chloroform	Trichloromethane
Crotonaldehyde	But-2-enal
Dichloroethylene	1,2-Dichloroethene
Diethyl ether (Ether)	Ethoxyethane
Dimethylol propionic acid	2,2-bis(Hydroxymethyl)propionic acid
Dipentene	p-Mentha-1,8-diene
Eleostearic acid	Octadeca-9,11,13-trienoic acid
Epichlorohydrin	1-Chloro-2,3-epoxypropane
Ethanolamine	1-Aminoethan-2-ol
Ethyl acetate	Ethyl ethanoate
Ethyl acetoacetate	Ethyl 3-oxo-butanoate
Ethyl acrylate	Ethyl propanoate
Ethyl alcohol	Ethanol
Ethyl propionate	Ethyl propanoate
Ethylene	Ethene
Ethylene carbonate	1,3-Dioxa-2-oxo cyclopentane
Ethylene dichloride	1,2-Dichloroethane
Ethylene glycol	Ethane-1,2-diol
Ethylene glycol mono butyl ether	2-Butoxyethanol
Ethylene oxide	1,2-Epoxyethane
Formaldehyde	Methanal
Formic acid	Methanoic acid
Fumaric acid	trans-Butenedioic acid
Glycerol	Propane-1,2,3-triol
Hydroquinone	Benzene-1,4-diol
Isoamyl acetate	3-Methyl-1-butyl ethanoate
Isoamyl alcohol	3-Methylbutan-1-ol
Isobutane	2-Methylpropane

Trivial name	Systematic name
Isobutyl alcohol (Isobutanol)	2-Methylpropan-1-ol
Isobutylene	2-Methylpropene
Isobutyl propionate	1-Methylpropyl propanoate
Isobutyric acid	2-Methylpropanoic acid
Isopentane	2-Methylbutane
Isophthalic acid	Benzene-1,3-dicarboxylic acid
Isoprene	2-Methylbuta-1,3-diene
Isopropyl alcohol (Isopropanol)	Propan-2-ol or 1-Methylethanol
Lactic acid	2-Hydroxypropanoic acid
Lauric acid	Dodecanoic acid
Lauroyl peroxide	Di(dodecanoyl) peroxide
Lauryl alcohol	Dodecanol
Linoleic acid	Octadeca-9,12-dienoic acid
Linolenic acid	Octadeca-9,12,15-trienoic acid
Maleic anhydride	cis-Butenedioic anhydride
Melamine	sym-Triaminotriazine
Methacrylic acid	2-Methyl-propenoic acid
Methyl acetate	Methyl ethanoate
Methyl alcohol	Methanol
Methyl ethyl ketone (MEK)	Butanone
Methyl formate	Methyl methanoate
Methyl methacrylate	Methyl 2-methyl-propenoate
Methyl propionate	Methyl propanoate
Methyl propyl ether	Methoxypropane
Methyl isobutyl ketone (MIBK)	2-Methylpentan-4-one
Methylene chloride	Dichloromethane
Oleic acid	Octadeca-9-enoic acid
Oxalic acid	Ethanedioic acid
Pentaerythritol	2,2-bis(Hydroxymethyl)-propane-1,3-diol
Neopentane	2,2-Dimethylpropane
Phosgene	Carbonyl chloride
Phthalic anhydride (ortho-Phthalic anhydride)	Benzene-1,2-dicarboxylic anhydride
Pinolenic acid	Octadeca-5,9,12-trienoic acid
Propionaldehyde	Propanal
Propionic acid	Propanoic acid
Propyl alcohol	Propan-1-ol
Propylene	Propene
Propylene chlorohydrin	1-Chloropropan-2-ol
Propylene glycol monoethyl ether	1-Ethoxypropan-2-ol
Propylene glycol monomethyl ether	1-Methoxypropan-2-ol
Propylene oxide	1,2-Epoxypropane
Propylene carbonate	1,3-Dioxa-2-oxo-4-methyl cyclopentane
Ricinoleic acid	12-Hydroxyoctadeca-9-enoic acid
Stearic acid	Octadecanoic acid
Styrene	Phenylethene (Ethenyl-benzene)
Succinic acid	Butanedioic acid
Terephthalic acid	Benzene-1,4-dicarboxylic acid
Tetrahydrofuran (THF)	Oxacyclopentane
Toluene	Methylbenzene
Trichloroacetic acid	Trichloroethanoic acid
Urea	Carbamide
Vinyl acetate	Ethanoyloxy-ethene
Vinyl acetate	Ethenyl ethanoate
Vinyl alcohol	Hydroxyethene
Vinyl chloride	Chloroethene
Vinyl fluoride	Fluoroethene
Vinylidene chloride	1,1-Dichloroethene
Vinylidene fluoride	1,1-Difluoroethene
meta-Xylene	1,3-Dimethylbenzene
ortho-Xylene	1,2-Dimethylbenzene
para-Xylene	1,4-Dimethylbenzene

Systematic name	Trivial name
1-Aminobutane	Butylamine
1-Aminoethan-2-ol	Ethanolamine
1-Chloro-2,3-epoxypropane	Epichlorohydrin
1-Chloropropan-2-ol	Propylene chlorohydrin
1-Ethoxypropan-2-ol	Propylene glycol monoethyl ether
1-Methoxypropan-2-ol	Propylene glycol monomethyl ether
1-Methylethanol	Isopropyl alcohol (Isopropanol)
1-Methylpropan-1-ol	sec-Butyl alcohol (sec-Butanol)
1-Methylpropyl propanoate	Isobutyl propionate
1-Pentyl ethanoate	Amyl acetate
2-Butoxyethanol	Ethylene glycol monobutyl ether
2-Hydroxypropanoic acid	Lactic acid
2-Methyl-propenoic acid	Methacrylic acid
2-Methylbuta-1,3-diene	Isoprene
2-Methylbutane	Isopentane
2-Methylpentan-4-one	Methyl isobutyl ketone (MIBK)
2-Methylpropan-1-ol	Isobutyl alcohol (Isobutanol)
2-Methylpropan-2-ol	tert-Butyl alcohol (tert-Butanol)
2-Methylpropane	Isobutane
2-Methylpropanoic acid	Isobutyric acid
3-Methyl-1-butyl ethanoate	Isoamyl acetate
3-Methylbutan-1-ol	Isoamyl alcohol
3-Methylbutan-2-ol	tert-Amyl alcohol
3-Oxo-butanoic acid	Acetoacetic acid
1,1-Dichloroethene	Vinylidene chloride
1,1-Difluoroethene	Vinylidene fluoride
1,2-Dichloroethane	Ethylene dichloride
1,2-Dichloroethene	Dichloroethylene
1,2-Dimethylbenzene	ortho-Xylene
1,2-Epoxyethane	Ethylene oxide
1,2-Epoxypropane	Propylene oxide
1,3-Dimethylbenzene	meta-Xylene
1,3-Dioxa-2-oxo cyclopentane	Ethylene carbonate
1,3-Dioxa-2-oxo-4-methyl cyclopentane	Propylene carbonate
1,4-Dimethylbenzene	para-Xylene
1,6-Hexanoide	Caprolactone
12-Hydroxyoctadeca-9-enoic acid	Ricinoleic acid
2,2-bis(4-Hydroxyphenyl)propane	Bisphenol A
2,2-bis(Hydroxymethyl)-propane-1,3-diol	Pentaerythritol
2,2-bis(Hydroxymethyl)propionic acid	Dimethylol propionic acid
2,2-Dimethylpropane	Neopentane
2-Methylpropene	Isobutylene
Benzene-1,2-dicarboxylic anhydride	Phthalic anhydride (ortho-Phthalic anhydride)
Benzene-1,2-diol	Catechol
Benzene-1,3-dicarboxylic acid	Isophthalic acid
Benzene-1,4-dicarboxylic acid	Terephthalic acid
Benzene-1,4-diol	Hydroquinone
Benzenecarboxylic acid	Benzoic acid
But-1-ene	Butylene
But-2-enal	Crotonaldehyde
Buta-1,3-diene	Butadiene (1,3-Butadiene)
Butan-1-ol	Butyl alcohol (n-Butanol)
Butan-2-ol	sec-Butyl alcohol (sec-Butanol)
Butanal	Butyraldehyde
Butanedioic acid	Succinic acid
Butanoic acid	Butyric acid
Butanone	Methyl ethyl ketone (MEK)
Carbamide	Urea
Carbonyl chloride	Phosgene
Chloroethene	Vinyl chloride
Di(dodecanoyl) peroxide	Lauroyl peroxide
Dibenzoyl peroxide	Benzoyl peroxide
Dichloromethane	Methylene chloride

Systematic name	Trivial name
Dodecanoic acid	Lauric acid
Dodecanol	Lauryl alcohol
Ethanal	Acetaldehyde
Ethanamide	Acetamide
Ethane-1,2-diol	Ethylene glycol
Ethanedioic acid	Oxalic acid
Ethanenitrile	Acetonitrile
Ethanoic acid	Acetic acid
Ethanoic anhydride	Acetic anhydride
Ethanol	Ethyl alcohol
Ethanoyl chloride	Acetyl chloride
Ethanoyloxy-ethene	Vinyl acetate
Ethene	Ethylene
Ethenyl ethanoate	Vinyl acetate
Ethoxyethane	Diethyl ether (Ether)
Ethyl 3-oxo-butanoate	Ethyl acetoacetate
Ethyl ethanoate	Ethyl acetate
Ethyl propanoate	Ethyl propionate
Ethyl propenoate	Ethyl acrylate
Ethyne	Acetylene
Fluoroethene	Vinyl fluoride
Hexanedioic acid	Adipic acid
Hydroxyethene	Vinyl alcohol
Methanal	Formaldehyde
Methanoic acid	Formic acid
Methanol	Methyl alcohol
Methoxypropane	Methyl propyl ether
Methyl 2-methyl-propenoate	Methyl methacrylate
Methyl ethanoate	Methyl acetate
Methyl methanoate	Methyl formate
Methyl propanoate	Methyl propionate
Methylbenzene	Toluene
Nonane-1,7-dioic acid	Azelaic acid
Octadeca-5,9,12-trienoic acid	Pinolenic acid
Octadeca-9,11,13-trienoic acid	Eleostearic acid
Octadeca-9,12,15-trienoic acid	Linolenic acid
Octadeca-9,12-dienoic acid	Linoleic acid
Octadeca-9-enoic acid	Oleic acid
Octadecanoic acid	Stearic acid
Oxacyclopentane	Tetrahydrofuran (THF)
Pentan-1-ol	Amyl alcohol
Phenyl-methanol	Benzyl alcohol
Phenylamine	Aniline
Phenylethanone	Acetophenone
Phenylethene (Ethenyl-benzene)	Styrene
Propan-1-ol	Propyl alcohol
Propan-2-ol	Isopropyl alcohol
Propanal	Propionaldehyde
Propane-1,2,3-triol	Glycerol
Propanoic acid	Propionic acid
Propanone	Acetone
Propenal	Acrolein
Propenamide	Acrylamide
Propene	Propylene
Propenenitrile	Acrylonitrile
Propenoic acid	Acrylic acid
Tetrachloromethane	Carbon tetrachloride
Trichloroethanoic acid	Trichloroacetic acid
Trichloromethane	Chloroform
cis-Butenedioic anhydride	Maleic anhydride
p-Mentha-1,8-diene	Dipentene
sym-Triaminotriazine	Melamine
trans-Butenedioic acid	Fumaric acid